Optics of Liquid Crystal
Displays

WILEY SERIES IN PURE AND APPLIED OPTICS

Founded by Stanley S. Ballard, University of Florida

EDITOR: Joseph W. Goodman, Stanford University

BEISER · *Holographic Scanning*
BERGER-SCHUNN · *Practical Color Measurement*
BOYD · *Radiometry and The Detection of Optical Radiation*
BUCK · *Fundamentals of Optical Fibers*
CATHEY · *Optical Information Processing and Holography*
CHUANG · *Physics of Optoelectronic Devices*
DELONE AND KRAINOV · *Fundamentals of Nonlinear Optics of Atomic Gases*
DERENIAK AND BOREMAN · *Infrared Detectors and Systems*
DERENIAK AND CROWE · *Optical Radiation Detectors*
DE VANY · *Master Optical Techniques*
GASKILL · *Linear Systems, Fourier Transform, and Optics*
GOODMAN · *Statistical Optics*
HOBBS · *Building Electro-Optical Systems: Optics That Work*
HUDSON · *Infrared System Engineering*
JUDD AND WYSZECKI · *Color in Business, Science, and Industry,* Third Edition
KAFRI AND GLATT · *The Physics of Moire Metrology*
KAROW · *Fabrication Methods for Precision Optics*
KLEIN AND FURTAK · *Optics,* Second Edition
MALACARA · *Optical Shop Testing,* Second Edition
MILONNI AND EBERLY · *Lasers*
NASSAU · *The Physics and Chemistry of Color*
NIETO-VESPERINAS · *Scattering and Diffraction in Physical Optics*
O'SHEA · *Elements of Modern Optical Design*
SALEH AND TEICH · *Fundamentals of Photonics*
SCHUBERT AND WILHELMI · *Nonlinear Optics and Quantum Electronics*
SHEN · *The Principles of Nonlinear Optics*
UDD · *Fiber Optic Sensors: An Introduction for Engineers and Scientists*
UDD · *Fiber Optic Smart Structures*
VANDERLUGT · *Optical Signal Processing*
VEST · *Holographic Interferometry*
VINCENT · *Fundamentals of Infrared Detector Operation and Testing*
WILLIAMS AND BECKLUND · *Introduction to the Optical Transfer Function*
WYSZECKI AND STILES · *Color Science: Concepts and Methods, Quantitative Data and Formulae,* Second Edition
XU AND STROUD · *Acousto-Optic Devices*
YAMAMOTO · *Coherence, Amplification, and Quantum Effects in Semiconductor Lasers*
YARIV AND YEH · *Optical Waves in Crystals*
YEH · *Optical Waves in Layered Media*
YEH · *Introduction to Photorefractive Nonlinear Optics*
YEH AND GU · *Optics of Liquid Crystal Displays*

Optics of Liquid Crystal Displays

POCHI YEH
University of California, Santa Barbara

CLAIRE GU
University of California, Santa Cruz

A Wiley Interscience Publication

John Wiley & Sons, Inc.

New York / Chichester / Weinheim / Brisbane / Singapore / Toronto

This book is printed on acid-free paper. ∞

Library of Congress Cataloging-in-Publication Data:

Yeh, Pochi, 1948–
    Optics of liquid crystal displays   /   Pochi Yeh, Claire Gu.
       p.   cm. — (Wiley series in pure and applied optics)
    Includes index.
    ISBN 0-471-18201-X (alk. paper)
    1. Liquid crystal displays.   2. Liquid crystals—Optical properties. I. Gu, Claire. II. Series.
TK7872.L56Y44   1999
621.3815'422—dc21   98-53672

Printed in the United States of America

10 9 8 7 6 5 4 3 2 1

Contents

Preface

This book introduces the basic principles and presents a systematic and self-contained treatment of the optics of liquid crystal displays (LCDs). It also describes the practical operation of various liquid crystal display systems. The book is intended as a textbook for students in electrical engineering and applied physics, as well as a reference book for engineers and scientists in the area of research and development of display technologies. It has two primary objectives: to present a clear physical picture of the fundamental principles of LCDs and to teach the reader how to analyze and design new components and subsystems for LCDs. The choice of the subject matter and the organization of the book follow closely a course on optics of LCDs that Pochi Yeh taught at the National Chiao Tung University in Hsinchu, Taiwan and at the University of California at Santa Barbara during the past 3 years.

The subject of liquid crystal displays has now grown to become an exciting interdisciplinary field of research and development, involving optics, materials, and electronics. New materials, device configurations, and processing technologies are being developed rapidly. To meet the rapid development of this display technology, this book emphasizes the fundamental principles that can be employed for understanding the limitation of various LCD systems, and to analyze and design high performance display systems. A significant effort is made to bridge the gap between theory and practice through the use of numerical examples based on real liquid crystal materials and information display systems. The book covers a very wide range of topics, including the basic physical properties of liquid crystals, the polarization of optical waves, the propagation of plane waves in liquid crystal media, the concept of phase retardation, Jones matrix method, operations of various LCD systems, passive and active matrix addressing, colors, and birefringence compensators. Near the end of the book, an advanced analytic technique known as the *extended Jones matrix method* is introduced to treat the optical transmission of LCDs at large viewing angles. In writing this book, we have assumed that the student has been introduced to basic electromagnetic plane waves in an undergraduate course in electricity and magnetism. It is further expected that the student has some background in elementary optics and linear matrix algebra.

The authors are grateful to Ragini Saxena, S. T. Wu, Don Taber, John Eblen, Zhiming Zhuang, Zili Li, Len Hale, W. Gunning, M. Khoshnevisan, Fang Luo, Hiap Ong, and K. H. Yang for helpful technical discussions, and to Miao Yang for the help in compiling the liquid crystal data. The authors also wish to thank their colleagues and students at National Chiao Tung University and the University of California for helpful remarks and discussions. Special thanks are given to the K. T. Li and K. Y. Jin Foundation of Taiwan for their sponsorship and to Professors Sien Chi and Ken Y. Hsu who provided the opportunity for teaching such a course during the summer of 1995 and 1997 at the Institute of Electro-Optical Engineering in Hsinchu, Taiwan. The authors also wish to thank Michele Brown and Florina Carvalho for their help in preparing the references. Their thanks are also extended to S. H. Lin and M. L. Hsieh for grading the homework problems and commenting on the manuscript.

Santa Barbara, California   POCHI YEH

Santa Cruz, California   CLAIRE GU

May 1999

Optics of Liquid Crystal
Displays

1

Preliminaries

In an information-dominated age, the display of information is becoming increasingly essential in many aspects of daily life. This requires display systems utilizing cathode ray tubes (CRTs), electroluminescence devices, plasma display devices, field emission devices, flat CRT, vacuum fluorescence and liquid crystal displays (LCDs). Among these systems, the LCD is one of the most important optical display systems for high performance flat panel display of information. The importance arises from many of its advantages, including flat panel, light weight, high definition, low driving voltages, and low power consumption. Generally speaking, a display panel consists of a two-dimensional array of pixels. Each pixel can be turned ON or OFF independently for the display of two-dimensional images. Figure 1.1 shows a schematic drawing of a basic element (pixel; picture element) of a LCD panel.

Referring to the figure, we note that the basic components of a LCD consist of a thin layer of liquid crystal sandwiched between a pair of polarizers. To control the optical transmission of the display element electronically, the liquid crystal layer is placed between transparent electrodes [e.g., indium tin oxide (ITO)]. The polarizers and the electrodes are cemented on the surfaces of the glass plates. A minimum thickness of a few millimeters is needed to maintain the structural integrity of the panel. The thickness of the liquid crystal layer is kept uniform by using spacers that are made of glass fibers or plastic microspheres. By applying a voltage across the electrodes, an electric field inside the liquid crystal can be obtained to control the transmission of light through the liquid crystal cell. To achieve the display of information, we need a two-dimensional array of these electrodes. These electrodes can be driven electrically for data input by using two sets (x,y) of parallel array of electrodes. In what follows, we will briefly describe each of the optical components.

1.1. BASIC LCD COMPONENTS

1.1.1. Polarizers

An optical arrangement that produces a beam of polarized light from a beam of unpolarized light is called a *polarizer*. Polarizers are key components in many LCDs. Virtually all light sources used in optical displays are unpolarized. A beam of polarized light is described by its direction of propagation, frequency,

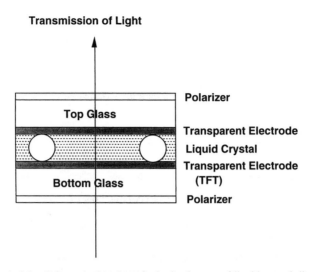

Figure 1.1. Schematic drawing of a basic element of liquid crystal display.

and a vector amplitude (electric field vector). The vector amplitude is related to the intensity of the beam and is perpendicular to the direction of propagation. Given a direction of propagation, there are two independent (orthogonal) components of the vector amplitude.

A beam of light is regarded as unpolarized provided two conditions are met: (1) the time-averaged intensity of the transmitted beam through a polarizer is independent of the orientation of the polarizer and (2) any two orthogonal components of the vector amplitude are totally uncorrelated. Thus, if we decompose the vector amplitude of a beam of unpolarized light into two mutually orthogonal transverse components (e.g., E_x, E_y for light propagating along z axis). These two components will have a random relative phase relationship between them. The random relative phase leads to a resultant amplitude vector that varies rapidly in time in a random manner.

Polarizers consist of anisotropic media that transmit one component and redirect or absorb the other component. It is known that anisotropic crystal prisms provide methods of obtaining plane polarized light by double refraction where one of the components can be removed by reflection or mechanical obstruction. Although prism polarizers can provide excellent extinction ratios, they are not suitable for flat panel applications.

In most LCDs, the polarizers consist of thin sheets of materials that transmit one polarization component and absorb the other component. This effect is known as *dichroism*. Natural dichroism of a material is, like natural birefringence, due to its anisotropic molecular structure. The ratio of the two absorption coefficients for two linear orthogonal polarizations is called the *dichroic ratio*. One of the best known materials of this kind is tourmaline—a naturally occurring mineral that exhibits uniaxial birefringence and a dichroic

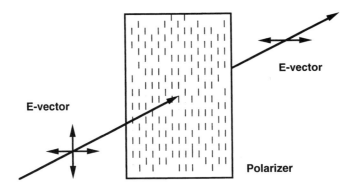

Figure 1.2. Schematic drawing of a sheet polarizer that transmits polarization component with **E** vector perpendicular to the direction of alignment and absorbs polarization component with **E** vector parallel to the direction of alignment.

ratio of about 10. In this crystal the ordinary polarization component (electric field vector perpendicular to c axis) is much more strongly absorbed (about 10 times) than the extraordinary polarization component (electric field vector parallel to c axis). A single crystal plate cut parallel to the axis can be used as a polarizer. Most sheet polarizers are, however, synthesized by stretching films that contain ultrafine rodlike or needle-like dichroic materials. The stretching provides an uniaxial alignment of the crystals or molecules (see Fig. 1.2). Dichroic ratios of greater than 100 can be easily obtained. A beam of polarized light with the vector amplitude parallel to the molecular axis may suffer more material absorption. Thus, given enough thickness, the stretched films transmit one polarization component that is perpendicular to the direction of alignment and absorb the other polarization component that is parallel to the direction of alignment. These sheet polarizers provide excellent extinction ratios while simultaneously accommodating a large angular aperture and an enormously large linear aperture. These are desirable features for large flat panel applications.

Another method of producing dichroic materials is to embed oriented ultrafine metallic needles that are not dichroic by themselves in a medium. The composite material then is said to have *form dichroism*, just as a composite medium containing oriented dielectric needles exhibits form birefringence [1].

Generally speaking, 50% of the incident energy of a beam of unpolarized light is absorbed by an ideal polarizer. This energy loss of about 3 dB accounts for the absorption of one of the polarization components resulting in a beam of plane polarized light. When two identical sheet polarizers are in series, the intensity transmission is approximately proportional to $\cos^2\theta$, where θ is the angle between the directions of alignment of the two polarizers. Thus a pair of cross-polarizers $(\theta = 90°)$ can block the transmission of light regardless of the polarization state of the incident beam (see Fig. 1.3).

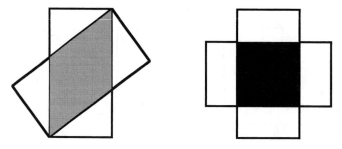

Figure 1.3. Transmission of unpolarized light through two polarizers in series.

1.1.2. Transparent Electrodes

To allow the transmission of light, the electrodes must be transparent in the spectral regime of interest. This requires transparent materials with a good electrical conductivity. For example, tin oxide (SnO_2) is a transparent material which also exhibits a good electrical conductivity. Generally speaking, all transparent semiconductors exhibit a finite electrical conductivity. The electrical conductivity in semiconductors can also be increased by an appropriate doping of impurity atoms. In recent LCD applications, ITO (indium tin oxide) is usually used because of its higher electrical conductivity. All transparent conductors exhibit a small absorption of light. Thus, the electrodes must be thin enough to allow adequate light transmission, yet thick enough to provide adequate electrical conductivity. As a result of the index mismatch between the glass substrate and the electrode materials, a Fabry–Perot cavity is formed in each of the electrodes which are sandwiched between glass and the liquid crystal. A proper choice of the thickness can ensure a constructive interference to maximize the transmission of light. In typical electrodes, the ITO thickness is in the range of 100–300 nm.

1.1.3. Liquid Crystal Cell

The space between the electrodes is filled with liquid crystal (LC) material. The thickness of the LC layer is kept uniform by using glass fibers or plastic balls as spacers. Typical thicknesses of the cell are in the range of a few micrometers. There are several different LC configurations that can be employed for display applications. For the purpose of illustrating the principle of operation of a basic element of LCD, we consider the case of a planar nematic LC cell where all the rodlike LC molecules are aligned parallel to the glass plates.

In an LC cell without any external field, the ordering of the molecules is usually determined by the anisotropy of the boundary. The surfaces of the electrodes are usually coated with a thin alignment layer (e.g., polyimide). Rubbing the surface of the alignment layer has been widely employed to align the liquid crystals in LCD panel fabrication. The LC molecules are aligned on the rubbed surface with their axes parallel to the rubbing direction, especially for

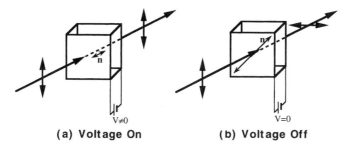

(a) Voltage On   **(b) Voltage Off**

Figure 1.4. LC cell can be controlled electrically to modify the polarization state of an incoming beam (**n** is a unit vector indicating the direction of alignment of the liquid crystal molecules): (*a*) voltage ON; (*b*) voltage OFF.

those molecules immediately next to the boundary. Depending on the boundary conditions, a long range ordering of the molecules may exist throughout the LC cell. By virtue of its electrical anisotropy, the ordering and the orientation of the molecules can be controlled electrically by applying an external field. In the presence of an applied electric field, rodlike molecules are aligned parallel to the applied electric field to minimize the electrostatic energy.

As a result of the ordering of the molecules, nematic liquid crystals exhibit a strong optical birefringence. In other words, there are two modes of optical propagation, each with a unique phase velocity. The difference in the phase velocity leads to a phase retardation between these two modes. As a consequence, the polarization state of an incoming beam of polarized light can be modified. An example is shown in Figure 1.4*b*, where an incoming beam of vertically polarized light is converted into a beam of horizontally polarized light as a result of a properly chosen phase retardation. If the axes of all the molecules are aligned parallel to the propagation direction of the incoming beam by applying an external electric field, the polarization state will remain unchanged. This is illustrated in Figure 1.4*a*. When such an LC cell is sandwiched between a pair of cross-polarizers (or parallel polarizers), the intensity of the transmitted beam can be controlled electrically.

1.2. PROPERTIES OF LIQUID CRYSTALS

Liquid crystal is a state of matter that is intermediate between the crystalline solid and the amorphous liquid. It may also be viewed as a liquid in which an ordered arrangement of molecules exists. Liquid crystals arise under certain conditions in organic substances having sharply anisotropic molecules, that is, highly elongated (rodlike) molecules or flat (disklike) molecules. A direct consequence of the ordering of the anisotropic molecules is the anisotropy of mechanical, electric, magnetic, and optical properties. This intermediate state was first observed in 1888 in cholesteryl benzoate, a crystalline solid. It becomes

Table 1.1. Liquid Crystal Materials

Name	Formula	Nematic Range (°C)
BCH-5	C_5H_{11}—⬡—⬡—⬡—CN	96–219
PAA	CH_3O—⬡—N=N—⬡—OCH_3 (with O below N)	118–135.5
EBBA	CH_3CH_2O—⬡—C=N—⬡—C_4H_9 (with O below C)	35–77
MBBA	CH_3O—⬡—C=N—⬡—C_4H_9 (with O below C)	22–47
CCH-501	C_5H_{11}—⬡—⬡—OCH_3	29–36.8
5CB	C_5H_{11}—⬡—⬡—CN	24–35
6CB	C_6H_{13}—⬡—⬡—CN	15–29

a turbid cloudy liquid, or liquid crystals, when heated to 145°C; on further heating to 179°C the liquid becomes isotropic and clear. The sequence is reversed when the substance is cooled. The cloudy intermediate phase contains domains that seem to have a crystal-like molecular structure. Color changes occur on both heating and cooling. Many organic compounds, such as hexylcyanobiphenyl (6CB) and sodium benzoate, exhibiting this behavior are known and used extensively in electric and electronic displays, electronic clocks, calculators, and similar devices dependent on temperature determination. Table 1.1 lists a number of LC materials and their nematic ranges.

Hexylcyanobiphenyl (6BC) is a nematic LC in the temperature range $15°C < T < 29°C$. When 6CB is heated at room temperature, a nematic–isotropic transition occurs at $T = 29°C$. The material "then becomes suddenly completely clear." 6CB has the following rodlike molecular structure:

$$C_6H_{13}\text{—⬡—⬡—CN}$$

which consists of two benzene rings and two terminal groups. In the nematic phase, only the long axes of the molecules have a statistical preferential orientation.

Referring to Table 1.1, we note that a typical rodlike LC molecule has the following general molecular structure:

$$X-\langle\ \rangle-A-\langle\ \rangle-Y$$

where X and Y are terminal groups and A is a linking group between two or more ring systems. We discuss the contribution of each group:

Ring Systems. The rings can be either benzene rings (unsaturated) or cyclohexanes (saturated), or a combination of both. The presence of the ring system is essential in most LCs that contain at least one ring (either phenyl or cyclohexyl). It is important to note that only σ electrons exist in cyclohexyl rings, whereas π electrons exist in phenyl rings. The presence of the rings provides the short range intermolecular forces needed to form the nematic phase. They also affect the absorption, dielectric anisotropy, birefringence, elastic constants, and viscosity. The $\sigma \rightarrow \sigma^*$ electronic transitions occur at vacuum ultraviolet regime ($\lambda < 180$ nm). Thus, the absorption of a saturated LC compound is negligible in the visible spectral regime. In unsaturated LC compound such as 5CB, the $\pi \rightarrow \pi^*$ transitions occur at $\lambda \sim 210$ nm, 280 nm with a strong dichroic ratio at 280 nm. The strong dichroic ratio is also responsible for the large dielectric anisotropy and optical birefringence in a nematic phase.

Terminal Group X (Also Called Side Chain). There are three common X-terminal groups: (1) alkyl chain C_nH_{2n+1}, (2) alkoxy chain $C_nH_{2n+1}O$, and (3) alkenyl chain that contains a double bond. The length of the chain can strongly influence the elastic constants of the nematic phase as well as the phase transition temperature. For short chains with one or two carbon atoms ($n = 1,2$), the molecules are too rigid to exhibit LC phases. Terminal groups with medium chain length (e.g., $n = 3$–8) are most suitable for nematic phases (which will be explained later in this section). Compounds with even longer chain length can exhibit smectic phases. Generally speaking, the melting point decreases as the chain length increases. However, some compounds are irregular and unpredictable. The nematic–isotropic transition temperature (known as the *clearing point* or N-I point) is a smooth function of n for both odd carbon number n and even carbon number n. Generally speaking, the N-I points (clearing points) for LC compounds with an even number of carbon atoms are lower than those of compounds with an odd number of carbon atoms. This is known as the "odd–even effect." Although LC compounds with longer chain length can offer a lower melting point, they also exhibit a higher viscosity. In LCDs, it is desirable to have a low viscosity in order to have a higher frame rate.

Linking Group A. The linking group can be a single bond between the two rings. The generic name of these LCs is biphenyl. In case of an additional ring as the linking group, the generic name of these LCs is terphenyl. Some common linking groups are C_2H_4 (diphenylethane), C_2H_2

(stilbene), $-C\equiv C-$ (tolane), $-N=N-$ (azobenzene), $-CH=N-$ (Schiff's base), and COO (phenyl benzoate, ester).

Terminal Group Y. The terminal group Y plays an important role in determining the dielectric constants ε and its anisotropy $\Delta\varepsilon$. In LCD, the operation voltage is usually a multiple of the threshold voltage, which is the minimum voltage required to cause a reorientation of the LC molecules. It is also known that the threshold voltage is inversely proportional to the square root of the dielectric anisotropy. Thus it is desirable to have a large $\Delta\varepsilon$ for low voltage operations. The terminal group Y can be either polar or nonpolar. Nonpolar terminal group Y, such as the alkyl chain C_nH_{2n+1}, has very little effect on the dielectric anisotropy. On the other hand, a polar terminal group such as CN can contribute significantly to the dielectric anisotropy. Some common polar terminal groups Y include CN, F, and Cl. Among them, CN has the highest polarity, leading to a high dielectric anisotropy and optical birefringence. On the other hand, LC compounds containing CN terminal group Y also exhibit high viscosity, insufficient resistivity, and stability problems under UV illumination. These physical properties can affect the operation of thin film transistor (TFT) LCD. For example, LC compounds with CN terminal group Y are not suitable for high temperature operations such as those used in projection displays. In addition, LC molecules with CN terminal group Y can be dissociated under the illumination of UV. On the contrary, LC molecules with F terminal group Y exhibit low viscosity, high resistivity and relatively high stability under UV illumination. As they are relatively lower in polarity, these LC materials exhibit relatively smaller dielectric anisotropy and optical birefringence.

Eutectic. A single LC compound that can fulfill all LC specifications for most display applications is not available at the moment. An example is 5CB, which has a melting point of 24°C and a clearing point of 35.3°C. It is clear that such a small temperature window for nematic phase is not adequate for most industrial applications with operation temperature from -20 to 80°C. It is known that the melting point of a binary mixture of compounds is less than either of its constituent compounds. The melting point of the mixture depends on the mixture ratio. At the eutectic point, the melting point reaches its minimum. The clearing point of the LC mixture is usually the linear average of the composition. Thus, a mixture of two LC compounds can offer a much larger temperature range that exhibits the nematic phase for LCD applications. For example, the eutectic mixture of MBBA and EBBA can provide a nematic range from 0 to 60°C. In addition to the melting point, many other physical properties, such as dielectric constants, dielectric anisotropy, elastic constants, birefringence, and viscosity, also depend on the mixture ratio. Many of the LC mixtures are designed to provide optimum physical parameters for LCD applications.

LC Phases. Generally speaking, there are three phases of liquid crystals, known as *smectic phase*, *nematic phase*, and *cholesteric phase*. For the sake of clarity, we assume that the liquid crystals are made of rodlike molecules. Figure 1.5*a* illustrates the smectic phase in which one-dimensional translational order as well as orientational order exist. A unit vector **n** in the direction of the preferred orientation of the molecular axes is known as the *director*. Figure 1.5*b* illustrates the nematic phase in

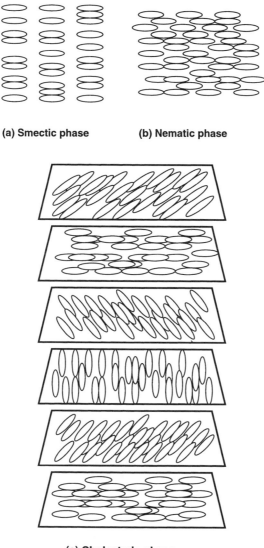

(a) Smectic phase   **(b) Nematic phase**

(c) Cholesteric phase

Figure 1.5.   Phases of liquid crystals: (*a*) smectic phase; (*b*) nematic phase; (*c*) cholesteric phase.

which only a long range orientational order of the molecular axes exists. The cholesteric phase is also a nematic type of LC except that it is composed of chiral molecules. As a consequence, the structure acquires a spontaneous twist about a helical axis normal to the director. The twist may be right-handed or left-handed depending on the molecular chirality. Figure 1.5c illustrates the cholesteric phase of LC by viewing the distribution of molecules at several planes that are perpendicular to the helical axis. Note that the LC is in a nematic phase in each perpendicular plane.

A smectic LC is closest in structure to solid crystals. It is interesting to note that in substances that form both a nematic phase and a smectic phase, the sequence of phase changes on rising temperature is as follows:

Solid crystal → smectic liquid crystal → nematic liquid crystal
→ isotropic liquid

Although the smectic phase possesses the highest degree of order, it is the nematic and cholesteric phases that have the greatest number of electrooptical applications. In the nematic phase, the medium may appear milky if the orientation order exists in many different domains. The nematic liquid crystal is clear only when a long range order exists in the whole medium. At the nematic–isotropic transition temperature, the medium becomes isotropic and looks clear and transparent. Thus the temperature is also known as the *clearing point*. In the following sections, we discuss various important physical properties that are the results of the orientational order.

1.2.1. Orientational Order Parameter

In the nematic phase, the molecules are rodlike with their long axes aligned approximately parallel to one another. Thus at any point in the medium, we can define a vector **n** to represent the preferred orientation in the immediate neighborhood of the point. This vector is known as the *director*. In a homogeneous nematic LC, the director is a constant throughout the medium. In an inhomogeneous nematic liquid crystal, the director **n** can change from point to point and is, in general, a function of space (x,y,z). If we define a unit vector to represent the long axis of each molecule, then the director **n** is the statistical average of the unit vectors over a small volume element around the point.

The order parameter S of a LC is defined as

$$S = \tfrac{1}{2} \langle 3 \cos^2\theta - 1 \rangle \qquad (1.2\text{-}1)$$

where θ is the angle between the long axis of an individual molecule and the director **n** and the angular brackets denote a statistical average. For perfectly parallel alignment, $S = 1$, while for totally random orientations, $S = 0$. In the nematic phase, the order parameter S has an intermediate value that is strongly temperature dependent. It is evident that $S = 0$ at the clearing point. Typical values of the order parameter S are in the range between 0.6–0.4 at low temperatures. As the clearing point is approached, the order parameter S drops abruptly to zero. The values of the order parameter S also depend on the structure of the molecules. Experimental observation indicates that liquid crystals based on cyclohexane rings exhibit higher values of S than do aromatic systems.

1.2.2. Dielectric Constants

Because of the orientational ordering of the rodlike molecules, the smectic and nematic liquid crystals are uniaxially symmetric, with the axis of symmetry parallel to the axes of the molecules (director $\mathbf{n}$). As a result of the uniaxial symmetry, the dielectric constants differ in value along the preferred axis ($\varepsilon_{\parallel}$) and perpendicular to this axis ($\varepsilon_{\perp}$). The dielectric anisotropy is defined as

$$\Delta\varepsilon = \varepsilon_{\parallel} - \varepsilon_{\perp} \tag{1.2-2}$$

The sign and magnitude of the dielectric anisotropy $\Delta\varepsilon$ are of the utmost importance in the applicability of the LC material in LCDs using one of the various electrooptic effects. To illustrate this, we consider an applied electric field along the z axis in a homogeneous nematic liquid crystal. As a result of the anisotropy, the induced dipole moment of the molecules is not parallel to the applied electric field, except when the molecular axis is parallel or perpendicular to the electric field. This creates a net torque that tends to align the molecules along the direction of the electric field for most rodlike molecules. The macroscopic electrostatic energy can be written

$$U = \tfrac{1}{2}\mathbf{D} \cdot \mathbf{E} \tag{1.2-3}$$

where $\mathbf{E}$ is the electric field vector and $\mathbf{D}$ is the displacement field vector. In a homogeneous medium the displacement field vector $\mathbf{D}$ is independent of the orientation of the liquid crystal. Letting θ be the angle between the director and the z axis, we can express z component of the displacement field vector as

$$D_z = (\varepsilon_{\parallel}\cos^2\theta + \varepsilon_{\perp}\sin^2\theta)E \tag{1.2-4}$$

Thus the electrostatic energy can be written (see Appendix A)

$$U = \frac{1}{2}\frac{D_z^2}{\varepsilon_{\parallel}\cos^2\theta + \varepsilon_{\perp}\sin^2\theta} \tag{1.2-5}$$

Table 1.2. Properties of Liquid Crystals

	$T(°C)$	λ (nm)	n_e	n_o	$\varepsilon_\parallel$	$\varepsilon_\perp$	k_1	k_2 $(10^{-12}\,N)$	k_3	Nematic	Range (°C)	Reference
MBBA[a]	22	589	1.769	1.549	4.7	5.4	6.2	3.8	8.6	20	47	4
PCH-5[b]	30.3	589	1.604	1.4875	17.1	5	8.5	5.1	16.2	30	55	5,12
		633	1.600	1.4851						30	55	
	38.5	589	1.5956	1.4863	16.6	15.3	7.3	4.5	13.2			
		633	1.5919	1.4840								
	46.7	589	1.5849	1.4860	15.9	5.7	5.9	3.9	9.9			
		633	1.5812	1.4836								
K15 (5CB)[c]	25	515	1.736	1.5442	19.7	6.7	6.4	3	10	24	35.3	18
K21 (7CB)[d]	37	577	1.6815	1.5248	15.7	6	5.95	4	6.6	30	42.8	8,15,16,17
M15 (5OCB)[e]	50	589	1.7187	1.5259	17.9	6.7	6.1	3.74	8.4	48	68	9,11,15,16
M21 (7OCB)[f]	60	589	1.6846	1.5139	16.3	6.5	6.9		7.7	54	74	15,16
M24 (8OCB)[g]	70	589	1.6639	1.5078	14.7	6.2	7.3		9.0	67	80	15,16
E5[h]	20	589	1.736	1.5228	19	5.9				−8	50.5	12,15
E7[i]	20	577	1.75	1.5231	19.6	5.1	12	9	19.5	−10	60.5	15
ZLI-1646[j]	20	589	1.558	1.478	10.6	4.6	7.7	4.0	12.2	−20	60	12,15
ZLI-4792[k]	20	589	1.573	1.479	8.3	3.1	13.2	6.5	18.3	−40	92	12

[a] p-Methoxybenzylidene-p'-n-butylaniline
[b] 4-(trans-4-Pentylcyclohexyl)benzonitrile
[c] Pentylcyanobiphenyl.
[d] Heptylcyanobiphenyl.
[e] Pentyloxycyanobiphenyl.
[f] Heptyloxycyanobiphenyl.
[g] Octyloxycyanobiphenyl.
[h] 45%K15 + 24%K21 + 10%M15 + 9%M21+12%M24.
[i] 47%K15 + 25%K21 + 18%M24 + 10%T15.
[j] Phenylcyclohexane and biphenylcyclohexane mixtures.
[k] SFM-TFT mixture.

For liquid crystals with positive dielectric anisotropy ($\varepsilon_\perp < \varepsilon_\parallel$), lowest electrostatic energy occurs at $\theta = 0$ when the director is parallel to the applied electric field.

In classical dielectric theory, the macroscopic dielectric constant is often proportional to the molecular polarizability. For rodlike molecules, the longitudinal polarizability (parallel to molecular axis) is often greater than the transverse polarizability (perpendicular to axis). So even in nonpolar liquid crystals with rodlike molecules, the dielectric anisotropy is positive ($0 < \Delta\varepsilon$). In the case of polar LC compounds, there is an additional contribution to the dielectric constant due to the permanent dipole moment. Depending on the angle between the dipole moment and the molecular axis, the dipole contribution can cause an increase or a decrease of $\Delta\varepsilon$, eventually leading to a negative value of $\Delta\varepsilon$ (e.g., MBBA). In practice, dielectric anisotropy varies between $-2\varepsilon_0$ and $+15\varepsilon_0$. Liquid crystal compounds with a large anisotropy can be synthesized by substitution of a strongly polar group (e.g., cyanide group) in specific positions. Table 1.2 lists properties of LC compounds with various dielectric anisotropy. The dielectric anisotropy also depends on the temperature and approaches zero abruptly at the clearing point. Beyond the clearing point, the dielectric constant becomes the mean dielectric constant

$$\bar{\varepsilon} = \frac{\varepsilon_\parallel + 2\varepsilon_\perp}{3} \qquad (1.2\text{-}6)$$

1.2.3.   Refractive Index

In a glass flask, nematic LCs often appear as an opaque milky fluid. The scattering of light is due to the random fluctuation of the refractive index of the sample. With no proper boundaries to define the preferred orientation, the sample consists of many domains of nematic LC. The discontinuity of the refractive index at the domain boundaries is the main cause of the scattering leading to the milky appearance. Under the proper treatment (e.g., rubbing the alignment layer on the glass substrate), a slab of nematic LC can be obtained with a uniform alignment of the director. Such a sample exhibits uniaxial optical symmetry with two principal refractive indices n_o and n_e. The ordinary refractive index n_o is for light with electric field polarization perpendicular to the director and the extraordinary refractive index n_e is for light with electric field polarization parallel to the director. The birefringence (or optical anisotropy) is defined as

$$\Delta n = n_e - n_o \qquad (1.2\text{-}7)$$

If $n_o < n_e$, the LC is said to be positive birefringent, whereas if $n_e < n_o$, it is said to be negative birefringent. In classical dielectric theory, the macroscopic refractive index is related to the molecular polarizability at optical frequencies.

The existence of the optical anisotropy is due mainly to the anisotropic molecular structures. Most LCs with rodlike molecules exhibit positive birefringence ranging from 0.05 to 0.45. The optical polarizability is due mainly to the presence of delocalized electrons not participating in chemical bonds and of π electrons. This is the reason that LC molecules composed of benzene rings have higher values of Δn than do the respective cyclohexane counterparts. As mentioned earlier, the substitution of a cyano terminal group (CN) can cause an increase in the optical anisotropy. Another effective way of increasing Δn or n_e is to employ triple bonds (tolane). For most LCs, the ordinary refractive index is around 1.5. The various components employed in the synthesis of LCs do not markedly affect n_o. On the other hand, the extraordinary refractive index can be higher by as much as 0.45 for some diphenyldiacetylenic compounds or only by 0.06 for bicyclohexanes. The optical anisotropy plays an essential role in changing the polarization state of light in liquid crystals.

1.2.4. Elastic Constants

Like most liquids and solids, LCs exhibit curvature elasticity. The elastic constants of LC determine the restoring torques that arise when the system is perturbed from its equilibrium configuration. These restoring torques are usually very weak compared with those of solids. In LCDs, an electric field is often applied to cause a reorientation of the molecules. It is the balance between the electric torque and the elastic restoring torque that determines the LC's static deformation pattern. Any static deformation of LCs can be divided into a combination of three basic deformations. These are splay, twist, and bend, as illustrated in Figure 1.6.

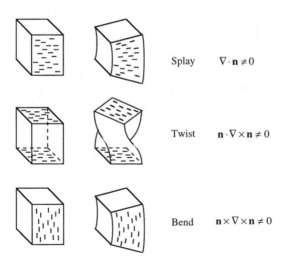

Splay $\nabla \cdot \mathbf{n} \neq 0$

Twist $\mathbf{n} \cdot \nabla \times \mathbf{n} \neq 0$

Bend $\mathbf{n} \times \nabla \times \mathbf{n} \neq 0$

Figure 1.6. Schematic drawing of splay, twist, and bend in LC.

It can be shown that for an isothermal deformation in an incompressible fluid, the free energy (elastic energy) can be written as a quadratic function of the curvature strain tensor. Following the notation of Oseen–Frank theory, the elastic energy density of a deformed LC can be written [2,3]

$$F = \tfrac{1}{2}k_1(\nabla \cdot \mathbf{n})^2 + \tfrac{1}{2}k_2(\mathbf{n} \cdot \nabla \times \mathbf{n})^2 + \tfrac{1}{2}k_3(\mathbf{n} \times \nabla \times \mathbf{n})^2 \qquad (1.2\text{-}8)$$

where k_1, k_2, and k_3 are the splay, twist, and bend elastic constants, respectively.

Like many other physical properties, the elastic constants are strongly temperature dependent. For most LC compounds, the elastic constants are in the range of 3–25 piconewtons (10^{-12} N). The ratio for the elastic constants varies from 0.7–1.8 for k_3/k_1, and varies from 1.3 to 3.2 for k_3/k_2. Table 1.2 lists the elastic constants for some typical LC materials.

1.2.5. Viscosity—Rotational Viscosity

The viscosity of fluid is an internal resistance to flow, defined as the ratio of shearing stress to the rate of shear. It arises from the intermolecular forces in the fluid. The viscous behavior of liquid crystals has a profound effect on the dynamical behavior of LCD systems. Like most liquids, the viscosity increases at low temperatures as a result of lower molecular kinetic energy. This can severely limit the operations of LCDs. An important parameter is the rotational viscosity coefficient γ_1, which provides a resistance to the rotational motion of the LC molecules. In most LCDs, the directors are reoriented by the application of an electric field. The switching time is approximately proportional to $\gamma_1 d^2$, where d represents the cell spacing. For most nematic LCs used in displays, the magnitude of the rotational viscosity is in the range of 0.02–0.5 Pa·s (comparable to light machine oils). As a reference, water at 20°C has a viscosity of 1.002 mPa·s. The viscosity unit of Pa·s (pascal-second) corresponds to $1 \text{N} \cdot \text{s}/\text{m}^2$ in SI (Système International) units, or 10 poise (1 poise = 1 dyn·s/cm^2 in cgs (centimeter–gram–second) units). It is known experimentally that molecules with a higher number of rings or longer alkyl chains are characterized by an increasing viscosity. In addition, LCs with high values of $\Delta\varepsilon$ usually exhibit higher viscosities. This is possibly due to the stronger polar interaction between the molecules.

1.2.6. Surface Alignment and Rubbing

As a result of the rodlike or disklike nature, the distribution of the orientation of LC molecules plays an essential role in the optical properties of liquid crystal in display applications. Uniform or well prescribed orientation of the LC molecules is required in most LCDs. The role of the surface is to ensure a single domain. Without special treatments of the surfaces (either physical or chemical), the LC will generally have many domains and many disclinations, or discontinuities

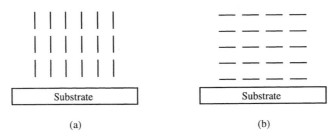

Figure 1.7. (*a*) Homeotropic (vertical) alignment; (*b*) parallel homogeneous alignment.

in orientations. These domains and discontinuities can cause a severe scattering of light, leading to a cloudy appearance.

The alignment of a LC by the surface treatment of a substrate has been one of the least understood aspects of LC behavior. It depends on both the nature of the LC and the surface. The most important factors include dipolar interactions, chemical and hydrogen bonding, van der Waals interactions, steric factors, surface topography, and the elasticity of LC molecules. For example, it is possible to deposit molecules known as *silane coupling agents* on to a glassy surface. These molecules may promote the adhesion of LC molecules, in a vertical manner, onto the surface; this is known as *homeotropic alignment* when the LC director is normal to the surface (see Fig. 1.7).

To achieve a "parallel homogeneous alignment" (see Fig. 1.7) where the LC director is uniformly parallel to the surface of the substrate, one must provide a preferred orientation to the alignment by physically or chemically treating the surface. Rubbing the surface has been a simple and effective way of achieving a preferred orientation. Typical rubbing materials include linen cloth and lens paper. Although the technique of rubbing to achieve parallel homogeneous alignment is still very much an art, it is generally believed that physically rubbing the surface produces a uniform and unidirectional tilt of the dangling bonds or side chains on the surface (see Fig. 1.8). The unidirectional tilt of these surface molecules may lead to the parallel homogeneous alignment of the LC molecules. Notice that the LC director usually tilts up in the direction of rubbing.

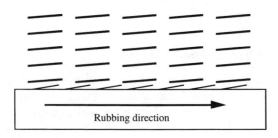

Figure 1.8. Schematic drawing of the anchoring of rodlike LC molecules near the surface of a rubbed substrate.

Pretilt Angle

As a result of the surface rubbing, a small tilt angle exists for the LC director. The pretilt angle is very important in electrooptic applications when an electric field is applied to the cell to reorient the LC director. Having no polarity, the rodlike LC molecules can turn in two ways following the application of an electric field, if the electric field is initially perpendicular to the molecular axis. A small tilt angle of the molecular axis would greatly facilitate the turning of the molecule under the application of an electric field, and ensure a single way of turning. A single way of turning for all molecules in the cell will ensure a single domain, which is essential in most display applications. Referring to Figure 1.9, we consider the distribution of director in a parallel cell of nematic LC. A uniform tilt angle can exist in a parallel cell provided the rubbing directions in the inner surfaces are antiparallel (see Fig. 1.9a). If, on the other hand, the rubbing directions are parallel as shown in Figure 1.9b,c, a nonuniform distribution of the director orientation exists with a zero tilt or 90° angle at midlayer. These are the splay or bend cells. For small pretilt angles, the splay cell has the lowest elastic energy density. This is, however, an undesirable config- uration for electrooptic applications, as multiple domains may occur when an electric field is applied. The bend cell is also known as the *Pi cell*, which was originally developed by Bos et al. for an electrically controllable wave plate [13].

Twisted Nematic Liquid Crystal Cell

The pretilt angle also plays an important role in twisted nematic liquid crystal (TN-LC) cells. Referring to Figure 1.10, we consider a liquid crystal cell with a pair of orthogonal rubbing directions. The rubbing directions are important to prepare a twisted nematic liquid crystal (TN-LC) cell with a total twist angle of 90°. Having no polarity and no chirality, the rodlike molecules can twist in either direction in the absence of a small pretilt angle. In other words, both

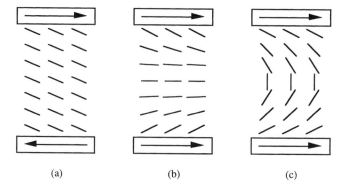

(a)   (b)   (c)

Figure 1.9.   Tilt of LC molecular axes due to rubbing of inner surfaces of the LC cell. The arrows indicate the direction of rubbing. (*a*) Parallel alignment occurs when the rubbings are in opposite directions. Splay cell (*b*) or bend cell (*c*) can occur when the rubbings are in the same direction.

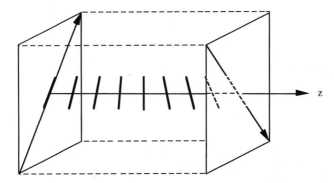

Figure 1.10. Distribution of liquid crystal molecular axes in a 90° TN-LC cell with a right-handed twist. The arrows indicate the rubbing directions at the boundaries. The rods indicate the distribution of director orientations in the cell.

right-handed and left-handed twist of 90° are possible to match the boundary conditions imposed by the rubbing. When a small pretilt angle exists, only the right-handed twist is possible in the cell described in Figure 1.10. In the cell shown in Figure 1.10, the upper parts of LC molecular axes tilt toward the right side (+ z direction).

REFERENCES

1. See, for example, P. Yeh, *Optical Waves in Layered Media*, Wiley, New York, 1988.
2. C. W. Oseen, *Trans. Faraday Soc.* **29**, 883 (1933).
3. F. C. Frank, *Disc. Faraday Soc.* **25**, 19 (1958).
4. P. G. de Gennes, *The Physics of Liquid Crystals*, Clarendon Press. Oxford, 1974.
5. U. Finkenzeller et al., "Liquid-crystalline reference compounds," *Liq. Cryst.* **5**, 313 (1989).
6. Shila Garg, K. A. Crandall, and A. A. Khan, "Bend and splay elastic constants of diheptylazoxybenzene," *Phys. Rev. E* **48**, 1123 (1993).
7. R. D. Polak, G. P. Crawford, B. C. Kostival, J. W. Doane, and S. Zumer, "Optical determination of the saddle-splay elastic constant K24 in nematic liquid crystals," *Phys. Rev. E* **49**, R978 (1994).
8. P. P. Karat and N. V. Madhusudana, *Mol. Cryst. Liq. Cryst.* **40**, 239 (1977).
9. Fu-Lung Chen, A. M. Jamieson et al., *J. Polym. Sci.* (Part B), **33**, 1213 (1995).
10. A. Buka, P. G. Owen, and A. H. Price "Dielectric relaxation in the nematic and isotropic phases of *n*-heptyl- and *n*-heptoxy-cyanobiphenyl," *Molec. Cryst. Liq. Cyst.* **51**, 273 (1979).
11. Fu-Lung Chen and A. M. Jamieson, "Odd–even effect in the viscoelastic properties of main-chain liquid crystal polymer-low molar mass nematogen mixtures," *Macromolecules* **26**, 6576 (1993).
12. Data sheet from EM Industries (7 Skyline Drive, Hawthorne, NY 10532).

13. P. J. Bos, K. R. Koehler/Beran, *Mol. Cryst. Liq. Cryst.* **113**, 329–339 (1984).

14. L. Pohl, G. Weber, R. Eidenschink, G. Baur, and W. Fehrenbach, *Appl. Phys. Lett.* **38**, 497 (1981).

15. S. D. Jacobs, "Liquid crystals for laser applications," in *CRC Handbook of Laser Science and Technology*, Vol. III, *Optical Materials*, Section 2, CRC Press, 1986.

16. M. J. Bradshaw, E. P. Raynes, J. D. Bunning, and T. E. Faber, *J. Phys. (Paris)* **46**, 1513 (1985).

17. D. A. Dunmur, M. R. Manterfield, W. H. Miller, and J. K. Dunleavy, *Mol. Cryst. Liq. Cryst.* **45**, 127 (1978).

18. L. M. Blinov and V. G. Chigrinov, *Electrooptic Effects in Liquid Crystal Materials*, Springer-Verlag, 1994.

SUGGESTED READINGS

P. G. de Gennes and J. Prost, *The Physics of Liquid Crystals*, Clarendon Press-Oxford, 1993.

B. Bahadur, ed., *Liquid Crystals—Applications and Uses*, Vol. 1, World Scientific, 1990.

L. Pohl and U. Finkenzeller, "Physical properties of liquid crystals," in *Liquid Crystals—Applications and Uses*, Vol. 1, B. Bahadur, ed., World Scientific, 1990, 139–170.

T. Scheffer and J. Nehring, "Twisted nematic and supertwisted nematic mode LCDs, " in *Liquid Crystals—Applications and Uses*, Vol. 1, B. Bahadur, ed., World Scientific, 1990, Chapter 10, pp. 231–274.

S. Chandrasekhar, *Liquid Crystals*, 2nd ed., Cambridge Univ. Press, 1992.

H. J. Deuling, "Elasticity of nematic liquid crystals," in *Liquid Crystals*, Solid State Physics Series, Vol. 14, L. Liebert, ed., Academic Press, 1978, pp. 77–107.

E. B. Priestley, P. J. Wojtowicz, and P. Sheng, *Introduction to Liquid Crystals*, Plenum Press, 1979.

I. C. Khoo, *Liquid Crystals*, Wiley, 1995.

S. D. Jacobs, K. L. Marshall, and A Schmid, "Liquid crystals," in *CRC Handbook of Laser Science and Technology*, Vol. II, *Optical Materials*, Section 14, CRC Press, 1995.

S. T. Wu, "Liquid crystals," in *Handbook of Optics*, 2nd ed., McGraw-Hill, 1995, Chapter 14.

J. L. Fergason, "Liquid crystals," *Sci. Am.* **211**(2), 77–82, 85 (1964).

PROBLEMS

1.1. (a) Show that 50% of the energy of a beam of unpolarized light transmits through an ideal polarizer regardless of the orientation of the polarizer.

   (b) As discussed in this chapter, the transmitted intensity through two polarizers in series is given by $\frac{1}{2}\cos^2\theta$. Show that the fractional error in the intensity measurement due to an error in $\Delta\theta$ is given by $-2\Delta\theta\tan\theta$.

1.2. *Polarizers in series*:

   (a) It is known that a pair of crossed polarizers can stop the transmission of light. Show that optical transmission occurs when a third polarizer is inserted between the crossed polarizers with a transmission axis oriented at $45°$ with respect to those of the crossed polarizers. Find the transmission for unpolarized light, assuming that the orientations of the crossed polarizers are at $0°$ and $90°$.

   (b) Consider the insertion of two polarizers oriented at $30°$ and $60°$ between the two crossed polarizers. Find the transmission for unpolarized light.

   (c) By generalizing (a) and (b), we consider the insertion of N polarizers oriented at equal angular separations. Show that the transmission of unpolarized light approaches $\frac{1}{2}$ when N tends to infinity. This is a situation when the electric field vector follows the direction of the transmission axis of the polarizers as the light propagates through the whole set of polarizers in series.

1.3. Derive Eqs. (1.2-4) and (1.2-5).

1.4. Let $\alpha_\perp$ and $\alpha_\parallel$ be the molecular polarizability of LC molecule with $\alpha_\perp < \alpha_\parallel$. Under the application of an external electric field, the induced dipole moment of the liquid crystal molecule can be written $p_\perp = \alpha_\perp E_\perp$ and $p_\parallel = \alpha_\parallel E_\parallel$.

   (a) Show that, in general, the dipole moment is not parallel to the applied electric field.

   (b) Show that a net torque is exerted on the molecule that tends to turn the molecule along the direction of the electric field. Evaluate the torque in terms of the polarizability.

   (c) Show that the torque discussed in (b) tends to minimize the electrostatic energy in Eq. (1.2-5).

1.5. *Elastic energy density and electrostatic energy density*:

   (a) Consider a uniformly twisted nematic LC with a director given by $\mathbf{n} = (\sin\phi, \cos\phi, 0)$, where ϕ is the twist angle given by $\phi = \alpha z$. Show that the twist elastic energy density is given by $U_t = \frac{1}{2}k_2\alpha^2$. Evaluate the energy density in terms of J/m^3 (joules per cubic meter) for the case of a 5-μm cell of ZLI-1646 liquid crystal with a $90°$ twist. (Use $k_2 = 4 \times 10^{-12}\,\text{N}$.)

   (b) Using Eq. (1.2-4) or (1.2-5), show that the difference between the electrostatic energy density at $\theta = 0$ and $\pi/2$ can be written $\Delta U = \frac{1}{2}\Delta\varepsilon E^2$. Evaluate ΔU in terms of J/m^3 for the case of a 5-μm cell of ZLI-1646 LC with an applied voltage of 1 V. Show that the electrostatic energy density is much larger than the twist elastic energy density. (Use $\Delta\varepsilon = 6\varepsilon_0$.) This example illustrates the advantages of LCs that can be easily reoriented by applying a small voltage.

1.6. Prove that the bend term in Eq. (1.2-8) can also be written as $\frac{1}{2}k_3(\mathbf{n}\cdot\nabla\mathbf{n})^2$; in other words, $(\mathbf{n}\times\nabla\times\mathbf{n})^2 = (\mathbf{n}\cdot\nabla\mathbf{n})^2$.

1.7. *Director distribution in a parallel cell due to pretilt*:

   (a) Show that the elastic energy density due to a tilt angle distribution $\theta(z)$ is given by

$$U_{EL} = \frac{1}{2}(k_1\cos^2\theta + k_3\sin^2\theta)\left(\frac{d\theta}{dz}\right)^2$$

   where k_1, k_3 are elastic constants.

   (b) Using variational calculus, show that minimum elastic energy occurs when

$$U_{EL} = \frac{1}{2}(k_1\cos^2\theta + k_3\sin^2\theta)\left(\frac{d\theta}{dz}\right)^2 = \text{constant}$$

   The absolute minimum occurs when $\theta = \text{constant}$. This is the case shown in Figure 1.9a.

   (c) Show that, in general, the tilt angle distribution $\theta(z)$ for minimum elastic energy is given by

$$\frac{d\theta}{dz} = \frac{s\sqrt{k_1}}{\sqrt{k_1\cos^2\theta + k_3\sin^2\theta}}$$

   where s is a constant. In the antisymmetric case when $\theta(0) = \theta_0$, $\theta(d) = -\theta_0$, the tilt angle is zero at midlayer $(z = d/2)$. In this case s is the slope $d\theta/dz$ at midlayer.

   (d) Integrate the equation in (c) and obtain an expression for s. Show that for small pretilt angle $(\theta_0 \ll 1)$, the slope can be written $s = (2\theta_0/d)$ and the tilt angle distribution $\theta(z)$ can be written $\theta(z) = \theta_0 + sz$.

   (e) In a Pi cell, the distribution $\theta(z)$ for minimum elastic energy is again given by

$$\frac{d\theta}{dz} = \frac{s\sqrt{k_1}}{\sqrt{k_1\cos^2\theta + k_3\sin^2\theta}}$$

   where s is a constant. In this case, $\theta(0) = \theta_0, \theta(d) = \pi - \theta_0$, the tilt angle is $\pi/2$ at midlayer $(z = d/2)$. In this case s is proportional to the slope $d\theta/dz$ at midlayer. Show that

$$s = \sqrt{\frac{k_3}{k_1}}\left(\frac{d\theta}{dz}\right)_{z=d/2}$$

   (f) Explain why the splay cell has a lower elastic energy for cells with small pretilt angles.

2

Polarization of Optical Waves

In the electromagnetic theory of light, beams of light are represented by electromagnetic waves propagating in space. A beam of light is often represented by its electric field vector, which vibrates in time and space as the beam propagates. In an isotropic medium, the direction of vibration is always orthogonal to the direction of propagation. With a transverse wave, there are two independent directions of vibration. In an isotropic medium (e.g., glass, vacuum), the two mutually independent directions of vibration can be chosen arbitrarily. If these two components of vibration are totally uncorrelated, the resultant direction of vibration is random and the beam of light is said to be unpolarized. For all thermal sources in nature, the direction of vibration is random. A beam of light is said to be linearly polarized if its electric field vector vibrates in one particular direction. In what follows, we describe the polarization states of light.

2.1. MONOCHROMATIC PLANE WAVES AND THEIR POLARIZATION STATES

The main objective of the book is to provide a comprehensive coverage of the transmission of light through the LCD assembly, which consists of glass plates, transparent electrodes, color filters, polarizers, retardation plates (compensators), LC medium, and other components, some of which are glued together by using optical cement. For display applications, we can assume that the whole system is linear relative to the light intensity. The assumption of linearity allows us to treat the transmission of light independently for each frequency (or color) component. In addition, we can treat the transmission of light independently for each angle of incidence. In other words, the problem of transmission is reduced to a linear superposition of the transmission of monochromatic plane waves through the LCD assembly.

A monochromatic plane wave propagating in an isotropic and homogeneous medium can be represented by its electric field $\mathbf{E}(\mathbf{r}, t)$, which can be written

$$\mathbf{E} = \mathbf{A}\cos(\omega t - \mathbf{k}\cdot\mathbf{r}) \qquad (2.1\text{-}1)$$

where ω is the angular frequency, $\mathbf{k}$ is the wavevector, and $\mathbf{A}$ is a constant vector representing the amplitude. The magnitude of the wavevector k is related to the frequency by the equation

$$k = n\frac{\omega}{c} = n\frac{2\pi}{\lambda} \tag{2.1-2}$$

Where n is the index of refraction of the medium, c is the speed of light in vacuum, and λ is the wavelength of light in vacuum. For transparent materials the index of refraction is a real number that may depend on the wavelength. The dependence on wavelength is known as the *optical dispersion*, or *chromatic dispersion*. For materials with some absorption, the index of refraction is a complex number. Reflecting the transverse nature, the electric field vector is always perpendicular to the direction of propagation:

$$\mathbf{k} \cdot \mathbf{E} = 0 \tag{2.1-3}$$

For mathematical simplicity, the monochromatic plane wave in Eq. (2.1-1) is often written as

$$\mathbf{E} = \mathbf{A} \exp[i(\omega t - \mathbf{k} \cdot \mathbf{r})] \tag{2.1-4}$$

with the understanding that only the real part of the right side represents the actual electric field. This is known as the *analytic representation* and is widely used in electromagnetic waves. In most situations, the analytic representation of the field poses no problems when linear mathematical operations, such as integration, differentiation, and summation, are involved. The exceptions are cases that involve the product (or powers) of field vectors, such as the field energy density and the Poynting vector. In these cases, one must use the real form of the physical quantities.

The polarization state of a beam of monochromatic light is specified by its electric field vector $\mathbf{E}(\mathbf{r},t)$. The time evolution of the electric field vector is exactly sinusoidal; that is, the electric field must oscillate at a definite frequency. For the purpose of describing various representations of the polarization states, we consider a propagation along the z axis. Because it is a transverse wave, the electric field vector must lie in the xy plane. The two mutually independent components of the electric field vector can be written

$$\begin{aligned} E_x &= A_x \cos(\omega t - kz + \delta_x) \\ E_y &= A_y \cos(\omega t - kz + \delta_y) \end{aligned} \tag{2.1-5}$$

where we have used two independent and positive amplitudes A_x, A_y and added two independent phases δ_x, δ_y to reflect the mutual independence of the two components. Because the amplitudes are positive, the phase angles are defined in the range $-\pi < \delta \leq \pi$. Since the x and the y components of the electric field vector can oscillate independently at a definite frequency, one must consider the effect produced by the vector addition of these two oscillating orthogonal components. The problem of superposing two independent oscillations at right angles to each other and with the same frequency is well known and is completely analogous to the classic motion of a two-dimensional harmonic oscillator. The orbit of a general motion is an ellipse, which corresponds to oscillations in which the x and y components are not in phase. For optical waves, this corresponds to elliptic polarization states. There are, of course, many special cases that are important in optics. These include linear polarization states and circular polarization states.

We start with the discussion of two special cases of interest. Without loss of generality, we consider the time evolution of the electric field vector at the origin $z = 0$. According to Eq. (2.1-5), the electric field components are written

$$
\begin{aligned}
E_x &= A_x \cos\left(\omega t + \delta_x\right) \\
E_y &= A_y \cos\left(\omega t + \delta_y\right)
\end{aligned}
\tag{2.1-6}
$$

We define a relative phase as

$$
\delta = \delta_y - \delta_x
\tag{2.1-7}
$$

where, again, δ is limited in the region $-\pi < \delta \leq \pi$.

2.1.1. Linear Polarization States

A beam of light is said to be linearly polarized if the electric field vector vibrates in a constant direction (in the xy plane). This occurs when the two components of oscillation are in phase ($\delta = \delta_y - \delta_x = 0$) or π out of phase ($\delta = \delta_y - \delta_x = \pi$):

$$
\delta = \delta_y - \delta_x = 0 \quad \text{or} \quad \pi
\tag{2.1-8}
$$

In this case, the electric field vector vibrates sinusoidally along a constant direction in the xy plane defined by the ratio of the two components:

$$
\frac{E_y}{E_x} = \frac{A_y}{A_x} \quad \text{or} \quad -\frac{A_y}{A_x}
\tag{2.1-9}
$$

Since the amplitudes A_x, A_y are independent, the electric field vector of linearly polarized light can vibrate along any direction in the xy plane. Linearly polarized light is often called *plane polarized light*. If we examine the space evolution of the electric field vector at a fixed point in time (say, $t = 0$), we can express the components of the electric field vector as

$$E_x = A_x \cos\left(-kz + \delta_x\right)$$
$$E_y = A_y \cos\left(-kz + \delta_y\right)$$

(2.1-10)

with $\delta = \delta_y - \delta_x = 0$, or π. We note that the sinusoidal curve traced by the components in space is confined in a plane defined by Eq. (2.1-9). The vibration of the electric field vector is confined in this plane. Thus, the beam of light is said to be plane polarized. The terms *plane polarized light* and *linearly polarized light* are interchangeable. Linear polarization states are most widely used in optics because of their simplicity.

2.1.2. Circular Polarization State

The other special case of importance is that of the circular polarization state. A beam of light is said to be circularly polarized if the electric field vector undergoes uniform rotation in the xy plane. This occurs when $A_x = A_y$ and

$$\delta = \delta_y - \delta_x = \pm \tfrac{1}{2}\pi$$

(2.1-11)

Table 2.1.   Circular Polarization States

Electric Field of a Beam of Light		Angular Momentum per Photon $(\hbar\omega)L_z$		Helicity	Sense of Rotation
$E_x = A\cos\left(\omega t - kz\right)$ $E_y = A\cos\left(\omega t - kz - \tfrac{1}{2}\pi\right)$	z = constant	$\hbar$	$k > 0$ $k < 0$	$\hbar$ $-\hbar$	Right-handed Left-handed
$E_x = A\cos\left(\omega t - kz\right)$ $E_y = A\cos\left(\omega t - kz + \tfrac{1}{2}\pi\right)$	z = constant	$-\hbar$	$k > 0$ $k < 0$	$-\hbar$ $\hbar$	Left-handed Right-handed

According to our convention, the beam of light is right-hand circularly polarized when $\delta = -\pi/2$, which corresponds to a counterclockwise rotation of the electric field vector in the xy plane and left-hand circularly polarized when $\delta = \pi/2$, which corresponds to a clockwise rotation of the electric field vector in the xy plane. Our convention for labeling right-hand and left-hand polarization is consistent with the terminology of modern physics in which a photon with a right-hand circular (RHC) polarization has a positive angular momentum along the direction of propagation (see Table 2.1 and Problem 2.4). However, some optics books adopt the opposite convention. The opposite convention arises from the description of the evolution of the electric field vector in space (see Problem 2.10).

It is interesting to note that the conditions of equal amplitude and $\pm\pi/2$ phase shift for circular polarization states are valid in any set of perpendicular coordinates in xy plane. In other words, when the electric field vector of a circularly polarized light is decomposed into any two mutually perpendicular components, the amplitudes are always equal and the phase shift is always $\pm\pi/2$.

2.1.3. Elliptic Polarization States

A beam of light is said to be elliptically polarized if the curve traced by the endpoint of the electric field vector is an ellipse (in xy plane). This is the most general case of a polarized light. Both linear polarization states and circular polarization states are special cases of elliptic polarization states. At a given point in space (say, $z = 0$), Eq. (2.1-5) is a parametric representation of an ellipse traced by the endpoint of the electric field vector. The equation of the ellipse can be obtained by eliminating ωt in Eq. (2.1-6). After several steps of elementary algebra, we obtain

$$\left(\frac{E_x}{A_x}\right)^2 + \left(\frac{E_y}{A_y}\right)^2 - 2\frac{\cos\delta}{A_x A_y}E_x E_y = \sin^2\delta \qquad (2.1\text{-}12)$$

This equation is an equation of conic. From Eq. (2.1-6). It is obvious that this conic is confined in a rectangular region with sides parallel to the coordinate axes and whose lengths are $2A_x, 2A_y$. Therefore, the curve must be an ellipse. Thus, we find that the polarization states of light are, in general, elliptical. A complete description of an elliptical polarization state includes the orientation of the ellipse relative to the coordinate axes, the shape and sense of revolution of the electric field vector. In general, the principal axes of the ellipse are not in the x and y directions. By using a transformation (rotation) of the coordinate system, we are able to diagonalize Eq. (2.1-12). Let x' and y' be the new set of axes along the principal axes of the ellipse. Then the equation of the ellipse in this new

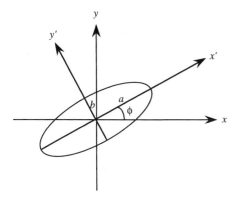

Figure 2.1.   Polarization ellipse.

coordinate system becomes

$$\left(\frac{E_{x'}}{a}\right)^2 + \left(\frac{E_{y'}}{b}\right)^2 = 1 \tag{2.1-13}$$

where a and b are the length of the principal semi-axes of the ellipse and $E_{x'}$ and $E_{y'}$ are the components of the electric field vector in this principal coordinate system.

Let ϕ be the angle between the x' axis and x axis (see Fig. 2.1). Then the length of the principal axes are given by

$$a^2 = A_x^2 \cos^2\phi + A_y^2 \sin^2\phi + 2A_x A_y \cos\delta \cos\phi \sin\phi$$
$$b^2 = A_x^2 \sin^2\phi + A_y^2 \cos^2\phi - 2A_x A_y \cos\delta \cos\phi \sin\phi \tag{2.1-14}$$

The angle ϕ can be expressed in terms of A_x, A_y, and $\cos\delta$ as

$$\tan 2\phi = \frac{2A_x A_y}{A_x^2 - A_y^2} \cos\delta \tag{2.1-15}$$

It is important to note that $\phi + \pi/2$ is also a solution, if ϕ is a solution of the equation. The sense of revolution of an elliptical polarization is determined by the sign of $\sin\delta$. The endpoint of the electric vector will revolve in a clockwise direction if $\sin\delta > 0$ and in a counterclockwise direction if $\sin\delta < 0$. Figure 2.2 illustrates how the polarization ellipse changes with varying phase difference δ.

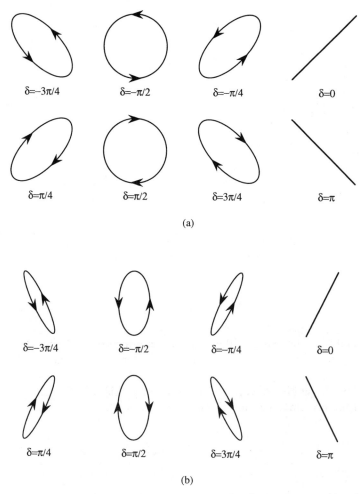

Figure 2.2. Polarization ellipses at various phase angles δ: (a) $E_x = \cos(\omega t - kz)$, $E_y = \cos(\omega t - kz + \delta)$; (b) $E_x = 0.5\cos(\omega t - kz)$, $E_y = \cos(\omega t - kz + \delta)$.

The ellipticity of a polarization ellipse is defined as

$$e = \pm\frac{b}{a} \tag{2.1-16}$$

where a and b are the half length of the principal axes. The ellipticity is taken as positive when the rotation of the electric field vector is right-handed and negative otherwise. With this definition, $e = \pm 1$ for circularly polarized light.

An elliptic polarization state can always be decomposed into two mutually orthogonal components. The relative phase shift between these two components can be anywhere between $-\pi$ and π. However, in the principal coordinate system, the relative phase shift between the two orthogonal components is always $-\pi/2$ or $\pi/2$, depending on the sense of revolution.

In summary, light is linearly polarized when the tip of the electric field vector **E** moves along a straight line. When it describes an ellipse, the light is elliptically polarized. When it describes a circle, the light is circularly polarized. If the endpoint of the electric field vector is seen to move in a counterclockwise direction by an observer facing the approaching wave, the field is said to possess right-handed polarization. Figure 2.2 also illustrates the sense of revolution of the ellipse.

2.2. COMPLEX-NUMBER REPRESENTATION

From the discussion in the previous section we found how the polarization state of a beam of light can be described in terms of the amplitudes and the phase angles of the x and y components of the electric field vector. In fact, all the information about the polarization state of a wave is contained in the complex amplitude **A** of the plane wave [see Eq. (2.1-4)]. Therefore, a complex number χ defined as

$$\chi = e^{i\delta} \tan \psi = \frac{A_y}{A_x} e^{i(\delta_y - \delta_x)} \qquad (2.2\text{-}1)$$

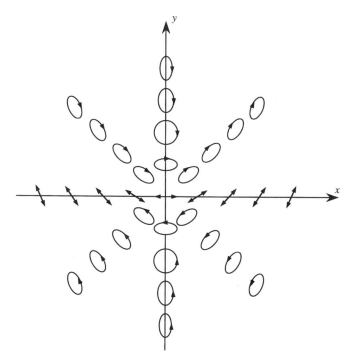

Figure 2.3. Complex-number representation of polarization states.

is sufficient to describe the polarization states. The angle ψ is defined to be between 0 and $\pi/2$. A complete description of the ellipse of polarization, which includes the orientation, sense of revolution, and ellipticity [see Eq. (2.1-16)], can be expressed in terms of δ and ψ. Figure 2.3 illustrates various different polarization states in the complex plane. It can be seen from the figure that all the right-handed elliptic polarization states are in the lower half of the plane, whereas the left-handed elliptic polarization states are in the upper half of the plane. The origin corresponds to a linear polarization state with direction of oscillation parallel to the x axis. Thus, each point on the complex plane represents a unique polarization state. Each point on the x axis represents a linearly polarized state with different azimuth angles of oscillation. Only two points $(0, \pm 1)$ correspond to circular polarization. Each point of the rest of the complex plane corresponds to a unique elliptic polarization state.

The inclination angle ϕ and the ellipticity angle $\theta (\theta \equiv \tan^{-1} e)$ of the polarization ellipse correspond to a given complex number χ are given by

$$\tan 2\phi = \frac{2 \, \mathrm{Re}[\chi]}{1 - |\chi|^2} = \tan 2\psi \cos \delta \qquad (2.2\text{-}2)$$

and

$$\sin 2\theta = -\frac{2 \, \mathrm{Im}[\chi]}{1 + |\chi|^2} = -\sin 2\psi \sin \delta \qquad (2.2\text{-}3)$$

2.3. JONES VECTOR REPRESENTATION

The Jones vector, introduced in 1941 by R. C. Jones [1], is very convenient for describing the polarization state of a plane wave. In this representation, the plane wave in Eq. (2.1-4) is expressed in terms of its complex amplitudes as a column vector

$$\mathbf{J} = \begin{pmatrix} A_x e^{i\delta_x} \\ A_y e^{i\delta_y} \end{pmatrix} \qquad (2.3\text{-}1)$$

Note that the Jones vector is a complex vector (i.e., its elements are complex numbers); $\mathbf{J}$ is not a vector in the real physical space—rather, it is a vector in an abstract mathematical space. To obtain, as an example, the real x component of the electric field, we must perform the operation $E_x(t) = \mathrm{Re}[J_x e^{i\omega t}] = \mathrm{Re}[A_x e^{i(\omega t + \delta_x)}]$.

The Jones vector contains complete information about the amplitudes and the phases of the electric field vector components. It thus specifies the polarization state of the wave uniquely. If we are interested only in the polarization state of the wave, it is convenient to use the normalized Jones vector that satisfies the condition that

$$\mathbf{J}^* \cdot \mathbf{J} = 1 \qquad (2.3\text{-}2)$$

where the asterisk (*) denotes complex conjugation. Thus, a beam of linearly polarized light with the electric field vector oscillating along a given direction can be represented by the Jones vector

$$\begin{pmatrix} \cos \psi \\ \sin \psi \end{pmatrix} \qquad (2.3\text{-}3)$$

where ψ is the azimuth angle of the oscillation direction with respect to the x axis. The state of polarization which is orthogonal to the state represented by Eq. (2.3-3) can be obtained by the substitution of ψ by $\psi + \frac{1}{2}\pi$, leading to a Jones vector

$$\begin{pmatrix} -\sin \psi \\ \cos \psi \end{pmatrix} \qquad (2.3\text{-}4)$$

The special case, when $\psi = 0$, represents linearly polarized waves whose electric field vector oscillates along the coordinate axes. These Jones vectors are given by

$$\mathbf{x} = \begin{pmatrix} 1 \\ 0 \end{pmatrix}, \qquad \mathbf{y} = \begin{pmatrix} 0 \\ 1 \end{pmatrix} \qquad (2.3\text{-}5)$$

Jones vectors for the right- and left-hand circularly polarized light waves are given by

$$\mathbf{R} = \frac{1}{\sqrt{2}} \begin{pmatrix} 1 \\ -i \end{pmatrix} \qquad (2.3\text{-}6)$$

$$\mathbf{L} = \frac{1}{\sqrt{2}} \begin{pmatrix} 1 \\ i \end{pmatrix} \qquad (2.3\text{-}7)$$

These two states of circular polarizations are mutually orthogonal in the

Table 2.2. Various Representation of Polarization States

Polarization Ellipse	Jones Vector	(δ, ψ)	(ϕ, θ)	Stokes
——	$\begin{pmatrix} 1 \\ 0 \end{pmatrix}$	$(0,0)$	$(0,0)$	$\begin{pmatrix} 1 \\ 1 \\ 0 \\ 0 \end{pmatrix}$
(vertical line)	$\begin{pmatrix} 0 \\ 1 \end{pmatrix}$	$(0,\pi/2)$	$(\pi/2,0)$	$\begin{pmatrix} 1 \\ -1 \\ 0 \\ 0 \end{pmatrix}$
(diagonal /)	$\dfrac{1}{\sqrt{2}}\begin{pmatrix} 1 \\ 1 \end{pmatrix}$	$(0,\pi/4)$	$(\pi/4,0)$	$\begin{pmatrix} 1 \\ 0 \\ 1 \\ 0 \end{pmatrix}$
(diagonal \)	$\dfrac{1}{\sqrt{2}}\begin{pmatrix} 1 \\ -1 \end{pmatrix}$	$(\pi,\pi/4)$	$(-\pi/4,0)$	$\begin{pmatrix} 1 \\ 0 \\ -1 \\ 0 \end{pmatrix}$
(circle, CCW)	$\dfrac{1}{\sqrt{2}}\begin{pmatrix} 1 \\ -i \end{pmatrix}$	$(-\pi/2,\pi/4)$	$(0,\pi/4)$	$\begin{pmatrix} 1 \\ 0 \\ 0 \\ -1 \end{pmatrix}$
(circle, CW)	$\dfrac{1}{\sqrt{2}}\begin{pmatrix} 1 \\ i \end{pmatrix}$	$(\pi/2,\pi/4)$	$(0,-\pi/4)$	$\begin{pmatrix} 1 \\ 0 \\ 0 \\ 1 \end{pmatrix}$
(ellipse, vertical)	$\dfrac{1}{\sqrt{5}}\begin{pmatrix} 1 \\ 2i \end{pmatrix}$	$(\pi/2, \tan^{-1}2)$	$(\pi/2, -\tan^{-1}\tfrac{1}{2})$	$\begin{pmatrix} 1 \\ -3/5 \\ 0 \\ 4/5 \end{pmatrix}$
(ellipse, horizontal)	$\dfrac{1}{\sqrt{5}}\begin{pmatrix} 2 \\ -i \end{pmatrix}$	$(-\pi/2, \tan^{-1}\tfrac{1}{2})$	$(0, \tan^{-1}\tfrac{1}{2})$	$\begin{pmatrix} 1 \\ 3/5 \\ 0 \\ -4/5 \end{pmatrix}$
(ellipse, tilted)	$\dfrac{1}{\sqrt{10}}\begin{pmatrix} 2+i \\ 2-i \end{pmatrix}$	$(-\tan^{-1}\tfrac{4}{3}, \pi/4)$	$(\pi/4, \tan^{-1}\tfrac{1}{2})$	$\begin{pmatrix} 1 \\ 0 \\ 3/5 \\ -4/5 \end{pmatrix}$

sense that

$$\mathbf{R}^* \cdot \mathbf{L} = 0 \qquad (2.3\text{-}8)$$

Since the Jones vector is a column matrix of rank 2, any pair of orthogonal Jones vectors can be used as a basis of the mathematical space spanned by all the Jones vectors. Any polarization state can be represented as a superposition of two mutually orthogonal polarization states $\mathbf{x}$ and $\mathbf{y}$, or $\mathbf{R}$ and $\mathbf{L}$. In particular, we can resolve the basic linear polarization states $\mathbf{x}$ and $\mathbf{y}$ into two circular polarization states $\mathbf{R}$ and $\mathbf{L}$ and vice versa. These relations are given by

$$\mathbf{R} = \frac{1}{\sqrt{2}}(\mathbf{x} - i\mathbf{y}) \qquad (2.3\text{-}9)$$

$$\mathbf{L} = \frac{1}{\sqrt{2}}(\mathbf{x} + i\mathbf{y}) \qquad (2.3\text{-}10)$$

$$\mathbf{x} = \frac{1}{\sqrt{2}}(\mathbf{R} + \mathbf{L}) \qquad (2.3\text{-}11)$$

$$\mathbf{y} = \frac{i}{\sqrt{2}}(\mathbf{R} - \mathbf{L}) \qquad (2.3\text{-}12)$$

Circular polarization states are seen to consist of linear oscillations along the x and y directions with equal amplitude $1/\sqrt{2}$, but with a phase difference of $\frac{1}{2}\pi$. Similarly, a linear polarization state can be viewed as a superposition of two oppositely sensed circular polarization states.

Thus far we have discussed only the Jones vectors of some simple special cases of polarization. It is easy to show that a general elliptic polarization state can be represented by the following Jones vector:

$$\mathbf{J}(\psi, \delta) = \begin{pmatrix} \cos \psi \\ e^{i\delta} \sin \psi \end{pmatrix} \qquad (2.3\text{-}13)$$

This Jones vector represents the same polarization state as the one represented by the complex number $\chi = e^{i\delta} \tan \psi$. Table 2.2 shows the Jones vectors of some typical polarization states.

The most important application of Jones vectors is in conjunction with the Jones calculus. This is a powerful technique used for studying the propagation of plane waves with arbitrary states of polarization through an arbitrary sequence of birefringent elements and polarizers. This topic will be considered in some detail in Chapter 4.

2.4. PARTIALLY POLARIZED AND UNPOLARIZED LIGHT

A monochromatic plane wave inherently must be polarized; that is, the end point of its electric field vector at each point in space must trace out periodically an ellipse or one of its special forms, such as a circle or a straight line. However, if the light is not absolutely monochromatic, the amplitudes and relative phase δ between the x and y components can both vary with time, and the electric field vector will first vibrate in one ellipse and then in another. As a result, the polarization state of a polychromatic plane wave may be constantly changing. If the polarization state changes more rapidly than the speed of observation, we say that the light is partially polarized or unpolarized depending on the time-averaged behavior of the polarization state. In optics, one often deals with the light with oscillation frequencies of about $10^{14}\,s^{-1}$, whereas the polarization state may change in a time period of 10^{-8} s depending on the nature of the light source.

We will limit ourselves to the case of quasimonochromatic waves, whose frequency spectrum is confined to a narrow bandwidth $\Delta\omega$ (i.e., $\Delta\omega \ll \omega$). Such a wave can still be described by Eq. (2.1-4), provided we relax the constancy condition of the amplitude $\mathbf{A}$. Now ω denotes the center frequency, and the complex amplitude $\mathbf{A}$ is a function of time. Because the bandwidth is narrow, $\mathbf{A}(t)$ may change by only a relatively small amount in a time interval $1/\Delta\omega$, and in this sense it is a slowly varying function of time. However, if the time constant of the detector, τ_D, is greater than $1/\Delta\omega$, $\mathbf{A}(t)$ may change significantly in a time interval τ_D. Although the amplitudes and phases are irregularly varying functions of time, certain correlations may exist among them.

To describe the polarization state of this type of radiation, we introduce the following time-averaged quantities:

$$
\begin{aligned}
S_0 &= \langle\langle A_x^2 + A_y^2 \rangle\rangle \\
S_1 &= \langle\langle A_x^2 - A_y^2 \rangle\rangle \\
S_2 &= 2\langle\langle A_x A_y \cos\delta \rangle\rangle \\
S_3 &= 2\langle\langle A_x A_y \sin\delta \rangle\rangle
\end{aligned}
\tag{2.4-1}
$$

where the amplitudes A_x, A_y and the relative phase δ are assumed to be time dependent, and the double brackets denote averages performed over a time interval τ_D that is the characteristic time constant of the detection process. These four quantities are known as the *Stokes parameters* of a quasimonochromatic plane wave. Note that all four quantities have the same dimension of intensity. It can be shown that the Stokes parameters satisfy the relation

$$
S_0^2 \geq S_1^2 + S_2^2 + S_3^2
\tag{2.4-2}
$$

where the equality sign holds only for polarized waves.

It is a simple exercise to compute, from the definitions, various Stokes parameters of principal interest. Consider, for example, unpolarized light. There is no preference between A_x and A_y; consequently, $\langle A_x^2 + A_y^2 \rangle$ reduces to $2\langle A_x^2 \rangle$, and $\langle A_x^2 - A_y^2 \rangle$ reduces to zero. The other quantities also reduce to zero because δ is a random function of time. If the field is normalized such that $S_0 = 1$, the Stokes vector representation of an unpolarized light wave is $(1,0,0,0)$. Similar reasoning shows that a horizontally polarized beam can be represented by the Stokes vector $(1,1,0,0)$ and a vertically polarized beam can be represented by $(1, -1,0,0)$. Right-hand circularly polarized light $(\delta = -\frac{1}{2}\pi)$ is represented by $(1,0,0, -1)$, and the left-hand circularly polarized light $(\delta = \frac{1}{2}\pi)$ is represented by $(1,0,0,1)$. From the definition, none of the parameters can be greater than the first S_0, which is normalized to 1. Therefore, each of the others lies in the range form -1 or 1. If the beam is entirely unpolarized, $S_1 = S_2 = S_3 = 0$. If it is completely polarized, $S_1^2 + S_2^2 + S_3^2 = 1$. The degree of polarization is therefore defined as

$$\gamma = \frac{(S_1^2 + S_2^2 + S_3^2)^{1/2}}{S_0} \tag{2.4-3}$$

According to Eq. (2.4-2), this parameter γ is a real number between 0 and 1. It is thus very useful in describing the partially polarized light. The polarization preference of a partially polarized light can be seen directly from the sign of the parameters S_1, S_2, and S_3.

The parameter S_1 describes the linear polarization along the x or y axis; the probability that the light is linearly polarized along the x axis is $\frac{1}{2}(1 + S_1)$ and along the y axis, $\frac{1}{2}(1 - S_1)$. Thus, the values $S_1 = 1, -1$ correspond to complete polarization in these directions. The parameter S_2 describes the linear polarization along directions at angles $\phi = \pm 45°$ to the x axis. The probability that the light is linearly polarized along these directions is, respectively, $\frac{1}{2}(1 + S_2)$ and $\frac{1}{2}(1 - S_2)$. Thus, the values $S_2 = 1, -1$ correspond to complete polarization in these directions. Finally, the parameter S_3 represents the degree of circular polarization; the probability that the lightwave had right-hand circular polarization is $\frac{1}{2}(1 - S_3)$, and left-hand circular polarization, $\frac{1}{2}(1 + S_3)$.

The Stokes parameters for a polarized light with a complex representation $\chi = e^{i\delta} \tan \psi$ are given by [according to (2.4-1)]

$$S_0 = 1, \qquad S_1 = \cos 2\psi, \qquad S_2 = \sin 2\psi \cos \delta, \qquad S_3 = \sin 2\psi \sin \delta \tag{2.4-4}$$

According to our convention, a positive S_3 corresponds to left-hand elliptical polarization $(\sin \delta > 0$, clockwise revolution).

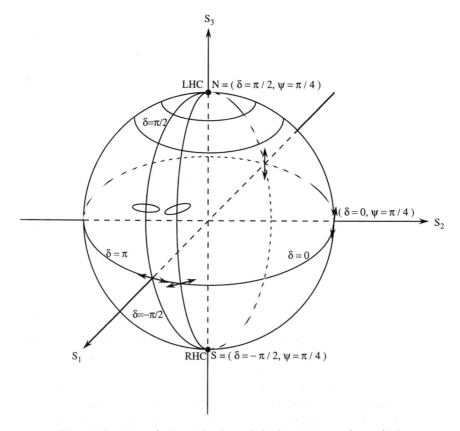

Figure 2.4. Poincaré sphere (showing polarization states at various points).

2.5. POINCARÉ SPHERE

Although the Stokes parameters are introduced for describing partially polarized light, they are also convenient parameters to describe the polarization states of polarized light. For polarized light, the Stokes parameters S_1, S_2, S_3 can also be employed to represent the polarization states. Since $S_0 = 1$, all points with coordinate (S_1, S_2, S_3) are confined on the surface of a unit sphere in three-dimensional (3D) space. This sphere is known as the *Poincaré sphere* (see Figure 2.4). Each point on the surface of the sphere represents a unique polarization state. For example, the north pole (0,0,1) corresponds to left-handed circular (LHC) polarization state, whereas the south pole $(0,0,-1)$ corresponds to a right-handed circular (RHC) polarization state. Point (1,0,0) corresponds to a linear polarization state parallel to horizontal direction, whereas point $(-1,0,0)$ corresponds to a linear polarization state parallel to vertical direction. In fact, all points on the equator correspond to a unique linear polarization state. The remaining points correspond to elliptical polarization states. It is interesting to note that any pair of antipodal points on the Poincaré sphere (two points on the opposite side of the center of the sphere) correspond to states with orthogonal polarization.

According to Eqs. (2.2-2) and (2.2-3), we have

$$\tan 2\phi = \frac{S_2}{S_1} \qquad \text{and} \qquad \sin 2\theta = -S_3 \qquad (2.5\text{-}1)$$

where ϕ is the inclination angle of the polarization ellipse and θ is the ellipticity angle defined as $\tan^{-1}e$. Generally speaking, $S_2/S_1 = \text{constant}$ represents a vertical plane containing the poles. Since both S_1 and S_2 are confined on the surface of the sphere for polarized light, $S_2/S_1 = \text{constant}$ actually represents a meridian—a half circle connecting the north and south poles. According to Eq. (2.5-1), ϕ is a constant on the meridian. Thus, each meridian represents a class of elliptic polarization states with the same inclination angle ϕ, but with different ellipticities. In addition, $S_3 = \text{constant}$ represents a circle on the sphere parallel to the equatorial plane. According to Eq. (2.5-1), θ is a constant on this circle (parallel or latitude). Thus, each parallel (latitude) represents a class of elliptic polarization states with the same ellipticity $e = \tan^{-1}\theta$, but with different inclination angles. The Poincaré sphere is particularly useful in optical birefringent networks where wave plates are employed to change the polarization state of light. Consider the following three examples:

1. *Polarization Conversion Using Quarter-Wave Plates.* An elliptic polarization state (with inclination angle ϕ and ellipticity angle θ) can be converted into a circular polarization state by using two quarter-wave plates. This is done by aligning the first quarter-wave plate with its c axis parallel to the major axis of the polarization ellipse. This leads to a linear polarization state along the direction either at $(\phi + \theta)$ or at $(\phi - \theta)$, depending on the handedness of the elliptical polarization state as well as the sign of the birefringence of the wave plate. The linear polarization state can then be converted into a circular polarization state by using the second quarter-wave plate with its c axis oriented at either $+45°$ or $-45°$ with respect to the linear polarization direction. The handedness of the circular polarization state is determined by the sign of the birefringence of the wave plate as well as whether the orientation angle is $-45°$ or $+45°$. In other words, given a quarter-wave plate, it is possible to obtain circular polarization of either handedness, regardless of the sign of the birefringence of the wave plate. By reversing this procedure, it is possible to convert a circular polarization state into an arbitrary elliptical polarization state by using two properly oriented quarter-wave plates in sequence. By combining the preceding two procedures, we are able to convert any elliptic polarization state into any other elliptic polarization state by using a total of four properly oriented quarter-wave plates in sequence. This is illustrated in Figure 2.5. Points A and B on the Pincaré sphere represent two arbitrary elliptical polarization states. Using two properly oriented quarter-wave plates, we are able to move

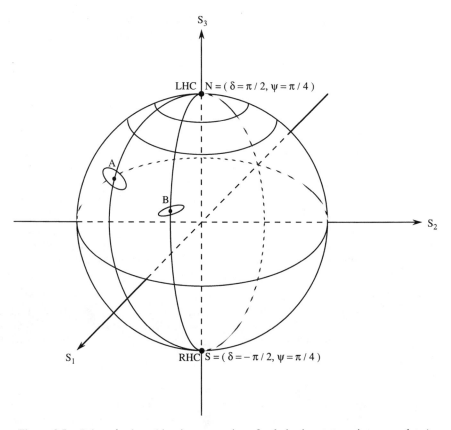

Figure 2.5. Poincaré sphere (showing conversion of polarization states using wave plates).

from point A to the north pole N which represents the left-handed circular polarization state. A similar procedure with two properly oriented quarter-wave plates can be employed to move from point N to point B.

2. *Polarization Conversion Using Two Quarter-Wave Plates and a Half-Wave Plate.* Consider the conversion of an elliptic polarization state (ϕ_1, θ_1) to another elliptic polarization state (ϕ_2, θ_2). As mentioned above, an elliptic polarization state (with inclination angle ϕ_1 and ellipticity angel θ_1) can be converted into a linear polarization state by using a quarter-wave plate with its c axis parallel to the major axis (azimuth ϕ_1) of the polarization ellipse. This leads to a linear polarization state along the direction either at $(\phi_1 + \theta_1)$ or at $(\phi_1 - \theta_1)$, depending on the handedness of the elliptic polarization state as well as the sign of the birefringence of the wave plate. The angle between the linear polarization state and the c axis of the first quarter-wave plate is either θ_1 or $-\theta_1$. A half-wave plate is then employed to obtain a linear polarization state with an inclination angle of ϕ_2 for the direction of polarization. This linear polarization state can then be converted into the final elliptical polarization (ϕ_2, θ_2), by using

the second quarter-wave plate with its c axis aligned at either θ_2 or $-\theta_2$ relative to the direction of polarization of the linear state (inclination ϕ_2).

3. *Polarization Conversion Using a General Birefringent Wave Plate.* In the examples above, we described the points of polarization states on the Poincaré sphere before and after the quarter- or half-wave plates. We now consider the effect of a general homogeneous wave plate that is described by its phase retardation Γ and the azimuth angle ψ of the slow axis (e.g., c axis of an a plate) measured from the x axis.

Case 1. $\psi = 0$
This is the case when the slow axis of the plate is oriented along the x axis. Consider now the effect of this plate on an arbitrary input polarization state of

$$P = \begin{pmatrix} \cos\theta \\ e^{i\delta}\sin\theta \end{pmatrix}$$

where θ and δ are real constants. The output polarization state in this case is given by

$$Q_1 = \begin{pmatrix} \cos\theta \\ e^{i\Gamma}e^{i\delta}\sin\theta \end{pmatrix}$$

where Γ is the phase retardation of the wave plate. These two polarization states are represented as points on the Poincaré sphere in Figure 2.6. According to the definition of the Stokes parameters, these two polarization states have the same S_1 coordinate of $\cos 2\theta$ for arbitrary values of δ and Γ. It can be easily shown that the output polarization state Q_1 on the Poincaré sphere can be obtained from the initial polarization state P on the Poincaré sphere by a rotation around the S_1-axis (OA) with an angle of Γ.

Case 2. $\psi \neq 0$
In the case when the slow axis (e.g., c axis of an a plate) of the wave plate is oriented at an azimuth angle ψ relative to the x axis, the output state can be obtained by a similar rotation on the Poincaré sphere. As discussed earlier in the paragraph after Eq. (2.5-1), the effect of a rotation of the polarization state by an angle α without changing the polarization ellipse is represented on the Poincaré sphere by a rotation of an angle of 2α around the polar axis (S_3 axis). To obtain the output polarization state in this case by using the result in Case 1, we follow the following three steps:

1. We first rotate the input polarization state by an angle $-\psi$ relative to the wave plate. This moves from P to a new point (say, P') on the Poincaré sphere by a rotation of -2ψ around the polar axis.

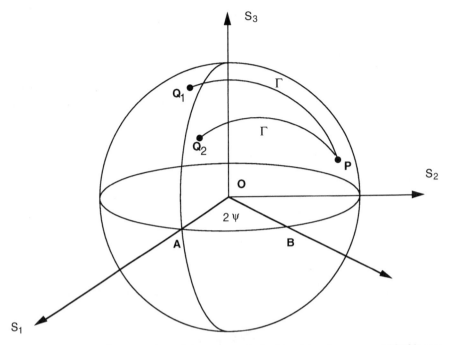

Figure 2.6. Poincaré construction of the output polarization states for a general birefringent wave plate with a phase retardation Γ and an azimuth angle of ψ. The output state Q_2 is obtained by rotating the input state P through an angle Γ around the OB axis which is oriented at an angle of 2ψ from the S_1 axis.

2. The effect of the wave plate is to move from P' to a new point (say, Q'_1) by a rotation of Γ around the S_1 axis (OA), according to the result in Case 1.

3. We then rotate the output polarization state Q'_1 by an angle ψ which is represented on the Poincaré sphere by a rotation of 2ψ around the polar axis. This leads to the final output state Q_2 for this case.

It can be shown that the three steps described above are equivalent to a single rotation of Γ around the axis OB. In other words, the output polarization state Q_2 on the Poincaré sphere can be obtained by a rotation around the axis OB with an angle of Γ. The axis of rotation OB is oriented at an angle of 2ψ from the S_1 axis, and lies in the equatorial plane. Figure 2.6 shows the construction of the output polarization states Q_1, Q_2 using the Poincaré sphere.

Although most liquid crystal cells in LCDs are optically inhomogeneous due to the spatial variation of the director orientation in the cells, the method described above can still be employed for the analysis of polarization evolution in the cells. This is often achieved by subdividing the cells into a large number of thin layers. Each of these layers can be considered as a homogeneous wave plate.

The method described above can then be employed several times to obtain the final output polarization state [3, 4]. In addition to the analysis of LCD cells, the Poincaré sphere can also be applied to the case when the birefringent plate exhibits both linear birefringence and optical rotation (circular birefringence, or Faraday effect) [5].

2.5.1. Pancharatnam Phase

As discussed earlier in this chapter, the polarization state of a beam of light is defined with an arbitrary overall phase. The polarization ellipse depends only on the relative phase between the two transverse components of the electric field vector (say, E_x, E_y). In 1956 Pancharatnam [2] introduced a convention for comparing the phases of two beams of light in different states of polarization. On the basis of his suggestion, a relative phase between two different polarization states $\mathbf{E}_A, \mathbf{E}_B$ can be defined as follows. We bring the two beams of light to interfere with each other. The intensity of the resultant signal can be written

$$I = (\mathbf{E}_A + \exp(i\chi)\mathbf{E}_B)^* \cdot (\mathbf{E}_A + \exp(i\chi)\mathbf{E}_B) \qquad (2.5\text{-}2)$$

where χ is a relative phase that can be due to a propagation path difference. The resultant intensity is obviously dependent on the relative phase χ. In general, a unique phase shift χ_{AB} (within 2π), defined as the relative phase between the two polarization states $\mathbf{E}_A, \mathbf{E}_B$, exists such that the resultant intensity reaches its maximum. In other words, we can say that $\mathbf{E}_A$ and $\exp(i\chi_{AB})\mathbf{E}_B$ are in phase.

On the basis of definition of relative phase, we can assign an overall phase to each polarization state so that all states of polarization are in phase with a reference polarization state, say, the left-handed circular (LHC) polarization state (the north pole N on Poincaré's sphere). Note that this additional phase does not affect the polarization ellipse. This additional phase is called the *Pancharatnam phase*. It is important to note that the assignment of a Pancharatnam phase depends on the reference state chosen. We now discuss Pancharatnam's theorem:

> Given three polarization states A,B,C such that A is in phase with B and B is in phase with C. Then, in general, C is not in phase with A. The phase difference between states C and A is equal to half the area of the geodesic triangle ABC on the Poincaré sphere.

To illustrate this theorem, we consider the following example. According to the definition given above, the linear polarization state $A = (1,0)$ is in phase with the LHC polarization state $N = (1, i)/\sqrt{2}$. In addition, the LHC polarization state N is in phase with a linear polarization state $B = \exp(-i\pi/4)\,(1,1)/\sqrt{2}$. When we compare the relative phase between A and B, we find that, according to the definition above, states A and B have a relative phase difference of $\pi/4$ that is exactly half the area of the geodesic triangle ABN.

As we know, birefringent wave plates are needed to convert the conventional polarization state from one polarization ellipse to another. If we include a Pancharatnam phase to each of the polarization states so that all states are in phase with a given polarization state (say, the LHC state), then converting one polarization state to another state would require birefringent wave plates and an isotropic phase plate. The isotropic phase plate (e.q., a glass plate) with the proper optical pathlength is essential to provide the Pancharatnam phase needed for each state to maintain the in-phase requirement with the reference state. With this definition for the polarization state, it can be shown that if the polarization state of a beam of light is to trace out a geodesic triangle on the Poincaré sphere, the net isotropic phase change needed (to maintain the Pancharatnam phase) would be exactly half the solid angle subtended by the geodesic triangle at the center of the Poincaré sphere. Let $\mathbf{a},\mathbf{b},\mathbf{c}$ be the unit vectors from the center to the three points forming the geodesic triangle on the Poincaré sphere; then the solid angle Ω is given by the following formula [6]:

$$\tan\left(\Omega/2\right) = |\mathbf{a}\cdot\mathbf{b}\times\mathbf{c}|/(1 + \mathbf{a}\cdot\mathbf{b} + \mathbf{b}\cdot\mathbf{c} + \mathbf{c}\cdot\mathbf{a}) \qquad (2.5\text{-}3)$$

REFERENCES

1. R. C. Jones, *J. Opt. Soc. Am.* **31**, 488 (1941).
2. S. Pancharatnam, *Proc. Ind. Acad. Sci. A* **44**, 247 (1956).
3. H. G. Jerrard, *J. Opt. Soc. Am.* **44**, 634–640 (1954).
4. J. E. Bigelow and R. A. Kashnow, *Appl. Opt.* **16**, 2090–2096 (1977).
5. G. N. Ramachandran and S. Ramaseshan, *J. Opt. Soc. Am.* **42**, 49–56 (1952).
6. P. K. Aravind, *Opt. Commun.* **94**, 191–196 (1992).

PROBLEMS

2.1. Derive Eq. (2.1-12).

2.2. Derive Eqs. (2.1-14) and (2.1-15).

2.3. Show that the endpoint of the electric vector of an elliptically polarized light will revolve in a clockwise direction if $\sin\delta>0$ and in a counterclockwise direction if $\sin\delta<0$.

2.4. (a) A RHC polarized wave ($\sin\delta<0$) propagating in the z direction has a finite extent in the x and y directions. Assuming that the amplitude modulation is slowly varying (the wave is many wavelengths broad), show that the electric and magnetic fields are given approximately by

$$\mathbf{E}(x, y, z, t) \approx \left[E_0(x, y)(\mathbf{x} - i\mathbf{y}) + \frac{-i}{k}\left(\frac{\partial E_0}{\partial x} - i\frac{\partial E_0}{\partial y} \right)\mathbf{z} \right] e^{i(\omega t - kz)}$$

$$\mathbf{H}(x, y, z, t) \approx i\frac{k}{\omega\mu} \mathbf{E}(x, y, z, t)$$

(b) Calculate the time-averaged component of angular momentum along the direction of propagation $(+z)$. Show that this component of the angular momentum is $\hbar$ provided the energy of the wave is normalized to $\hbar\omega$. This shows that a RHC polarized photon carries a positive angular momentum $\hbar$ along the direction of its momentum vector (see helicity in Table 2.1).

(c) Show that the transverse components of the angular momentum vanish.

2.5. *Orthogonal polarization states*:

(a) Find a polarization state that is orthogonal to the polarization state

$$\mathbf{J}(\psi, \delta) = \begin{pmatrix} \cos\psi \\ e^{i\delta}\sin\psi \end{pmatrix}$$

Answer:

$$\begin{pmatrix} \sin\psi \\ e^{i(\pi+\delta)}\cos\psi \end{pmatrix}$$

(b) Show that the major axes of the ellipses of two mutually orthogonal polarization states are perpendicular to each other and the senses of revolution are opposite.

2.6. (a) An elliptically polarized beam propagating in the z direction has a finite extent in the x and y directions:

$$\mathbf{E}(x, y, z, t) \approx E_0(x, y)(\alpha\mathbf{x} + \beta\mathbf{y})e^{i(\omega t - kz)}$$

where $\alpha = \cos\psi$, $\beta = e^{i\delta}\sin\psi$. Show that the electric field must have a component in the z direction [see Problem 2.4(a)], and derive the expressions for the electric field and the magnetic field.

(b) Calculate the z component of the angular momentum, assuming that the total energy of the wave is $\hbar\omega$. Answer: $L_z = -\hbar\sin 2\psi\sin\delta$.

(c) Decompose elliptically polarized light into a linear superposition of right-hand and left-hand polarized states $\mathbf{R}$ and $\mathbf{L}$, that is, if $\mathbf{J}$ is the Jones vector of the polarization state in (a), find r and l such that $\mathbf{J} = r\mathbf{R} + l\mathbf{L}$.

(d) If r and l are the probability amplitudes that the photon is RHC and LHC polarized, respectively, show that the angular momentum can be obtained by evaluating $|r|^2 - |l|^2$:

$$L_z = \hbar(|r|^2 - |l|^2)$$

(e) Express the angular momentum in terms of the ellipticity angle θ.

2.7. Derive Eqs. (2.2-2) and (2.2-3).

2.8. *Orthogonal polarization states*:
   Consider two monochromatic plane waves of the form

   $$\mathbf{E}_a(z, t) = \text{Re}[\mathbf{A}e^{i(\omega t - kz)}] \quad \text{and} \quad \mathbf{E}_b(z, t) = \text{Re}[\mathbf{B}e^{i(\omega t - kz)}]$$

   The polarization states of these two waves are orthogonal, that is, $\mathbf{A}^* \cdot \mathbf{B} = 0$.

   (a) Let δ_a, δ_b be the phase angles defined in Eq. (2.1-5). Show that $\delta_a - \delta_b = \pm\pi$.

   (b) Since δ_a, δ_b are all in the range $-\pi < \delta \le \pi$, show that $\delta_a\delta_b \le 0$.

   (c) Let χ_a, χ_b be the complex numbers representing the polarization states of these two waves. Show that $\chi_a^*\chi_b = -1$.

   (d) Show that the major axes of the polarization ellipses are mutually orthogonal and the ellipticities are of the same magnitude with opposite signs.

2.9. Show that any polarization state can be converted into a linearly polarized state by using a quarter-wave plate. Describe your approach.

2.10. Draw the endpoint of the electric field vector in space for a RHC polarized light at a given point in time. Show that the locus is a left-handed helix. (This has been the traditional way of defining the handedness of circular polarization states. The advantage of this convention is that the handedness of the space helix is independent of the direction of viewing.)

2.11. (a) Show that with a linear combination of two beams of circularly polarized light with equal amplitude and opposite handedness, a beam of plane polarized light results. What determines the plane of polarization of the resultant?

   (b) Show that the combination of two beams of elliptically polarized light is in general another beam of elliptically polarized light. Write down the conditions that the resultant shall be plane polarized, or circularly polarized.

2.12. *Analysis of polarization states*:
   (a) Show that it is impossible to distinguish between a beam of unpolarized light and a beam of circularly polarized light by using linear polarizers. Generally speaking, wave plates and polarizers are needed to determine the degree of polarization.

(b) Given a beam of completely polarized light (say, elliptically polarized), show that the ellipticity and the orientation of the principal axes can be measured by using a single linear polarizer.

2.13. Superposition of orthogonal elliptically polarized states. Use the following basis

$$\mathbf{E}_1 = (a, ib) \qquad \text{and} \qquad \mathbf{E}_2 = (b, -ia)$$

where a and b are real. Let $\mathbf{E} = (\cos \psi, \sin \psi) = c_1 \mathbf{E}_1 + c_2 \mathbf{E}_2$. Find c_1 and c_2.

3

Electromagnetic Propagation in Anisotropic Media

As we mentioned earlier in Chapter 1, liquid crystals exhibit birefringent optical properties by virtue of the orientational order of anisotropic molecules. Like all crystals, the optical properties of LCs depend on the direction of the propagation as well as the polarization state of the lightwaves relative to the orientation of the LC director. The anisotropic optical phenomena in LCs include double refraction, polarization effects, optical rotation, and electrooptical and magnetooptical effects. In addition to LCDs, many optical devices are made of anisotropic materials, such as, sheet polarizers, prism polarizers, wave plates, birefringent filters, and electrooptic modulators. A thorough understanding of light propagation in anisotropic media, especially LCs, is thus important if these phenomena are to be utilized for practical applications such as LCD. The present chapter is devoted to the study of the propagation of electromagnetic radiation in these media. Specifically, we consider only the propagation of plane waves in homogeneous media (e.g., nematic LCs).

3.1. MAXWELL EQUATIONS AND DIELECTRIC TENSOR

We begin by a brief review of the Maxwell equations and the material equations. The most fundamental equations in electrodynamics are Maxwell's equations, which are given in the following in rationalized mks (meter–kilogram–second) units:

$$\nabla \times \mathbf{E} + \frac{\partial \mathbf{B}}{\partial t} = 0 \qquad (3.1\text{-}1)$$

$$\nabla \times \mathbf{H} - \frac{\partial \mathbf{D}}{\partial t} = \mathbf{J} \qquad (3.1\text{-}2)$$

$$\nabla \cdot \mathbf{D} = \rho \qquad (3.1\text{-}3)$$

$$\nabla \cdot \mathbf{B} = 0 \qquad (3.1\text{-}4)$$

In these equations, **E** and **H** are the electric field vector (in volts per meter) and magnetic field vector (in amperes per meter), respectively. These two field vectors are often used to describe an electromagnetic field, especially the propagation of optical waves in various media. The quantities **D** and **B** are called the *electric displacement* (in coulombs per square meter) and the *magnetic induction* (in webers per square meter), respectively. These two quantities are introduced to include the effect of the field on matter. The quantities ρ and **J** are the *electric charge density* (in coulombs per cubic meter) and *current density* (in amperes per square meter), respectively, and may be considered as the sources of the fields **E** and **H**. These four Maxwell equations completely determine the electromagnetic field and are the fundamental equations of the theory of such field, that is, of electrodynamics.

In optics of liquid crystals, one often deals with propagation of electromagnetic radiation in regions of space where both charge density and current density are zero. In fact, if we set ρ = 0 and **J** = 0 in Maxwell's equations, we find that nonzero solutions exist. This means that an electromagnetic field can exist even in the absence of any charges and currents. Electromagnetic fields occurring in media in the absence of charges are called *electromagnetic waves*. In this book we treat the propagation of optical waves in liquid crystals and its application in displays.

Maxwell's equations [Eqs. (3.1-1)–(3.1-4)] consist of 8 scalar equations that relate a total of 12 variables, 3 for each of the 4 vectors **E**, **H**, **D** and **B**. They cannot be solved uniquely unless the relationships between **B** and **H** and that between **E** and **D** are known. To obtain a unique determination of the field vectors, Maxwell's equations must be supplemented by the so-called constitutive equations (or material equations):

$$\mathbf{D} = \varepsilon\mathbf{E} = \varepsilon_0\mathbf{E} + \mathbf{P} \tag{3.1-5}$$

$$\mathbf{B} = \mu\mathbf{H} = \mu_0\mathbf{H} + \mathbf{M} \tag{3.1-6}$$

where the constitutive parameters ε and μ are tensors of rank 2 and are the dielectric tensor (or permittivity tensor) and the permeability tensor, respectively; **P** and **M** are electric and magnetic polarizations, respectively. When an electromagnetic field is present in matter, the electric field can perturb the motion of electrons and produce a dipole polarization **P** per unit volume. Similarly, the magnetic field can also induce a magnetization **M** in materials having a permeability that is different from μ_0. The constant ε_0 is called the *permittivity of vacuum* and has a value of 8.854×10^{-12} F/m. The constant μ_0 is known as the *permeability of vacuum*. It has, by definition, the exact value of $4\pi \times 10^{-7}$ H/m. If the medium is isotropic, both ε and μ tensors reduce to scalars. In many cases, the quantities ε and μ can be assumed to be independent of the field strengths. However, if the fields are sufficiently strong, such as obtained, for example, by focusing a laser beam or applying a strong dc (direct-current) electric field to an electrooptic crystal, the dependence of these quantities on **E** and **H** must be considered.

In anisotropic media such as nematic LCs, calcite, quartz, and lithium niobate, plane-wave propagation is determined by the dielectric tensor ε_{ij} that links the displacement vector and the electric field vector

$$D_i = \varepsilon_{ij}E_j \qquad (3.1\text{-}7)$$

where the convention of summation over repeated indices is observed. In nonmagnetic and transparent materials, this tensor is real and symmetric:

$$\varepsilon_{ij} = \varepsilon_{ji} \qquad (3.1\text{-}8)$$

The magnitude of these nine tensor elements depends, of course, on the choice of the x, y, and z axes relative to the crystal structure. Because of its real and symmetric nature, it is always possible to find three mutually orthogonal axes in such a way that the off-diagonal elements vanish, leaving

$$\varepsilon = \varepsilon_0 \begin{pmatrix} n_x^2 & 0 & 0 \\ 0 & n_y^2 & 0 \\ 0 & 0 & n_z^2 \end{pmatrix} = \begin{pmatrix} \varepsilon_x & 0 & 0 \\ 0 & \varepsilon_y & 0 \\ 0 & 0 & \varepsilon_z \end{pmatrix} \qquad (3.1\text{-}9)$$

where $\varepsilon_x, \varepsilon_y,$ and ε_z are the principal dielectric constants and $n_x, n_y,$ and n_z are the principal indices of refraction. These directions (x, y, and z) are called the *principal dielectric axes* of the crystal. According to Eqs. (3.1-7) and (3.1-9), a plane wave propagating along the z axis can have two phase velocities, depending on its state of polarization. Specifically, the phase velocity is c/n_x for x-polarized light and c/n_y for y-polarized light. Generally speaking, there are two normal modes of polarization for each direction of propagation.

It is important to note that the dielectric tensor (or dielectric constants) is a function of the frequency (or wavelength) of the electromagnetic field. This is known as *dispersion*. In the regime of optical waves, the frequencies are in the range of 10^{14} s^{-1}, we often use the refractive indices to describe the propagation of waves in optical media. In the low frequency regime (0 Hz–1 GHz), the dielectric constants are responsible for the effect of applied electric field on the orientation of the LC molecules. It is the same dielectric tensor, but at different frequency regimes, that is affecting the optical transmission as well as the electric-field-induced reorientation of the LCs.

3.2. PLANE WAVES IN HOMOGENEOUS MEDIA AND NORMAL SURFACE

To study such propagation along a general direction, we assume a monochromatic plane wave with an electric field vector

$$\mathbf{E} \exp[i(\omega t - \mathbf{k} \cdot \mathbf{r})] \qquad (3.2\text{-}1)$$

and a magnetic field vector

$$\mathbf{H} \exp[i(\omega t - \mathbf{k}\cdot\mathbf{r})] \tag{3.2-2}$$

where $\mathbf{k}$ is the wavevector $\mathbf{k} = (\omega/c)n\mathbf{s}$, with $\mathbf{s}$ as a unit vector in the direction of propagation. The phase velocity c/n, or equivalently the refractive index n, is to be determined. Substituting $\mathbf{E}$ and $\mathbf{H}$ from Eqs. (3.2-1) and (3.2-2), respectively, into Maxwell's equations (3.1-1) and (3.1-2) gives

$$\mathbf{k} \times \mathbf{E} = \omega\mu\mathbf{H} \tag{3.2-3}$$

$$\mathbf{k} \times \mathbf{H} = -\omega\varepsilon\mathbf{E} = -\omega\mathbf{D} \tag{3.2-4}$$

By eliminating $\mathbf{H}$ from Eqs. (3.2-3) and (3.2-4), we obtain

$$\mathbf{k} \times (\mathbf{k} \times \mathbf{E}) + \omega^2\mu\varepsilon\mathbf{E} = 0 \tag{3.2-5}$$

This equation will now be used to solve for the eigenvectors $\mathbf{E}$ and the corresponding eigenvalues n.

In the principal coordinate system, the dielectric tensor ε is given by Eq. (3.1-9). Equation (3.2-5) can be written as

$$\begin{pmatrix} \omega^2\mu\varepsilon_x - k_y^2 - k_z^2 & k_x k_y & k_x k_z \\ k_y k_x & \omega^2\mu\varepsilon_y - k_x^2 - k_z^2 & k_y k_z \\ k_z k_x & k_z k_y & \omega^2\mu\varepsilon_z - k_x^2 - k_y^2 \end{pmatrix} \begin{pmatrix} E_x \\ E_y \\ E_z \end{pmatrix} = 0 \tag{3.2-6}$$

where we recall that $\varepsilon_x = \varepsilon_0 n_x^2$, $\varepsilon_y = \varepsilon_0 n_y^2$, and $\varepsilon_z = \varepsilon_0 n_z^2$.

For nontrivial solutions to exist, the determinant of the matrix in Eq. (3.2-6) must vanish. This leads to a relation between ω and $\mathbf{k}$:

$$\det \begin{vmatrix} \omega^2\mu\varepsilon_x - k_y^2 - k_z^2 & k_x k_y & k_x k_z \\ k_y k_x & \omega^2\mu\varepsilon_y - k_x^2 - k_z^2 & k_y k_z \\ k_z k_x & k_z k_y & \omega^2\mu\varepsilon_z - k_x^2 - k_y^2 \end{vmatrix} = 0 \tag{3.2-7}$$

At a given frequency ω, this equation represents a three-dimensional surface in $\mathbf{k}$ space (momentum space). This surface, known as the *normal surface*, consists of two shells. These two shells, in general, have four points in common. The two lines that go through the origin and these points are known as the optical axes. Figure 3.1 shows one octant of a general normal surface. In the case when

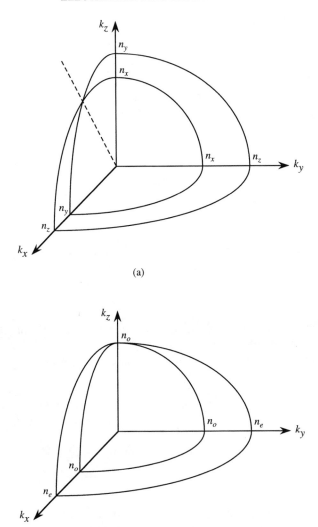

Figure 3.1. (*a*) One octant of a normal surface in momentum space with $n_x < n_y < n_z$. k_x, k_y, and k_z are in units of ω/c. (*b*) One octant of a normal surface in momentum space with $n_o = n_x = n_y < n_z = n_e$.

$n_x = n_y < n_z$, the normal surface consists of a sphere and an ellipsoid. Given a direction of propagation, there are in general two k values that are the intersections of the direction of propagation **s** and the normal surface. These two k values correspond to two different phase velocities (ω/k) of the waves propagating along the chosen direction. The directions of the electric field vector associated with these propagations can also be obtained from Eq. (3.2-6) and are given by

$$
\begin{pmatrix}
\dfrac{k_x}{k^2 - \omega^2 \mu \varepsilon_x} \\[2ex]
\dfrac{k_y}{k^2 - \omega^2 \mu \varepsilon_y} \\[2ex]
\dfrac{k_z}{k^2 - \omega^2 \mu \varepsilon_z}
\end{pmatrix}
\tag{3.2-8}
$$

provided the denominators do not vanish.

For propagation in the direction of the optic axes, there is only one value of k and consequently only one phase velocity. There are, however, two independent directions of polarization.

Equations (3.2-7) and (3.2-8) are often written in terms of the direction cosines of the wavevector. By using the relation $\mathbf{k} = (\omega / c)n\mathbf{s}$ for the plane wave given by Eq. (3.2-1), we can express Eqs. (3.2-7) and (3.2-8) as

$$
\frac{s_x^2}{n^2 - n_x^2} + \frac{s_y^2}{n^2 - n_y^2} + \frac{s_z^2}{n^2 - n_z^2} = \frac{1}{n^2}
\tag{3.2-9}
$$

and

$$
\begin{pmatrix}
\dfrac{s_x}{n^2 - n_x^2} \\[2ex]
\dfrac{s_y}{n^2 - n_y^2} \\[2ex]
\dfrac{s_z}{n^2 - n_z^2}
\end{pmatrix}
\tag{3.2-10}
$$

respectively, where we have used $\varepsilon_x = \varepsilon_0 n_x^2$, $\varepsilon_y = \varepsilon_0 n_y^2$, and $\varepsilon_z = \varepsilon_0 n_z^2$.

Equation (3.2-9) is known as *Fresnel's equation of wavenormals* and can be solved for the eigenvalues of index of refraction, and Eq. (3.2-10) gives the directions of polarization. Note that equation (3.2-9) is a quadratic equation in n^2. Therefore, for each direction of propagation (a set of s_x, s_y, s_z), two solutions for n^2 can be obtained from Eq. (3.2-9). To complete the solution of the problem, we use the values of n^2, one at a time, in Eq. (3.2-10). This gives us the polarizations (electric field vectors) of these waves. It can be seen that in a nonabsorbing medium these normal modes are linearly polarized since all the components are real in Eq. (3.2-10). Let $\mathbf{E}_1, \mathbf{E}_2$ be the electric field vectors and $\mathbf{D}_1, \mathbf{D}_2$ be the displacement vectors of the linearly polarized normal modes associated with n_1^2 and n_2^2, respectively. Maxwell's equation $\nabla \cdot \mathbf{D} = 0$ requires that $\mathbf{D}_1, \mathbf{D}_2$ be orthogonal to $\mathbf{s}$. Since $\mathbf{D}_1 \cdot \mathbf{D}_2 = 0$ (the proof of this orthogonal relation is left as a problem for students), the three vectors $\mathbf{D}_1, \mathbf{D}_2$ and $\mathbf{s}$ form an orthogonal triad. According to Eqs. (3.2-3) and (3.2-4), $\mathbf{D}$ and $\mathbf{H}$ are both perpendicular to the direction of propagation $\mathbf{s}$. Consequently, the direction of energy flow as given by the Poynting vector $\mathbf{E} \times \mathbf{H}$ is, in general, not collinear with the direction of propagation $\mathbf{s}$. Since $\mathbf{D}$, $\mathbf{E}$, and $\mathbf{k}$ are all orthogonal to $\mathbf{H}$, they must lie in the same plane.

Orthogonality of Normal Modes (Eigenmodes)

Using $\mathbf{k} = (\omega/c)n\mathbf{s}$, Eqs. (3.2-3) and (3.2-4) can be written as

$$\mathbf{D} = -\frac{n}{c}\mathbf{s} \times \mathbf{H} \tag{3.2-11}$$

and

$$\mathbf{H} = \frac{n}{\mu c}\mathbf{s} \times \mathbf{E} \tag{3.2-12}$$

By substituting Eq. (3.2-12) for $\mathbf{H}$ in Eq. (3.2-11) and using the vector identity $\mathbf{A} \times (\mathbf{B} \times \mathbf{C}) = \mathbf{B}(\mathbf{A} \cdot \mathbf{C}) - \mathbf{C}(\mathbf{A} \cdot \mathbf{B})$, we obtain the following expression:

$$\mathbf{D} = -\frac{n^2}{c^2\mu}\mathbf{s} \times (\mathbf{s} \times \mathbf{E}) = \frac{n^2}{c^2\mu}[\mathbf{E} - \mathbf{s}(\mathbf{s} \cdot \mathbf{E})]$$

$$= \frac{n^2}{c^2\mu}\mathbf{E}_{\text{transverse}} \tag{3.2-13}$$

and since $\mathbf{s} \cdot \mathbf{D} = 0$ and $n^2/c^2\mu = n^2\varepsilon_0$, we obtain

$$D^2 = \frac{n^2}{c^2\mu}\mathbf{E} \cdot \mathbf{D} = n^2\varepsilon_0\mathbf{E} \cdot \mathbf{D} \tag{3.2-14}$$

In other words, $\mathbf{D}$, $\mathbf{E}$, and $\mathbf{s}$ all lie in the same plane. It can be shown that these field vectors satisfy the following relations (see Problems 3.1):

$$\mathbf{D}_1 \cdot \mathbf{D}_2 = 0$$
$$\mathbf{D}_1 \cdot \mathbf{E}_2 = 0$$
$$\mathbf{D}_2 \cdot \mathbf{E}_1 = 0 \tag{3.2-15}$$
$$\mathbf{s} \cdot \mathbf{D}_1 = \mathbf{s} \cdot \mathbf{D}_2 = 0$$

$\mathbf{E}_1$ and $\mathbf{E}_2$ are in general not orthogonal. The orthogonality relation of the eigenmodes of propagation is often written as

$$\mathbf{s} \cdot (\mathbf{E}_1 \times \mathbf{H}_2) = 0. \qquad (3.2\text{-}16)$$

This latter relation shows that the power flow in an anisotropic medium along the direction of propagation is the sum of the power carried by each mode individually.

We now derive the orthogonality relation (3.2-16) between the two eigenmodes of propagation along a given directions $\mathbf{s}$. Using Eqs. (3.2-1) and (3.2-2) for the field vectors and the Lorentz reciprocity theorem, we obtain

$$\mathbf{s} \cdot (\mathbf{E}_1 \times \mathbf{H}_2) = \mathbf{s} \cdot (\mathbf{E}_2 \times \mathbf{H}_1). \qquad (3.2\text{-}17)$$

If we substitute Eq. (3.2-12) for $\mathbf{H}_1$ and $\mathbf{H}_2$ in Eq. (3.2-17), this becomes

$$\frac{n_2}{\mu c} \mathbf{s} \cdot [\mathbf{E}_1 \times (\mathbf{s} \times \mathbf{E}_2)] = \frac{n_1}{\mu c} \mathbf{s} \cdot [\mathbf{E}_2 \times (\mathbf{s} \times \mathbf{E}_1)] \qquad (3.2\text{-}18)$$

This equation can be further simplified by using the identity

$$\mathbf{A} \cdot (\mathbf{B} \times \mathbf{C}) = \mathbf{C} \cdot (\mathbf{A} \times \mathbf{B})$$

and becomes

$$\frac{n_2}{\mu c} (\mathbf{s} \times \mathbf{E}_1) \cdot (\mathbf{s} \times \mathbf{E}_2) = \frac{n_1}{\mu c} (\mathbf{s} \times \mathbf{E}_1) \cdot (\mathbf{s} \times \mathbf{E}_2) \qquad (3.2\text{-}19)$$

Since this equation must hold for any arbitrary direction of propagation $\mathbf{s}$ with $n_1 \neq n_2$, it can be satisfied only when both sides vanish. This proves, according to Eq. (3.2-12)

$$\mathbf{s} \cdot (\mathbf{E}_1 \times \mathbf{H}_2) = \mathbf{s} \cdot (\mathbf{E}_2 \times \mathbf{H}_1) = 0 \qquad (3.2\text{-}20)$$

To summarize, along an arbitrary direction of propagation $\mathbf{s}$, there can exist two independent plane-wave, linearly polarized propagation modes. These modes have phase velocities $\pm(c/n_1)$ and $\pm(c/n_2)$, where n_1^2 and n_2^2 are the two solutions of Fresnel's equation (3.2-9). The electric field vectors of these two normal modes are given by Eq. (3.2-8) or (3.2-10).

Although Eq. (3.2-8) provides a general explicit expression for the polarization states ($\mathbf{E}$ vector) of the normal modes in a general anisotropic medium, one must be careful in applying the equation for propagation along the principal axes or principal planes when the denominators in Eq. (3.2-8) can become zero. For propagation along these special directions, it is often easier to obtain the normal modes directly from the wave equation (3.2-6). To illustrate this, we consider the following special cases:

3.2.1. k in the xy Plane

For propagation along the principal xy plane, $k_z = 0$. The marix in Eq. (3.2-6) is block diagonalized. Thus, one of the normal modes is polarized along the z axis with its $\mathbf{E}$ vector and wavenumber given by

$$\mathbf{E}_1 = \begin{pmatrix} 0 \\ 0 \\ 1 \end{pmatrix} \tag{3.2-21}$$

$$k_1 = \frac{n_z \omega}{c} \tag{3.2-22}$$

respectively. The other normal mode is polarized in the xy plane. The $\mathbf{E}$ vector and the wavenumber can be obtained from the 2×2 portion of the matrix equation (3.2-6). A simple examination of the equation yields

$$\mathbf{E}_2 = \begin{pmatrix} \dfrac{s_x}{n^2 - n_x^2} \\ \dfrac{s_y}{n^2 - n_y^2} \\ 0 \end{pmatrix} \tag{3.2-23}$$

$$k_2 = \frac{\omega}{c} \left[\frac{n_x^2 n_y^2}{n_x^2 \cos^2\theta + n_y^2 \sin^2\theta} \right]^{1/2} = n\frac{\omega}{c} \tag{3.2-24}$$

where θ is the angle of the wavevector measured from the x axis. Substituting Eq. (3.2-24) for n in Eq (3.2-23) yields

$$\mathbf{E}_2 = \begin{pmatrix} \dfrac{s_y}{n_x^2} \\ \dfrac{-s_x}{n_y^2} \\ 0 \end{pmatrix} = \begin{pmatrix} \dfrac{\sin\theta}{n_x^2} \\ \dfrac{-\cos\theta}{n_y^2} \\ 0 \end{pmatrix} . \tag{3.2-25}$$

We notice that the electric field vector is in the xy plane and is tangent to the ellipse that is the intersection of the normal surface with the xy plane.

3.2.2. k in the yz Plane

Similarly, for propagation in the principal yz plane, $k_x = 0$. The matrix in Eq. (3.2-6) is block diagonalized. Thus, one of the normal modes is polarized along the x axis with its **E** vector and wavenumber given by

$$\mathbf{E}_1 = \begin{pmatrix} 1 \\ 0 \\ 0 \end{pmatrix} \tag{3.2-26}$$

$$k_1 = \frac{n_x \omega}{c} \tag{3.2-27}$$

respectively. The other normal mode is polarized in the yz plane. The **E** vector and the wavenumber can be obtained from the 2×2 portion of the matrix equation (3.2-6). A simple examination of the equation yields

$$\mathbf{E}_2 = \begin{pmatrix} 0 \\ \dfrac{s_y}{n^2 - n_y^2} \\ \dfrac{s_z}{n^2 - n_z^2} \end{pmatrix} \tag{3.2-28}$$

$$k_2 = \frac{\omega}{c} \left[\frac{n_y^2 n_z^2}{n_y^2 \cos^2\theta + n_z^2 \sin^2\theta} \right]^{1/2} = n\frac{\omega}{c} \tag{3.2-29}$$

where θ is the angle of the wavevector measured from the y axis. Substituting Eq. (3.2-29) for n in Eq. (3.2-28) yields

$$\mathbf{E}_2 = \begin{pmatrix} 0 \\ \dfrac{s_z}{n_y^2} \\ \dfrac{-s_y}{n_z^2} \end{pmatrix} = \begin{pmatrix} 0 \\ \dfrac{\sin\theta}{n_y^2} \\ \dfrac{-\cos\theta}{n_z^2} \end{pmatrix} \tag{3.2-30}$$

We notice that the electric field vector is in the yz plane and is tangent to the ellipse that is the intersection of the normal surface with the yz plane.

3.2.3. k in the zx Plane

Similarly, for propagation in the principal zx plane, $k_y = 0$. Thus, one of the normal modes is polarized along the y axis with its **E** vector and wavenumber given by

$$\mathbf{E}_1 = \begin{pmatrix} 0 \\ 1 \\ 0 \end{pmatrix} \tag{3.2-31}$$

$$k_1 = \frac{n_y \omega}{c} \tag{3.2-32}$$

respectively. The other normal mode is polarized in the zx plane. The **E** vector and the wavenumber can be obtained as

$$\mathbf{E}_2 = \begin{pmatrix} \dfrac{s_x}{n^2 - n_x^2} \\ 0 \\ \dfrac{s_z}{n^2 - n_z^2} \end{pmatrix} \tag{3.2-33}$$

$$k_2 = \frac{\omega}{c} \left[\frac{n_x^2 n_z^2}{n_x^2 \cos^2\theta + n_z^2 \sin^2\theta} \right]^{1/2} = n \frac{\omega}{c} \tag{3.2-34}$$

where θ is the angle of the wavevector measured from the x axis. Substituting Eq. (3.2-34) for n in Eq. (3.2-33) yields

$$\mathbf{E}_2 = \begin{pmatrix} \dfrac{s_z}{n_x^2} \\ 0 \\ \dfrac{-s_x}{n_z^2} \end{pmatrix} = \begin{pmatrix} \dfrac{\sin\theta}{n_x^2} \\ 0 \\ \dfrac{-\cos\theta}{n_z^2} \end{pmatrix} \tag{3.2-35}$$

We notice that the electric field vector is in the zx plane and is tangent to the ellipse that is the intersection of the normal surface with the zx plane.

3.2.4. Classification of Media

We have shown above that the normal surface contains a good deal of information about the wave propagation in anisotropic media. The normal surface is uniquely determined by the principal indices of refraction n_x, n_y, n_z. In the general case when the three principal indices n_x, n_y, n_z are all different, there are two optical axes. In this case, the medium is said to be biaxial. In many optical materials (e.g., nematic liquid crystals) it happens that two of the principal indices are equal, in which case the equation for the normal surface [Eq. (3.2-7) or (3.2-9)] can be factored according to

$$\left(\frac{k_x^2 + k_y^2}{n_e^2} + \frac{k_z^2}{n_o^2} - \frac{\omega^2}{c^2}\right)\left(\frac{k^2}{n_o^2} - \frac{\omega^2}{c^2}\right) = 0 \qquad (3.2\text{-}36)$$

where $n_o^2 = \varepsilon_x/\varepsilon_0 = \varepsilon_y/\varepsilon_0, n_e^2 = \varepsilon_z/\varepsilon_0$.

The normal surface in this case consists of a sphere and an ellipsoid of revolution (see Fig. 3.1b). These two sheets of the normal surface touch at two points on the z axis. The z axis is therefore the only optic axis, and the medium is said to be uniaxial. If all the three principal indices are equal, the two sheets of normal surface degenerate to a single sphere, and the medium is optically isotropic.

It is obvious that the optical symmetry of the materials is closely related to the point group of the materials. For example, in a cubic material, the three principal axes are physically equivalent. Therefore, we expect a cubic material to be optically isotropic. Table 3.1 lists the optical symmetry of solid crystals, liquid crystals, and the corresponding dielectric tensors.

In a biaxial medium, the principal coordinate axes are labeled in such a way that the three principal indices are in the following order:

$$n_x < n_y < n_z \qquad (3.2\text{-}37)$$

In this convention the optical axes lie in the xz plane. The cross section of the normal surfaces with the xz plane is shown in Figure 3.2a. In a uniaxial medium, the index of refraction that corresponds to the two equal elements, $n_o^2 = \varepsilon_x/\varepsilon_0 = \varepsilon_y/\varepsilon_0$, is called the *ordinary index* n_o; the other index, corresponding to ε_z, is called the *extraordinary index* n_e. If $n_o < n_e$, the medium is said to be positive, whereas if $n_o > n_e$, it is said to be negative. Most LCs with rodlike molecules are positive uniaxial media with $n_o < n_e$. The intersection of the normal surfaces with the xz plane is again shown in Figure 3.2b,c. The optic axis corresponds to the principal axis, which has a unique index of refraction. Table 3.2a lists some examples of solid crystals with their indices of refraction. Table 3.2b lists some examples of liquid crystals with their indices of refraction.

3.2.5. Power Flow in Anisotropic Media

As discussed earlier, the Poynting power flow $\mathbf{E} \times \mathbf{H}$ is, in general, not collinear with the direction of popagation $\mathbf{s}$. Consider a general propagation of a beam of

Table 3.1a.   Solid Crystals

Optical Symmetry	Crystal System	Point Groups	Dielectric Tensor
Isotropic	Cubic	$\overline{4}3m$ 432 $m3$ 23 $m3m$	$\varepsilon = \varepsilon_0 \begin{pmatrix} n^2 & 0 & 0 \\ 0 & n^2 & 0 \\ 0 & 0 & n^2 \end{pmatrix}$
Uniaxial	Tetragonal	4 $\overline{4}$ $4/m$ 422 $4mm$ $\overline{4}2m$ $4/mmm$	
	Hexagonal	6 $\overline{6}$ $6/m$ 622 $6mm$ $\overline{6}m2$ $6/mmm$	$\varepsilon = \varepsilon_0 \begin{pmatrix} n_o^2 & 0 & 0 \\ 0 & n_o^2 & 0 \\ 0 & 0 & n_e^2 \end{pmatrix}$
	Trigonal	3 $\overline{3}$ 32 $3m$ $\overline{3}m$	
Biaxial	Triclinic	1 $\overline{1}$	
	Monoclinic	2 m $2/m$	$\varepsilon = \varepsilon_0 \begin{pmatrix} n_x^2 & 0 & 0 \\ 0 & n_y^2 & 0 \\ 0 & 0 & n_z^2 \end{pmatrix}$
	Orthorhombic	222 $2mm$ mmm	

light in an anisotropic medium. Given a direction of propagation, the electric field vector can be written

$$\mathbf{E} = c_1\mathbf{E}_1 + c_2\mathbf{E}_2 \tag{3.2-38}$$

where $\mathbf{E}_1$ and $\mathbf{E}_2$ are the normal modes of propagation and c_1, c_2 are constants.

Table 3.1b. Liquid Crystals

Optical Symmetry	Liquid Crystal Phase	Dielectric Tensor
Isotropic	Isotropic Smectic D (cubic)	$\varepsilon = \varepsilon_0 \begin{pmatrix} n^2 & 0 & 0 \\ 0 & n^2 & 0 \\ 0 & 0 & n^2 \end{pmatrix}$
Uniaxial	Nematic Smectic A Smectic B	$\varepsilon = \varepsilon_0 \begin{pmatrix} n_o^2 & 0 & 0 \\ 0 & n_o^2 & 0 \\ 0 & 0 & n_e^2 \end{pmatrix}$
Biaxial	Smectic C (monoclinic) Smectic E (orthorhombic) Smectic F (monoclinic) Smectic G (monoclinic) Smectic H (monoclinic) Smectic I (monoclinic) Smectic J (monoclinic) Smectic K (monoclinic) Biaxial nematic N_6	$\varepsilon = \varepsilon_0 \begin{pmatrix} n_x^2 & 0 & 0 \\ 0 & n_y^2 & 0 \\ 0 & 0 & n_z^2 \end{pmatrix}$

Note: Only homogenous liquid crystals are listed. Chiral phases and twisted nematic phases are considered inhomogenous media.

To obtain the power flow, we must have the magnetic field, which is given by

$$\mathbf{H} = \frac{n_1}{\mu c}\mathbf{s} \times \mathbf{E}_1 + \frac{n_2}{\mu c}\mathbf{s} \times \mathbf{E}_2 \qquad (3.2\text{-}39)$$

where n_1, n_2 are the eigen refractive indices. The Poynting vector, which represents the power flow, is given by

$$\mathbf{S} = \mathbf{E} \times \mathbf{H} \qquad (3.2\text{-}40)$$

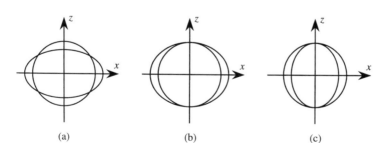

Figure 3.2. Intersection of the normal surface with the xz plane for (*a*) biaxial media, (*b*) positive uniaxial media, and (*c*) negative uniaxial media.

Table 3.2a.   Refractive Indices of Some Typical Solid Crystals

Isotropic		1.392		
	Fluorite, CaF_2	1.392		
	Sodium chloride, NaCl	1.544		
	Diamond, C	2.417		
	CdTe	2.69		
	GaAs	3.40		
	Ge	3.40		
	InP	3.61		
	GaP	3.73		
Uniaxial		n_o		n_e
Positive	Ice, H_2O	1.309		1.310
	MgF_2	1.378		1.390
	Quartz, SiO_2	1.544		1.553
	Beryllium oxide, BeO	1.717		1.732
	Zircon, $ZrSiO_4$	1.923		1.968
	SnO_2	2.01		2.10
	ZnS	2.354		2.358
	CdS	2.483		2.511
	Rutile, TiO_2	2.616		2.903
Negative	KDP, KH_2PO_4	1.507		1.467
	ADP, $(NH_4)H_2PO_4$	1.522		1.478
	Beryl, $Be_3Al_2(SiO_3)_6$	1.598		1.590
	Sodium nitrate, $NaNO_3$	1.587		1.366
	Calcite, $CaCO_3$	1.658		1.486
	Tourmaline	1.638		1.618
	Sapphire, Al_2O_3	1.768		1.760
	Lithium niobate, $LiNbO_3$	2.300		2.208
	Barium titanate, $BaTiO_3$	2.416		2.364
	Proustite, Ag_3AsS_3	3.019		2.739
Biaxial		n_x	n_y	n_z
	Gypsum, $CaSO_4 \cdot 2H_2O$	1.520	1.523	1.530
	Feldspar	1.522	1.526	1.530
	Mica	1.552	1.582	1.588
	Topaz, $Al_2(SiO_4)(OH,F)_2$	1.619	1.620	1.627
	Sodium nitrite, $NaNO_2$	1.344	1.411	1.651
	$YAlO_3$	1.923	1.938	1.947
	SbSI	2.7	3.2	3.8

Note: The refractive indices of most materials depend on the wavelength (dispersion). The listed numbers are typical values.

Table 3.2b.   Refractive Indices of Some Typical Nematic Liquid Crystals [1]

	$T(°C)$	Wavelength (nm)	n_e	n_o
MBBA	25	467.8	1.837	1.575
(Schiff base)		480	1.825	1.57
		508.6	1.802	1.563
		589	1.764	1.549
		643.8	1.749	1.544
RO-TN-601	25	467.8	1.718	1.515
		480	1.7116	1.5131
		508.6	1.7041	1.5098
		546.1	1.6937	1.506
Phase 4 Licristal (EM-Merck), azoxy				
	25	546.1	1.856	1.5606
		589.3	1.8291	1.553
K15 (5CB) (BDH, Ltd.), Cyanoalkylbiphenyl				
	25	436		
		509	1.7411	1.5443
		577	1.7201	1.5353
		644	1.7072	1.5292
	30	436	1.7648	1.5624
		509	1.725	1.5481
		577	1.7044	1.539
		644	1.6926	1.5323
K21 (7CB) (BDH, Ltd.), cyanoalkylbiphenyl				
	37	436	1.736	1.5443
		509	1.6998	1.5329
		577	1.6815	1.5248
		644	1.6702	1.5186
	41	436	1.714	1.5517
		509	1.6805	1.5389
		577	1.6632	1.5305
		644	1.6526	1.5236
M15(5OCB) (BDH, Ltd.), cyanoalkoxybiphenyl				
	50	589	1.7187	1.5259
M21(7OCB) (BDH, Ltd.), cyanoalkoxybiphenyl				
	60	589	1.6846	1.5139
M24(8OCB) (BDH, Ltd.), cyanoalkoxybiphenyl				
	70	589	1.6639	1.5078
E5 (BDH, Ltd.), K15−45% + K21−24% + M15−10% + M21−9% + M24−12%				
	20	436	1.8038	1.5447
		509	1.7584	1.5303
		577	1.736	1.5228
		644	1.7222	1.5169

Table 3.2b.   *(Continued)*

	$T(°C)$	Wavelength (nm)	n_e	n_o
	30	436	1.7856	1.5455
		509	1.7417	1.5318
		577	1.7208	1.5232
		644	1.7071	1.5172
E7 (BDH, Ltd.), K15−47% + K21−25% + M24−18% + T15−10%				
	20	436	1.8208	1.544
		509	1.7737	1.5311
		577	1.75	1.5231
		644	1.7354	1.5175
E8 (BDH, Ltd.), 3 cyanobiphenyls and 1 cyanoterphenyl				
	30	589	1.76	1.52
N10 (Merck), azoxy compounds + aromatic esters				
	30	589	1.77	1.56
PCH-1132 (Merck), 3 cyanophenyl-cyclohexanes + 1 cyanobiphenyl-cyclohexane				
	30	589	1.62	1.49
PCH-5	30.3	589	1.604	1.4875
		633	1.6001	1.4851
	38.5	589	1.5956	1.4863
		633	1.5919	1.484
	46.7	589	1.5849	1.486
		633	1.5812	1.4836
ZLI-1646	20	589	1.558	1.478
ZLI-4792	20	589	1.573	1.479

We note that, the Poynting vector, is in general, not a simple sum of the individual Poynting vectors. In other words

$$\mathbf{S} \neq \mathbf{E}_1 \times \mathbf{H}_1 + \mathbf{E}_2 \times \mathbf{H}_2 \qquad (3.2\text{-}41)$$

Using the orthogonality of the normal modes Eq. (3.2-16), we can write

$$\mathbf{S} \cdot \mathbf{s} = \mathbf{s} \cdot (\mathbf{E}_1 \times \mathbf{H}_1) + \mathbf{s} \cdot (\mathbf{E}_2 \times \mathbf{H}_2) \qquad (3.2\text{-}42)$$

We note that the total power flow along the direction of propagation is a sum of the individual mode power along the same direction. This is known as the *power orthogonality theorem*.

3.3. LIGHT PROPAGATION IN UNIAXIAL MEDIA

Most of LC materials with rodlike molecules are optically uniaxial when an orientational order exists. In addition, most LCDs involve the use of nematic LCs. A planar nematic slab is a good example of homogeneous uniaxial LC. Thus, the propagation of optic waves in uniaxially anisotropic media deserves special attention. By putting

$$\varepsilon_x = \varepsilon_y = \varepsilon_0 n_o^2, \qquad \varepsilon_z = \varepsilon_0 n_e^2 \qquad (3.3\text{-}1)$$

the normal surface becomes

$$\left(\frac{k_x^2 + k_y^2}{n_e^2} + \frac{k_z^2}{n_o^2} - \frac{\omega^2}{c^2}\right)\left(\frac{k^2}{n_o^2} - \frac{\omega^2}{c^2}\right) = 0 \qquad (3.3\text{-}2)$$

This is identical to Eq. (3.2-36). We note that the normal surface consists of two parts. The sphere gives the relation between ω and $\mathbf{k}$ of the ordinary (O) wave. The ellipsoid of revolution gives the similar relation for the extraordinary (E) wave. These two surfaces touch at two points on the z axis (see Figure 3.1(b)). The eigen refractive indices associated with these two modes of propagation are given by

$$O \text{ wave}: \qquad n = n_o \qquad (3.3\text{-}3)$$

$$E \text{ wave}: \qquad \frac{1}{n^2} = \frac{\cos^2\theta}{n_o^2} + \frac{\sin^2\theta}{n_e^2} \qquad (3.3\text{-}4)$$

where θ is the angle between the direction of propagation and the optic axis (the crystal c axis). For propagation along the optic axis (c axis), the eigen refractive indices of both modes are n_o, according to Eqs. (3.3-3) and (3.3-4).

The electric field vector of the O wave cannot be obtained directly from Eq. (3.2-10) because of the vanishing denominators. It can be easily obtained from Eq. (3.2-6). By using $\varepsilon_x = \varepsilon_y = \varepsilon_0 n_o^2$, $\varepsilon_z = \varepsilon_0 n_e^2$, and $\mathbf{k}_o = (\omega/c)n_o\mathbf{s}$, Eq. (3.2-6) can be written as

$$\begin{pmatrix} s_x^2 & s_x s_y & s_x s_z \\ s_y s_x & s_y^2 & s_y s_z \\ s_z s_x & s_z s_y & \left(\frac{n_e}{n_o}\right)^2 - (s_x^2 + s_y^2) \end{pmatrix}\begin{pmatrix} E_x \\ E_y \\ E_z \end{pmatrix} = 0 \qquad (3.3\text{-}5)$$

A simple inspection of this equation yields the following direction of polarization:

$$O \text{ wave}: \qquad \mathbf{E} = \begin{pmatrix} s_y \\ -s_x \\ 0 \end{pmatrix} \qquad (3.3\text{-}6)$$

where we recall that s_x, s_y, s_z are directional cosines of the direction of propagation. The electric field vector of the E wave can be obtained from Eq. (3.2-10) and is given by

$$E \text{ wave}: \qquad \mathbf{E} = \begin{pmatrix} \dfrac{s_x}{n^2 - n_o^2} \\[2mm] \dfrac{s_y}{n^2 - n_o^2} \\[2mm] \dfrac{s_z}{n^2 - n_e^2} \end{pmatrix} \qquad (3.3\text{-}7)$$

where n is as given by Eq. (3.3-4). The corresponding wavevector is $\mathbf{k}_e = (\omega/c)n\mathbf{s}$.

Note that the electric field vector of the O wave is perpendicular to the plane formed by the wave vector $\mathbf{k}_o$ and the c axis, whereas the electric field vector of the E wave is not exactly perpendicular to the wavevector $\mathbf{k}_e$. However, the deviation from $90°$ is very small. This small angle between the field vectors $\mathbf{E}$ and $\mathbf{D}$ is also the angle between the phase velocity and the group velocity (see Problem 3.4). Therefore, for practical purposes, we may assume that the electric field is transverse to the direction of propagation. The displacement vectors $\mathbf{D}$ of the normal modes are exactly perpendicular to the wavevectors $\mathbf{k}_o$ and $\mathbf{k}_e$, respectively, and can be written as

$$O \text{ wave}: \qquad \mathbf{D}_o = \frac{\mathbf{k}_o \times \mathbf{c}}{|\mathbf{k}_o \times \mathbf{c}|} \qquad (3.3\text{-}8)$$

$$E \text{ wave}: \qquad \mathbf{D}_e = \frac{\mathbf{D}_o \times \mathbf{k}_e}{|\mathbf{D}_o \times \mathbf{k}_e|} \qquad (3.3\text{-}9)$$

where $\mathbf{c}$ is a unit vector along the optic axis.

Let (θ, ϕ) be the angle of propagation in spherical coordinate. The unit vector $\mathbf{s}$ can be written

$$\mathbf{s} = \begin{pmatrix} \sin\theta \cos\phi \\ \sin\theta \sin\phi \\ \cos\theta \end{pmatrix} \qquad (3.3\text{-}10)$$

Using Eq. (3.3-10), the normal modes for $\mathbf{E}$ can be written

$$O \text{ wave}: \quad \mathbf{E}_o = \begin{pmatrix} \sin\phi \\ -\cos\phi \\ 0 \end{pmatrix} \tag{3.3-11a}$$

$$E \text{ wave}: \quad \mathbf{E}_e = \begin{pmatrix} n_e^2 \cos\theta \cos\phi \\ n_e^2 \cos\theta \sin\phi \\ -n_o^2 \sin\theta \end{pmatrix} \tag{3.3-11b}$$

We note that these two modes are mutually orthogonal. The normal modes for $\mathbf{D}$ can be written

$$O \text{ wave}: \quad \mathbf{D}_o = \begin{pmatrix} \sin\phi \\ -\cos\phi \\ 0 \end{pmatrix} \tag{3.3-12}$$

$$E \text{ wave}: \quad \mathbf{D}_e = \begin{pmatrix} \cos\theta \cos\phi \\ \cos\theta \sin\phi \\ -\sin\theta \end{pmatrix} \tag{3.3-13}$$

These two $\mathbf{D}$ vectors and $\mathbf{s}$ are also mutually orthogonal.

If a polarized light inside the uniaxial medium is generated that is to propagate along a direction $\mathbf{s}$, the displacement vector of this light can always be written as a linear combination of these two normal modes:

$$\mathbf{D} = C_o \mathbf{D}_o \exp(-i\mathbf{k}_o \cdot \mathbf{r}) + C_e \mathbf{D}_e \exp(-i\mathbf{k}_e \cdot \mathbf{r}) \tag{3.3-14}$$

where C_o and C_e are constants and $\mathbf{k}_o$ and $\mathbf{k}_e$ are the wavevectors that are, in general, different [Eq. (3.3-2)]. As the light propagates inside the medium, a phase retardation between these two components is built up as a result of the difference in their phase velocities. Such a phase retardation leads to a new polarization state. Thus, birefringent plates can be used to alter the polarization state of light. To illustrate this, we consider some special cases of interest.

3.3.1. Propagation Perpendicular to the c Axis

For propagation perpendicular to the c axis, the unit vector $\mathbf{s}$ in Eq. (3.3-10) can be written

$$\mathbf{s} = \begin{pmatrix} \cos\phi \\ \sin\phi \\ 0 \end{pmatrix} \tag{3.3-15}$$

where ϕ is the angle of the direction of propagation in the xy plane and $\theta = \pi/2$. According to Eqs. (3.3-3) and (3.3-4), the index of refractive for the O wave is $n = n_o$ and that for the E wave is $n = n_e$. The direction of polarization for the ordinary wave is obtained from Eq. (3.3-6):

$$O \text{ wave}: \quad \mathbf{E}_o = \begin{pmatrix} \sin\phi \\ -\cos\phi \\ 0 \end{pmatrix} \quad\quad\quad (3.3\text{-}16)$$

Note that the $\mathbf{E}$ vector of the ordinary wave is perpendicular to both the direction of propagation and the c axis. The direction of polarization for the extraordinary wave is obtained from Eq. (3.3-11):

$$E \text{ wave}: \quad \mathbf{E}_e = \begin{pmatrix} 0 \\ 0 \\ 1 \end{pmatrix} \quad\quad\quad (3.3\text{-}17)$$

Note that the $\mathbf{E}$ vector of the extraordinary wave is parallel to the c axis.

The electric field vector of a polarized light can be written as a linear combination of O wave and E wave:

$$\mathbf{E} = C_o \mathbf{E}_o \exp(-i\mathbf{k}_o \cdot \mathbf{r}) + C_e \mathbf{E}_e \exp(-i\mathbf{k}_e \cdot \mathbf{r})$$

where C_o and C_e are constants and $\mathbf{k}_o = (\omega/c)n_o\,\mathbf{s}$ and $\mathbf{k}_e = (\omega/c)n_e\,\mathbf{s}$ are the ordinary and extraordinary wavevectors, respectivley. To show the phase retardation and the polarization change during propagation, let us assume $\phi = \pi/2$. In this case, $\mathbf{s} = \mathbf{y}$, $\mathbf{E}_o = \mathbf{x}$, and $\mathbf{E}_e = \mathbf{z}$. The electric field vector is

$$\mathbf{E} = C_o\mathbf{x}\exp(-ik_o y) + C_e\mathbf{z}\exp(-ik_e y)$$

Suppose $C_o = C_e = 1$, that is, that the ordinary and extraordinary waves are in phase and have equal amplitude at $y = 0$. The corresponding electric field vector, $\mathbf{E} = \mathbf{x} + \mathbf{z}$, represents a linearly polarized lightwave. At the propagation distance $y = d_{\lambda/4}$, where $d_{\lambda/4}$ is given by $(\omega/c)(n_e - n_o)d_{\lambda/4} = \pi/2$, the electric field vector becomes $\mathbf{E} = \exp(-ik_e d_{\lambda/4})[i\mathbf{x} + \mathbf{z}]$ which represents a circular polarization state. At the propagation distance $y = d_{\lambda/2}$, where $d_{\lambda/2}$ is given by $d_{\lambda/2} = 2d_{\lambda/4}$, the electric field vector becomes $\mathbf{E} = \exp(-ik_e d_{\lambda/2})[-\mathbf{x} + \mathbf{z}]$, which represents a linear polarization state perpendicular to the original linear polarization state at $y = 0$. A birefringent plate with thickness $d_{\lambda/4}$ is known as a *quarter-wave plate* ($\lambda/4$), which is usually used to convert a linear polarization state to a circular polarization state or vice versa. A birefringent plate with thickness $d_{\lambda/2}$ is known as a *half-wave plate* ($\lambda/2$), which is usually used to change the direction of linear polarization states.

3.3.2. Propagation in the *xz* Plane

For propagation in the *xz* plane, $\phi = 0$ in Eq. (3.3-10), the unit vector **s** can be written

$$\mathbf{s} = \begin{pmatrix} \sin\theta \\ 0 \\ \cos\theta \end{pmatrix} \tag{3.3-18}$$

where θ is the angle of **s** measured from the *c* axis. According to Eqs. (3.3-3) and (3.3-4), the eigen refractive index of the ordinary wave is $n = n_o$ and the eigen refractive index of the extraordinary wave is

$$n_e(\theta) = \left[\frac{\cos^2\theta}{n_o^2} + \frac{\sin^2\theta}{n_e^2} \right]^{-1/2}$$

The corresponding wavevectors are given by $\mathbf{k}_0 = (\omega/c)n_o\mathbf{s}$ and $\mathbf{k}_e = (\omega/c)n_e(\theta)\mathbf{s}$, respectively. Note that, depending on θ, the eigen refractive index of the extraordinary wave can vary from $n_e(\theta = 0) = n_o$ to $n_e(\theta = \pi/2) = n_e$. The directions of polarization for the ordinary and extraordinary waves are obtained from Eqs. (3.3-6) and (3.3-11), respectively:

$$O\text{ wave:} \quad \mathbf{E}_o = \begin{pmatrix} 0 \\ 1 \\ 0 \end{pmatrix} \tag{3.3-19}$$

$$O\text{ wave:} \quad \mathbf{E}_e = \begin{pmatrix} n_e^2 \cos\theta \\ 0 \\ -n_o^2 \sin\theta \end{pmatrix} \tag{3.3-20}$$

Note that the **E** vector of the ordinary wave is in the *y* direction, which is perpendicular to both the direction of propagation and the *c* axis; the **E** vector of the extraordinary wave is in the *xz* plane but it is not exactly perpendicular to the direction of propagation. As the ordinary and extraordinary waves propagate in the medium for a distance *d*, there will be a phase difference $\Gamma = (\omega/c)(n_e(\theta) - n_o)d$. Such a phase retardation Γ results in a change of polarization state, provided the beam of light consists of both ordinary and extraordinary components.

3.3.3. Propagation along the c Axis

For propagation along the c axis, $\theta = 0$ in Eq. (3.3-10), the unit vector $\mathbf{s}$ can be written

$$\mathbf{s} = \begin{pmatrix} 0 \\ 0 \\ 1 \end{pmatrix} \qquad (3.3\text{-}21)$$

Equations (3.3-3) and (3.3-4) predict that the refractive indices associated with the two modes of propagation are equal, that is $n_1 = n_2 = n_o$. Their corresponding wave vectors are equal as well $[\mathbf{k}_1 = \mathbf{k}_2 = (\omega/c)n_o\mathbf{s}]$. In other words, the two eigenmodes are degenerate. The direction of polarization can be obtained directly by solving Eq. (3.3-5). Equation (3.3-5) shows that the $\mathbf{E}$ vector can be an arbitrary vector in the xy plane. For the two eigenmodes, we can choose

$$\mathbf{E}_1 = \begin{pmatrix} 1 \\ 0 \\ 0 \end{pmatrix} \qquad (3.3\text{-}22)$$

$$\mathbf{E}_2 = \begin{pmatrix} 0 \\ 1 \\ 0 \end{pmatrix}. \qquad (3.3\text{-}23)$$

Note that both $\mathbf{E}$ vectors are perpendicular to the direction of propagation, which is along the c axis. Since there is no difference in eigen refractive indices of the two waves, there is no phase retardation. Therefore, the polarization state does not change during the propagation.

3.4. DOUBLE REFRACTION AT A BOUNDARY

Consider a plane wave incident on the surface of an anisotropic medium. The refracted wave, in general, is a mixture of the two eigenmodes. In a uniaxial medium, the refracted wave, in general, is a mixture of the ordinary wave and the extraordinary wave. According to the arguments used in connection with reflection and refraction at a plane boundary, the boundary condition requires that all the wavevectors lie in the plane of incidence and that their tangential components along the boundary be the same. This kinematic condition remains true for refraction at a boundary of an anisotropic medium.

Let $\mathbf{k}_i$ be the wavevector of the incident wave, and let $\mathbf{k}_1, \mathbf{k}_2$ be the wave vectors of the refracted waves. Given a prescribed value of the projection of the

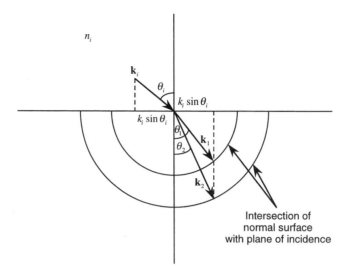

Figure 3.3. Double refraction at a boundary of an anisotropic medium and the graphic method of determining θ_1 and θ_2. Note that the tangential components of the wavevectors remain the same.

propagation vector $\mathbf{k}_i$ on the boundary, the two shells of the normal surface in general yields two propagation vectors, thus giving rise to two refracted waves as shown in Figure 3.3. The kinematic condition requires that

$$k_i \sin\theta_i = k_1 \sin\theta_1 = k_2 \sin\theta_2 \qquad (3.4\text{-}1)$$

for the refracted waves, where θ_i, θ_1, and θ_2 are angles of the wavevectors measured from the normal of the boundary.

Equation (3.4-1) resembles Snell's law. However, it is important to remember that k_1, k_2 are not, in general, constant; rather, they vary with the directions of the vectors $\mathbf{k}_1, \mathbf{k}_2$. The algebraic problem of determining θ_1, θ_2 involves solving a quadratic equation. Figure 3.3 illustrates a graphical method of determining the refraction angles.

In the case of uniaxial media, one shell of the normal surface is a sphere. The corresponding wavenumber k is therefore a constant for all directions of propagation. This wave is the ordinary wave and obeys Snell's law

$$n_i \sin\theta_i = n_1 \sin\theta_1 \qquad (3.4\text{-}2)$$

where n_i is the index of refraction of the incident medium and $n_1 = n_o$ is the ordinary refractive index of the medium. The other shell of the normal surface is an ellipsoid of revolution. Therefore, the corresponding wavenumber k_2 depends

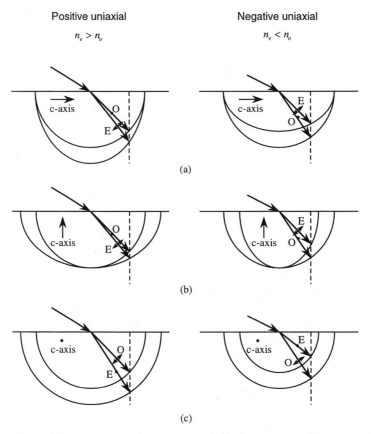

Figure 3.4. Wavevectors for double refraction in uniaxial media, with the optic axis (*a*) parallel to the boundary and parallel to the plane of incidence, (*b*) perpendicular to the boundary and parallel to the plane of incidence, and (*c*) parallel to the boundary and perpendicular to the plane of incidence. Symbols *O* and *E* stand for ordinary and extraordinary rays, respectively.

on the direction of propagation. This wave is the extraordinary wave. Some examples of double refraction are illustrated in Figure 3.4.

3.5. ANISOTROPIC ABSORPTION AND POLARIZERS

Certain materials exhibit strong anisotropic absorption in addition to the birefringence. Tourmaline, tin oxide crystals, and Polaroid sheets are some examples. Polaroid sheets consist of a thin layer of small needle-like crystals of herapathite (a salt of iodine and guinine) all aligned with their axes parallel. Such materials are characterized by complex refractive indices. If these materials are made of ordered arrangements of sharply anisometric molecules (e.g., needle-like), they are uniaxially anisotropic (e.g., Polaroid). These molecules may be absorptive for light with electric fields parallel to their absorption axes and transparent for light with electric fields perpendicular to these axes. Most sheet polarizers are uniaxially anisotropic in their absorption. These polarizers can be described by the complex ordinary and extraordinary indices of refraction, which can be written in their real parts and imaginary parts as

$$\hat{n}_o = n_o - i\kappa_o, \qquad \hat{n}_e = n_e - i\kappa_e \qquad (3.5\text{-}1)$$

where the imaginary parts κ_o, κ_e are known as the extinction coefficients.

A good polarizer should have a significant difference between the imaginary parts of the complex refractive indices so that one of the modes suffers strong attenuation in the bulk of the polarizer. Thus, in this model, polarizers can be classified into two types. The O-type polarizer, which transmits ordinary waves and attenuates extraordinary waves, has a real ordinary refractive index (i.e., $\kappa_o = 0$) and a complex extraordinary refractive index (i.e., $0 < \kappa_e$). The E-type polarizer, which transmits extraordinary waves and attenuates ordinary waves, has a real extraordinary refractive index $\hat{n}_e$ (i.e., $\kappa_e = 0$) and a complex ordinary refractive index $\hat{n}_o$ (i.e., $0 < \kappa_o$). Most of the commercially produced sheet-type polarizers are made by processes involving the stretching of large plastic sheets. A commonly used plastic is the polymer polyvinyl alcohol (PVA). The stretching creates a unidirectional parallelism for the anisometric molecules. If the absorption axis is parallel to the long axis, these stretched PVA sheet polarizers are of the O type (i.e., $0 \approx \kappa_o \leq \kappa_e$). Typical examples of these polarizers are HN-22 and HN-38. If the polarizers are prepared by a unidirectional compression, these compressed sheet polarizers are of the E type, provided the absorption axis of the molecules is parallel to the long axis. An ideal O-type polarizer would have a real ordinary refractive index and a complex extraordinary refractive index such that $|\exp(-ik_o d)| = 1$ and $|\exp(-ik_e d)| \approx 0$, where d is the thickness of the plate. In practice, most sheet polarizers are O-type with a small residual absorption for the ordinary wave. It is important to note that the transmission axis of an O-type polarizer is perpendicular to the c axis of the polarizer.

It is also important to note that E-type polarizers attenuate both E waves and O waves at a general angle of incidence. This can be understood by noting that the wavenumber of E wave depends on both n_e and n_o. Because n_o is complex, both O wave and E wave will suffer attenuation at a general angle of incidence. The attenuation of E wave diappears only when the wave propagates along a direction perpendicular to the c axis.

3.5.1. Extinction Ratio and Real Sheet Polarizers

On the basis of the preceding discussion, the principal transmittance of a polarizer can be written

$T_1 =$ transmission with polarization parallel to the transmission axis
$T_2 =$ transmission with polarization perpendicular to the transmission axis

An ideal polarizer would have $T_1 = 1$ and $T_2 = 0$. For real polarizers, T_1 is always less than 1, and T_2 is always greater than 0. Generally speaking, both T_1 and T_2 depend on the wavelength of light. For high performance applications, it is desirable to have polarizers with high T_1 and low T_2. A parameter known as the extinction ratio (or contrast ratio) is defined as

$$\text{Extinction ratio} = \frac{T_2}{T_1} \tag{3.5-2}$$

which represents a measure of the extinguishing efficiency of the polarizer. Thus, for a polarizer with $T_1 = 0.8, T_2 = 0.0008$, the extinction ratio is 10^{-3}. For a beam of unpoarized light, the transmittance through a polarizer is

$$T_0 = \tfrac{1}{2}(T_1 + T_2) \tag{3.5-3}$$

For a pair of parallel polarizers, the transmittance for unpolarized light is

$$T_p = \tfrac{1}{2}(T_1^2 + T_2^2) \tag{3.5-4}$$

For a pair of crossed polarizers, the transmittance for unpolarized light is

$$T_x = \tfrac{1}{2}T_1 T_2 \tag{3.5-5}$$

According to this equation, the transmittance of a pair of crossed ideal polarizers would be zero.

3.5.2. Field of View of Crossed Polarizers

Properties of polarizers at oblique incidence are described in this section. Equation (3.5-5) gives a simple formula for evaluation of the leakage of light through a pair of crossed polarizers due to a residual transmission T_2. According to this equation, the leakage is zero for ideal polarizers with $T_2 = 0$. It is important to note that this is true only at normal incidence. For large angles of incidence, the leakage of light is far more severe than that of Eq. (3.5-5). This is known for some time for a pair of crossed sheet polarizers. We will show in what follows that the leakage of light occurs at large angles of incidence even for a pair of ideal crossed polarizers. The understanding of the leakage requires the knowledge of the normal modes of propagation in polarizers.

Let us consider the leakage of light through a pair of crossed ideal polarizers. Since most polarizers are of O type, we assume that the polarizers are made of uniaxial materials that completely eliminate the extraordinary wave, and transmit the ordinary wave. Let c_1, c_2 be unit vectors representing the c axes of the two polarizers. In a crossed configuration, the two c axes are mutually perpendicular. From the discussion earlier, the polarization state of the transmitted light in the first polarizer is

$$o_1 = \frac{k \times c_1}{|k \times c_1|} \tag{3.5-6a}$$

and the polarization state of the transmitted light in the second polarizer is

$$o_2 = \frac{k \times c_2}{|k \times c_2|} \tag{3.5-6b}$$

where k is the wavevector of the light inside the medium of the polarizers. Thus, if o_1 and o_2 are mutually orthogonal, then no light can be transmitted through the crossed polarizers. A careful examination of Eqs. (3.5-6a) and (3.5-6b) indicates that o_1 and o_2 are, in general, not orthogonal even if c_1 and c_2 are orthogonal. In the case of crossed polarizers (c_1 and c_2 are mutually orthogonal), o_1 and o_2 are orthogonal only at normal incidence. The leakage of unpolarized light through a pair of crossed polarizers can be written

$$T_x = \tfrac{1}{2}|o_1 \cdot o_2|^2 \tag{3.5-7}$$

If we define the internal angle of propagation (θ, ϕ) such that the incident wavevector inside the polarizer medium can be written

$$k = k(\sin\theta \cos\phi, \sin\theta \sin\phi, \cos\theta) \tag{3.5-8}$$

then the leakage for a pair of crossed polarizers with $c_1 = x$ and $c_2 = y$ can be written

$$T_x = \frac{1}{2} \frac{\sin^4\theta \sin^2\phi \cos^2\phi}{2(1 - \sin^2\theta \sin^2\phi)(1 - \sin^2\theta \cos^2\phi)} \tag{3.5-9}$$

The angle θ is related to the external angle of incidence by Snell's law. In other words

$$\sin\theta_{ext} = n_o \sin\theta \tag{3.5-10}$$

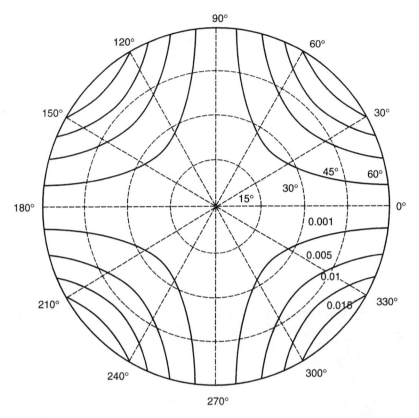

Figure 3.5. Contours of equitransmission curves in the viewing field showing the leakage of light through a pair of crossed O-type polarizers in air. The center corresponds to the viewing at normal incidence.

We note that the worst leakage occurs at large angles θ with $\phi=45°$. There is no leakage of light at $\phi = 0°$ or $90°$ when the plane of incidence is parallel to one of the transmission axes of the polarizers. According to Eq. (3.5-9), we find a leakage of $T_x = \frac{1}{98}$ at $\theta = 30°$, $\phi = 45°$. This is a significant leakage (1.02%) of light. For polarizer medium with $n_o = 1.5$, this corresponds to an angle of incidence in air of $\theta_{ext} = 48.6°$. Figure 3.5 is a contour plot showing the leakage as a function of the angle of incidence in air. We note that the leakage pattern exhibits a fourfold symmetry with a dark cross determined by the crossed transmission axes of the polarizers.

On the basis of the discussion above, we can prevent the leakage of light by using a pair of crossed polarizers that consists of an O-type polarizer and an E-type polarizer. It is important to note that E-type polarizers can cause an attenuation for the transmitted light at large angles of incidence.

3.6. OPTICAL ACTIVITY AND FARADAY ROTATION

Substances that rotate the plane of polarization of a beam of light traversing through them are said to be optically active. It was first observed in 1811 that quartz crystal rotates the plane of polarization of a beam traversing through it in the direction of the optic axis. The plane of polarization rotates 21.72° per mm for sodium light at 20°C. Both right-handed and left-handed rotation exist in different specimens. The optical activity of liquids was first observed in sugar solutions in 1815.

It is known that fused quartz is optically inactive. The optical rotatory power in solid is due to the crystalline structure and the arrangement of molecules in the crystal. In liquids, the molecular orientations are random. Thus the cause of optical rotatory power in liquids must lie in the structure of the molecules themselves. Virtually all molecules in optically active liquids possess an asymmetric atom of carbon, nitrogen, or sulfur. In addition, all these liquids have a twin substance with an opposite rotatory power.

Many LC materials consist of chiral molecules that are asymmetric with the existence of isomers of mirror images. The existence of mirror-image isomers is a result of the presence of one or more asymmetric carbon atoms in the compound. Thus, these molecular structures can have left- and right-hand (chiral) forms. These forms are conventionally designated as dextro (D) and levo (L) because they compare to each other structurally as do the right and left hands when the carbon atoms are lined up vertically. Optical rotatory power in liquids is due to the asymmetric structural arrangement (e.g., helical structure) of atoms in the randomly oriented molecules.

Figure 3.6 illustrates the rotation of the plane of polarization by an optically active medium. The amount of rotation is proportional to the pathlength of light in the medium. Conventionally, the rotatory power of a medium is given in degrees per centimeter; that is, the specific rotatory power is defined as the amount of rotation per unit length.

The sense of rotation bears a fixed relation to the propagation wavevector of the beam of light, so if the beam of light is made to traverse the medium once in

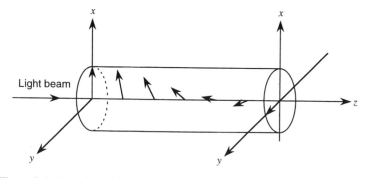

Figure 3.6. Rotation of the plane of polarization by an optically active medium.

each of two opposite directions, as in the case of reflection from the right end
face in Figure 3.6, the net rotation is zero. A substance is called *dextrorotatory*
or *right-handed* if the sense of rotation of the plane of polarization is
counterclockwise as viewed by an observer facing the approaching light beam. If
the sense of rotation is clockwise, the substance is called *levorotatory* or *left-
handed*. Quartz occurs in both right-handed and left-handed crystalline forms.
Many other substances are now known to exhibit optical activity; these include
cinnabar, sodium chlorate, turpentine, sugar, strychnine sulfate, tellurium,
selenium, and silver thiogallate ($AgGaS_2$). Many liquid crystal materials and
organic compounds also exhibit optical rotatory power. The specific rotatory
powers of some optically active media are given in Table 3.3a,b.

Fresnel first recognized in 1825 that optical activity arises from circular
double refraction, in which the eigenwaves of propagation (i.e., the independent
plane-wave solutions of Maxwell's equations) are right and left circularly
polarized waves.

Let n_r and n_l be the refractive indices associated with these two waves, and
assume that the waves are propagating in the $+z$ direction. The displacement
vector amplitudes for these two waves are

$$\mathbf{R} \exp\left[i\omega\left(t - \frac{zn_r}{c}\right)\right]$$

Table 3.3a. Optical Rotatory Powers of Some Solids

	λ (Å)	ρ (degrees/mm)
Quartz	4,000	49
	4,550	37
	5,000	31
	5,500	26
	6,000	22
	6,500	17
$AgGaS_2$	4,850	950
	4,900	700
	4,950	600
	5,000	500
	5,050	430
Se	7,500	180
	10,000	30
Te	(6 μm)	40
	(10 μm)	15
TeO_2	3,698	587
	4,382	271
	5,300	143
	6,328	87
	10,000	30

Table 3.3b. Optical Rotatory Powers of Some Liquid Crystals

Compound	Wavelength (nm)	T (°C)	ρ (degrees/mm)
CEEC	400	RT[a]	472
(Cholesteryl 2-2-ethoxy-	450		474
ethoxy ethyl carbonate)	500		315
	550		295
	600		290
	650		285
	700		275
COC	460	RT	4167
(Cholesteryl oleyl carbonate)	470		2500
	490		1111
	500		833
	530		500
	550		444
	600		222
	650		167
PBLG in chloroform	436	RT	820
(Poly γ-benzyl-L-glutamate)	509		585
	546		500
	589		425
	644		350
PBLG in 1,1,2,2, tetra-chloroethane	350	22	355
	400		270
	450		220
	500		170
	550		135
PBDG in 1,1,2,2, tetra-chloroethane	350	22	−430
(Poly γ-benzyl-D-glutamate)	400		−312
	450		−240
	500		−185
	550		−140
CEC	400	RT	472
(Cholesteryl ethyl carbonate)	433		394
	460		354
	500		315
	533		295
	566		295
	600		276
	633		276
	666		295
	700		295
Enantiotropic cholesteryl chloride	400	RT	−2,860
mixture	530		−10,000
	595		−23,570

[a] Room temperature.

and

$$\mathbf{L}\exp\left[i\omega\left(t - \frac{zn_l}{c}\right)\right]$$

where $\mathbf{R}$ and $\mathbf{L}$ are unit Jones vectors for the circular polarizations (2.3-9) and (2.3-10), respectively.

If a linearly polarized beam of amplitude D_0 that is polarized along the x direction enters the medium at $z = 0$, it will be represented by the sum of two such waves with mplitudes $D_0/\sqrt{2}$. At distance z in the medium, the resultant will be

$$\frac{D_0}{\sqrt{2}}e^{i\omega t}(\mathbf{R}e^{-i\omega zn_r/c} + \mathbf{L}e^{-i\omega zn_l/c}) \qquad (3.6\text{-}1)$$

This can further be written, if we use Eqs. (2.3-9) and (2.3-10), as

$$D_o\mathbf{p}\exp\left\{i\omega\left(t - \frac{z(n_r + n_l)}{2c}\right)\right\} \qquad (3.6\text{-}2)$$

with

$$\mathbf{P} = \mathbf{x}\cos\left[\frac{\omega(n_l - n_r)}{2c}z\right] + \mathbf{y}\sin\left[\frac{\omega(n_l - n_r)}{2c}z\right] \qquad (3.6\text{-}3)$$

The polarization of the resultant wave is represented by Eq. (3.6-3). The wave is linearly polarized with its plane of polarization turned in the counterclockwise sense from x through an angle $\omega z(n_l - n_r)/2c$. Thus the specific rotatory power is given by

$$\rho = \frac{\pi}{\lambda}(n_l - n_r) \qquad (3.6\text{-}4)$$

The optical rotation is right-handed (counterclockwise) if $n_r < n_l$. Thus the plane of polarization turns in the same sense as the circularly polarized wave, which travels with the greater phase velocity. It can be shown that the polarization ellipse of a beam of elliptically polarized light will rotate the same angle $\omega z(n_l - n_r)/2c$ with its shape remaining unchanged. (See Problems 3.15)

The specific rotatory power of the liquid crystal CEEC at $\lambda = 650$ nm is $285°/$ mm, giving $|n_r - n_l| = 1.03 \times 10^{-3}$. It is evident that the rotation of the plane of polarization is an extremely sensitive way of measuring very small amounts of circular double refraction.

The dielectric property described by the material equation (3.1-7) does not allow the existence of optical acitivity. To develop a theory of optical activity, we need to generalize the constitutive relations of materials. The electromagnetic theory of optical activity is due mainly to Born and his co-workers and has been well summarised by Condon [2]. We note that the rotatory power is strongly dependent on the wavelength due to the $1/\lambda$ factor in Eq. (3.6-4), as well as the material dispersion.

According to their theory, the optical activity originates in properties of a molecule represented by the following relationship

$$p = \alpha E - \beta \dot{H} \qquad (3.6\text{-}5)$$

in which p is the induced dipole moment of the molecule. For a linear molecule it is obvious that $\beta = 0$. A nonvanishing β arises from an intrinsic chiral structure in the molecule. When such a chiral molecule is in a changing magnetic field, the changing magnetic flux through the molecule sets up an induced current circulating around $\mathbf{H}$ in the sense given by Lenz law (3.1-1). This induced current will now give rise to a time-varying separation of charges in the direction of $\dot{\mathbf{H}}$, thus setting up an electric dipole moment that is represented by the term in β in Eq. (3.6-5). For plane-wave propagation in a homogeneous medium, the material equation for an optically active material can thus be written

$$\mathbf{D} = \varepsilon \mathbf{E} + i\varepsilon_0 \mathbf{G} \times \mathbf{E} \qquad (3.6\text{-}6)$$

where ε is the dielectric tensor without optical activity and $\mathbf{G}$ is a vector parallel to the direction of propagation and is called the gyration vector. The vector product $\mathbf{G} \times \mathbf{E}$ can always be represented by the product of an antisymmetric tensor $[G]$ with $\mathbf{E}$. The matrix elements of $[G]$ are given by

$$
\begin{aligned}
[G]_{23} &= -[G]_{32} = -G_x \\
[G]_{31} &= -[G]_{13} = -G_y \\
[G]_{12} &= -[G]_{21} = -G_z
\end{aligned}
\qquad (3.6\text{-}7)
$$

Thus Eq. (3.6-6) becomes

$$\mathbf{D} = (\varepsilon + i\varepsilon_o[G])\mathbf{E} \qquad (3.6\text{-}8)$$

where $[G]$ is the matrix representation of the gyration vector. It is often convenient to define a new dielectric tensor as

$$\varepsilon' = \varepsilon + i\varepsilon_0[G]. \qquad (3.6\text{-}9)$$

This new tensor is Hermitian, that is, $\varepsilon'_{ij} = \varepsilon'^{*}_{ji}$. We can now substitute ε' into Eq. (3.2-5) and solve for the normal modes of propagation. Again, let $\mathbf{s}$ be a unit vector in the direction of propagation, so that the gyration vector can be written

$$\mathbf{G} = G\mathbf{s}. \tag{3.6-10}$$

The resulting Fresnel equation for the eigen indices of refraction is given by

$$
\begin{aligned}
&\frac{s_x^2}{n^2 - n_x^2} + \frac{s_y^2}{n^2 - n_y^2} + \frac{s_z^2}{n^2 - n_z^2} - \frac{1}{n^2} \\
&= G^2 \frac{s_x^2 n_x^2 + s_y^2 n_y^2 + s_z^2 n_z^2}{n^2(n^2 - n_x^2)(n^2 - n_y^2)(n^2 - n_z^2)}
\end{aligned}
\tag{3.6-11}
$$

where n_x, n_y, and n_z are the principal refractive indices and s_x, s_y, and s_z are the components of $\mathbf{s}$ along the principal dielectric axes. Let n_1^2 and n_2^2 be the roots of the Fresnel equation with $G = 0$. Equation (3.6-11) can be written in terms of n_1^2 and n_2^2 as

$$(n^2 - n_1^2)(n^2 - n_2^2) = G^2 \tag{3.6-12}$$

For propagation along the optic axes we know that $n_1 = n_2 = \bar{n}$, say. Equation (3.6-12) then gives

$$n^2 = \bar{n}^2 \pm G$$

and since G is small, we obtain

$$n \cong \bar{n} \pm \frac{G}{2\bar{n}} \tag{3.6-13}$$

These correspond to two circularly polarized waves. From Eq. (3.6-4), the rotatory power is

$$\rho = \frac{\pi G}{\lambda \bar{n}} \tag{3.6-14}$$

In uniaxial crystals, $\bar{n} = n_o$. The parameter G in Eq. (3.6-12) varies with the direction of the wavevector and is a quadratic function of the direction cosines

s_x, s_y, s_z; thus we write

$$G = g_{11}s_x^2 + g_{22}s_y^2 + g_{33}s_z^2 + 2g_{12}s_xs_y + 2g_{23}s_ys_z + 2g_{31}s_zs_x \qquad (3.6\text{-}15)$$

or

$$G = g_{ij}s_is_j, \qquad i,j = x, y, z \qquad (3.6\text{-}16)$$

where g_{ij} are matrix elements of the gyration tensor that describe the optical activity of the crystal.

To study the polarization states of the normal modes of propagation, it is convenient to use the displacement vector $\mathbf{D}$, since $\mathbf{D}$ is always perpendicular to the direction of propagation ($\mathbf{D} \cdot \mathbf{s} = 0$). It is also more convenient to use the inverse tensor ε^{-1}. The material constitutive relation can now be written

$$\mathbf{E} = \frac{1}{\varepsilon'}\mathbf{D} \qquad (3.6\text{-}17)$$

where ε' is as given by Eq. (3.6-9). The inverse tensor $1/\varepsilon'$ is also a Hermitian tensor. The normal modes of propagation can be obtained from the following wave equation (3.2-5):

$$n^2\mathbf{s} \times \left(\mathbf{s} \times \frac{\varepsilon_0}{\varepsilon'}\right)\mathbf{D} + \mathbf{D} = 0 \qquad (3.6\text{-}18)$$

Since $[G]$ is small compared with $\varepsilon/\varepsilon_0$, the inverse tensor can be written

$$\frac{1}{\varepsilon'} = \frac{1}{\varepsilon} - i\varepsilon_0\frac{1}{\varepsilon}[G]\frac{1}{\varepsilon} \qquad (3.6\text{-}19)$$

Thus Eq. (3.6-18) can be written in terms of the impermeability tensor $\eta(= \varepsilon_0/\varepsilon)$ as

$$[s][s]\{\eta - i\eta[G]\eta\}\mathbf{D} = -\frac{1}{n^2}\mathbf{D} \qquad (3.6\text{-}20)$$

where $[s]$ is the antisymmetric tensor representation of $\mathbf{s} \times$ and is defined in a way similar to that for $[G]$ [see Eq. (3.6-7)]. Let $\mathbf{D}_1, \mathbf{D}_2$ be the normalized eigenmodes of propagation in the absence of optical activity ($G = 0$):

$$\left\{[s][s]\eta + \frac{1}{n_{1,2}^2}\right\}\mathbf{D}_{1,2} = 0 \qquad (3.6\text{-}21)$$

and

$$\mathbf{D}_i \cdot \mathbf{D}_j = \delta_{ij}$$

We will now solve the eigenvalue problem in the coordinate system formed by the triad $(\mathbf{D}_1, \mathbf{D}_2, \mathbf{s})$. In this coordinate system, Eq. (3.6-20) becomes

$$\begin{pmatrix} \dfrac{1}{n_1^2} & \dfrac{iG}{n_1^2 n_2^2} \\[2mm] -\dfrac{iG}{n_1^2 n_2^2} & \dfrac{1}{n_2^2} \end{pmatrix} \mathbf{D} = \frac{1}{n^2} \mathbf{D} \qquad (3.6\text{-}22)$$

The refractive indices n of the eigenmodes satisfy the follwing secular equation:

$$\left(\frac{1}{n_1^2} - \frac{1}{n^2}\right)\left(\frac{1}{n_2^2} - \frac{1}{n^2}\right) = \left(\frac{G}{n_1^2 n_2^2}\right)^2 \qquad (3.6\text{-}23)$$

The roots of Eq. (3.6-23) are given by

$$\frac{1}{n^2} = \frac{1}{2}\left(\frac{1}{n_1^2} + \frac{1}{n_2^2}\right) \pm \sqrt{\frac{1}{4}\left(\frac{1}{n_1^2} - \frac{1}{n_2^2}\right)^2 + \left(\frac{G}{n_1^2 n_2^2}\right)^2} \qquad (3.6\text{-}24)$$

The corresponding polarization states can be represented by the Jones vectors

$$\mathbf{J}_{\pm} = \begin{pmatrix} \dfrac{1}{2}\left(\dfrac{1}{n_1^2} - \dfrac{1}{n_2^2}\right) \pm \sqrt{\dfrac{1}{4}\left(\dfrac{1}{n_1^2} - \dfrac{1}{n_2^2}\right)^2 + \left(\dfrac{G}{n_1^2 n_2^2}\right)^2} \\[4mm] -\dfrac{iG}{n_1^2 n_2^2} \end{pmatrix} \qquad (3.6\text{-}25)$$

These Jones vectors represent two elliptically polarized waves that are orthogonal to each other. Since the first component is real and the second component is pure imaginary, the principal axes of the polarization ellipses are parallel to the "unperturbed" polarizations $\mathbf{D}_1$, $\mathbf{D}_2$ (see Fig 3.7). The two senses of rotation are opposite each other. The ellipticity (defined as the ratio of the

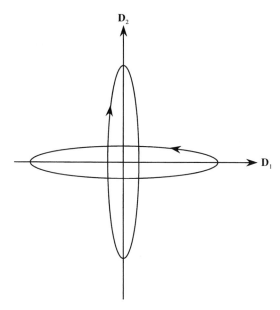

Figure 3.7. Polarization ellipses of the normal modes in the presence of both optical birefringence and rotatory power.

lengths of the principal axes) of the polarization ellipses is given by

$$e = \frac{-G}{\frac{1}{2}(n_2^2 - n_1^2) \pm \sqrt{\frac{1}{4}(n_2^2 - n_1^2)^2 + G^2}} \tag{3.6-26}$$

In the case of propagation along the optic axes of an anisotropic medium or propagation in an isotropic medium ($n_1 = n_2 = \bar{n}$), the refractive indices of the normal modes (3.6-24) become

$$\frac{1}{n^2} = \frac{1}{\bar{n}^2} \pm \frac{G}{\bar{n}^4} = \frac{1}{\bar{n}^2}\left(1 \pm \frac{G}{\bar{n}^2}\right) \tag{3.6-27}$$

which agrees with Eq. (3.6-13). The corresponding polarization states (3.6-25) become

$$\mathbf{J}_\pm = \begin{pmatrix} \pm 1 \\ -i \end{pmatrix} \tag{3.6-28}$$

which represent right and left circularly polarized light.

Table 3.4. Forms of the Gyration Tensor $[g_{ij}]$

Centrosymmetric ($\bar{1}$, $2/m$, mmm, $4/m$, $4/mmm$, $\bar{3}$, $\bar{3}m$, $6/m$, $6/mmm$, $m3$, $m3m$):

$$\begin{pmatrix} 0 & 0 & 0 \\ 0 & 0 & 0 \\ 0 & 0 & 0 \end{pmatrix}$$

Triclinic:

$$1$$

$$\begin{pmatrix} g_{11} & g_{12} & g_{13} \\ g_{12} & g_{22} & g_{23} \\ g_{13} & g_{23} & g_{33} \end{pmatrix}$$

Monoclinic:

$$2(2 \parallel x_2) \qquad\qquad 2(2 \parallel x_3)$$

$$\begin{pmatrix} g_{11} & 0 & g_{13} \\ 0 & g_{22} & 0 \\ g_{13} & 0 & g_{33} \end{pmatrix} \qquad \begin{pmatrix} g_{11} & g_{12} & 0 \\ g_{12} & g_{22} & 0 \\ 0 & 0 & g_{33} \end{pmatrix}$$

$$m(m \perp x_2) \qquad\qquad m(m \perp x_3)$$

$$\begin{pmatrix} 0 & g_{12} & 0 \\ g_{12} & 0 & g_{23} \\ 0 & g_{23} & 0 \end{pmatrix} \qquad \begin{pmatrix} 0 & 0 & g_{13} \\ 0 & 0 & g_{23} \\ g_{13} & g_{23} & 0 \end{pmatrix}$$

Orthorhombic:

$$222 \qquad\qquad 2mm$$

$$\begin{pmatrix} g_{11} & 0 & 0 \\ 0 & g_{22} & 0 \\ 0 & 0 & g_{33} \end{pmatrix} \qquad \begin{pmatrix} 0 & g_{12} & 0 \\ g_{12} & 0 & 0 \\ 0 & 0 & 0 \end{pmatrix}$$

Tetragonal:

$$4, 422 \qquad\qquad \bar{4} \qquad\qquad \bar{4}2m \quad (2 \parallel x_1)$$

$$\begin{pmatrix} g_{11} & 0 & 0 \\ 0 & g_{11} & 0 \\ 0 & 0 & g_{33} \end{pmatrix} \quad \begin{pmatrix} g_{11} & g_{12} & 0 \\ g_{12} & -g_{11} & 0 \\ 0 & 0 & 0 \end{pmatrix} \quad \begin{pmatrix} 0 & g_{12} & 0 \\ g_{12} & 0 & 0 \\ 0 & 0 & 0 \end{pmatrix}$$

Trigonal and hexagonal:

$$3, 32, 622$$

$$\begin{pmatrix} g_{11} & 0 & 0 \\ 0 & g_{11} & 0 \\ 0 & 0 & g_{33} \end{pmatrix}$$

Table 3.4. (*Continued*)

Cubic:

432, 23

$$\begin{pmatrix} g_{11} & 0 & 0 \\ 0 & g_{11} & 0 \\ 0 & 0 & g_{11} \end{pmatrix}$$

Isotropic (without center of symmetry):

$$\begin{pmatrix} g & 0 & 0 \\ 0 & g & 0 \\ 0 & 0 & g \end{pmatrix}$$

Others (4mm, $\bar{4}3m$, 3m, 6mm, $\bar{6}$, $\bar{6}m2$):

$$\begin{pmatrix} 0 & 0 & 0 \\ 0 & 0 & 0 \\ 0 & 0 & 0 \end{pmatrix}$$

In the case of a general propagation in anisotropic media, G is usually very small compared with $n_2^2 - n_1^2$, and the ellipticity of the polarization ellipse is extremely small (i.e., $e \ll 1$), so that the waves are almost linearly polarized (see Fig. 3.7). As an example, for the propagation of a light beam perpendicular to the optic axes of a quartz crystal at $\lambda = 5100$ Å, the G value according to Szivessy and Munster is 6×10^{-5}, and the ellipticity is 2×10^{-3} [3].

The gyration tensor is symmetric and in general has six independent components. Some components may vanish because of the effect of crystal symmetry. For example, a material with a center of symmetry cannot be optically active (see Table 3.4). The arrays of nonvanishing components of the gyration tensors of are shown in Table 3.4.

3.6.1. Faraday Rotation

The Faraday effect is a property of transparent substances that causes a rotation of the plane of polarization with distance when the material is placed in a magnetic field, for light propagated along the magnetic field. More accurately, the rotation is proportional to the component of the magnetic field along the direction of propagation of the light. The gyration vector, defined by Eq. (3.6-6), is proportional to the external magnetic field:

$$\mathbf{G} = \gamma \mathbf{B} \tag{3.6-29}$$

in which γ is the magnetogyration coefficient of the medium. In an optically active medium, the direction of rotation bears a fixed relation to the direction of propagation, so that if a beam of light is reflected back on itself, the net rotation is zero. In the Faraday effect, however, the rotation bears a fixed relation to the magnetic field **B**, so that reflection back on itself doubles the rotation. The specific rotation (i.e., rotation per unit length) is often written

$$\rho = VB \tag{3.6-30}$$

where V is the Verdet constant.

The Faraday effect originates from the effect of the static magnetic field on the motion of electrons. In the presence of the electric field of the optical beam, the electrons are displaced from their equilibrium position. This motion, coupled with the static magnetic field, creates a lateral displacement of electrons due to the Lorentz force $\mathbf{v} \times \mathbf{B}$. As a result, the induced dipole moment involves a term that is proportional to $\mathbf{B} \times \mathbf{E}$. The material relation becomes

$$\mathbf{D} = \varepsilon\mathbf{E} + i\varepsilon_0\gamma\mathbf{B} \times \mathbf{E} \tag{3.6-31}$$

The factor i accounts for the $\pi/2$ phase lag between the velocity and the electric field. Faraday rotation has been observed in many solids, liquids, and even gases. A few values of the Verdet constant are given in Table 3.5.

From the atomic point of view, the Faraday effect is related to the Zeeman effect. As a result of interaction between the orbiting electrons and the magnetic field, each electron energy level is split into several sublevels. By virtue of conservation of angular momentum, RHC polarized light and LHC polarized light interact with different sets of the sublevels. Thus, the medium exhibits circular birefringence in the presence of the magnetic field, leading to a rotation of the polarization vector.

Table 3.5. Value of the Verdet Constant at $\lambda = 5893\,\text{Å}$

Substance	T (°C)	V (degrees/G·mm)
Water	20	2.18×10^{-5}
Fluorite		1.5×10^{-6}
Diamond		2.0×10^{-5}
Glass (crown)	18	2.68×10^{-5}
Glass (flint)		5.28×10^{-5}
Carbon disulfide (CS_2)	20	7.05×10^{-5}
Phosphorus	33	2.21×10^{-4}
Sodium chloride		6.0×10^{-5}
MBBA	20	6.67×10^{-5}

3.7. LIGHT PROPAGATION IN BIAXIAL MEDIA

Strictly speaking, most of the LC molecules are biaxial in nature. In other words, these molecules are not cylindrically symmetric. In the unidirectional alignment of LC molecules, only the long axes are aligned. The other axes remain randomly oriented. Thus, the nematic LC phase exhibits optically uniaxial properties. When orientational order exists for two of the axes, the LC exhibits biaxial optical properties. Two order parameters are needed to describe the degree of alignment of the liquid crystal. In some advanced LC displays, biaxial thin films are employed as phase compensators to improve the contrast and gray levels at large viewing angles.

For light propagation in biaxial crystals, the analytic results obtained in Section 3.2 for the normal modes can be employed. We know that there are two independent modes of propagation. However, unlike uniaxial crystals, the polarization states of these normal modes cannot be easily constructed in terms of their directions. The general results, Eqs. (3.2-9) and (3.2-10), must be employed to obtain the eigenrefractive indices and the polarization states of the normal modes of propagation.

We now investigate the propagation of electromagnetic radiation in a biaxial medium. The normal surface [i.e., $\omega(\mathbf{k}) = $ constant] of a typical biaxial medium is shown in Fig 3.1. It will help us visualize this surface if we consider first its intersections with the three coordinate planes (see Fig. 3.8.a). If we set $k_y = 0$ in Eq. (3.2-7), the equation breaks into two factors, giving

$$\frac{k_x^2 + k_z^2}{n_y^2} = \left(\frac{\omega}{c}\right)^2, \qquad \frac{k_x^2}{n_z^2} + \frac{k_z^2}{n_x^2} = \left(\frac{\omega}{c}\right)^2 \tag{3.7-1}$$

where n_x, n_y, and n_z are the principal refractive indices. The intersection of the normal surface with the plane $k_y = 0$ consists of a circle with a radius of $n_y\omega/c$ and an ellipse with semiaxes $n_x\omega/c$ and $n_z\omega/c$. The intersection with each of the other two coordinate planes likewise consists of a circle and an ellipse. The circle and the ellipse intersect only in the plane $k_y = 0$, as a result of the choice of the coordinate axes (i.e., $n_x < n_y < n_z$). These four points of intersection define the two optic axes of the crystal.

We now consider the wave propagation in the coordinate plane $k_y = 0$. For a general direction of propagation, we draw a line OS along the direction of propagation. There are in general two points of intersection. The distances between the origin O and the points of intersection determine the length of the wavevectors. The modes associated with the circle (with $n = n_y$) are polarized perpendicular to the coordinate plane $k_y = 0$, whereas the modes associated with the ellipse are polarized in the plane of the ellipse. Wave propagation in each of the other two coordinate planes is similar except for the propagation along the optic axes. Light propagation in the direction of these optic axes has a unique phase velocity (c/n_y) regardless of the state of polarization. The group velocity

$\mathbf{v}_g$ defined by

$$\mathbf{v}_g = \nabla_\mathbf{k}\omega(\mathbf{k}) \qquad (3.7\text{-}2)$$

which represents the flow of electromagnetic energy, is however, undefined in this direction, because the two shells of the normal surface degenerate to a point. The refraction phenomena associated with the propagation in this direction are closely related to the nature of the singularity at these points.

If we draw unit normal vectors perpendicular to the normal surface at points in an infinitesimal neighborhood of the singular point, we obtain an infinite number of unit vectors corresponding to the direction of energy flow. These vectors form a surface of a cone. Therefore, we expect that the flow of electromagnetic energy will take the form of a cone. This phenomenon is known as *conical refraction*.

To study the properties of conical refraction, we need to examine the normal surface near the singularity point $\mathbf{k}_0$ (see Fig. 3.8a). For light propagation in the direction of the optic axis shown in Fig. 3.8a, the wavevector is given by

$$\mathbf{k}_0 = \hat{\mathbf{x}}k_{x0} + \hat{\mathbf{y}}k_{y0} + \hat{\mathbf{z}}k_{z0} \qquad (3.7\text{-}3)$$

with

$$k_{x0} = n_y \frac{\omega}{c}\sin\theta, \qquad k_{y0} = 0, \qquad k_{z0} = n_y \frac{\omega}{c}\cos\theta \qquad (3.7\text{-}4)$$

where θ is the angle between the optic axis and the z axis and is given by

$$\tan\theta = \frac{n_z}{n_x}\left(\frac{n_y^2 - n_x^2}{n_z^2 - n_y^2}\right)^{1/2} \qquad (3.7\text{-}5)$$

For a medium with $n_x = 1.5, n_y = 1.6$, and $n_z = 1.7$, Eq. (3.7-5) yields $\theta = 47.7°$. This angle becomes zero when $n_y = n_x$. To examine the normal surface near the point $\mathbf{k}_0$, we need to perform a Taylor series expansion around this point. Let

$$k_x = k_{x0} + \xi, \qquad k_y = k_{y0} + \eta, \qquad k_z = k_{z0} + \zeta \qquad (3.7\text{-}6)$$

and then substitute them into Eq. (3.2-7). We obtain, after neglecting higher order terms in ξ, η, ζ,

$$4(k_{x0}\xi + k_{z0}\zeta)(n_x^2 k_{x0}\xi + n_z^2 k_{z0}\zeta) + \eta^2(n_y^2 - n_x^2)(n_y^2 - n_z^2) = 0 \qquad (3.7\text{-}7)$$

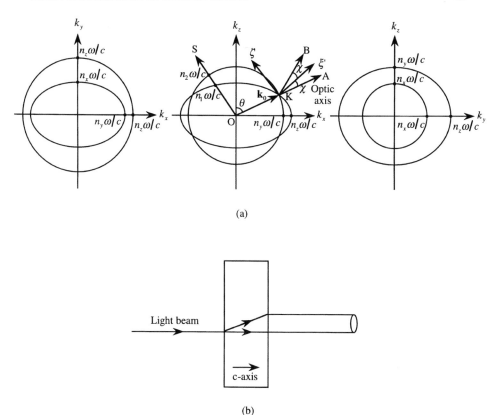

(a)

(b)

Figure 3.8. (a) Section of the normal surface of a biaxial crystal. (b) Conical refraction. The biaxial crystal plate is cut in such a way that the plate surfaces are perpendicular to one of the optic axes.

This equation of second degree represents a cone whose vertex lies at point $\mathbf{k}_0$ (i.e., $\xi = \eta = \zeta = 0$). It can be diagonalized by a rotation of the coordinates (See Fig. 3.8a) and becomes

$$\zeta'^2 + \frac{1}{1 - \tan^2\chi}\eta^2 = \xi'^2\cot^2\chi \qquad (3.7\text{-}8)$$

where χ is given by

$$\tan^2 2\chi = \frac{(n_z^2 - n_y^2)(n_y^2 - n_x^2)}{n_x^2 n_z^2} \qquad (3.7\text{-}9)$$

According to (3.7-8), the normal surface near the optic axis is a cone with a vertex at $\mathbf{k}_0$. Hence when the wavevector coincides with the optic axis, there are an infinite number of directions for energy flows (i.e., an infinite number of $\mathbf{v}_g$), which lie on the cone given by

$$\zeta'^2 + (1 - \tan^2\chi)\eta^2 = \xi'^2\tan^2\chi \qquad (3.7\text{-}10)$$

This is an elliptical cone with the ξ' axis as the cone axis and the point $\mathbf{k}_0$ as the vertex. This cone contains the optic axis OA and intersects any plane that is perpendicular to OA in a circle (see Problem 3.6). The aperture angle of this cone is 2χ in the xz plane (see Fig. 3.8a). Each unit vector originating from the vertex of the cone and lying in the cone represents a direction of energy flow for propagation with wavevector $\mathbf{k}_0$. Each such direction corresponds to a linear polarization state. For example, the direction KA represents the energy flow for a y-polarized wave, whereas the direction KB represents the energy flow for waves polarized in the xz plane (see Fig. 3.8a).

We now consider a plate of a biaxial material (e.g., mica) cut so that its two parallel surfaces are perpendicular to one of the optic axes. If this plate is illuminated by an unpolarized collimated beam of monochromatic light, such as laser light, incident normally on one of its faces, the energy will spread out in the plate in a hollow cone, and on emerging at the other side it will form a hollow cylinder, as shown in Figure 3.8. Thus, on a screen parallel to the crystal face, we expect to see a bright circular ring.

The eigenvalues of refractive indices and eigenmodes of polarization along a given direction of propagation can be found by using Eqs. (3.2-9) and (3.2-10), respectively. Figure 3.9 shows the $\mathbf{E}$ vectors on one octant of the normal surface illustrating the polarization.

3.7.1. Method of Index Ellipsoid

Here we describe the method of index ellipsoid, which can also be employed to determine the normal modes of propagation. We define the impermeability tensor η as

$$\eta = \varepsilon_0 \varepsilon^{-1} \qquad (3.7\text{-}11)$$

where ε^{-1} is the inverse of the dielectric tensor. In the principal coordinate, the impermeability tensor is given by

$$\eta = \varepsilon_0 \begin{pmatrix} \varepsilon_x^{-1} & 0 & 0 \\ 0 & \varepsilon_y^{-1} & 0 \\ 0 & 0 & \varepsilon_z^{-1} \end{pmatrix} = \begin{pmatrix} \dfrac{1}{n_x^2} & 0 & 0 \\ 0 & \dfrac{1}{n_y^2} & 0 \\ 0 & 0 & \dfrac{1}{n_z^2} \end{pmatrix}. \qquad (3.7\text{-}12)$$

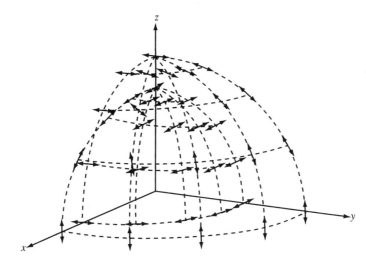

Figure 3.9. **E** vectors on one octant of the normal surface illustrating the polarization.

The index ellipsoid in the principal coordinate can be written

$$\frac{x^2}{n_x^2} + \frac{y^2}{n_y^2} + \frac{z^2}{n_z^2} = 1.$$

To find the normal modes of propagation along a general direction **s**, we draw the vector **s** from the origin. We then draw a plane through the origin and perpendicular to this vector **s**. An intersection ellipse is obtained between the plane and the ellipsoid. The two principal axes of the intersection ellipse are equal in length to $2n_1$ and $2n_2$, where n_1 and n_2 are the two eigen indices of refraction, that is, the solutions of the Fresnel equation. These two axes are parallel, respectively, to the directions of the vectors $\mathbf{D}_1, \mathbf{D}_2$ of the normal modes.

In what follows, we show that this method is equivalent to the method described in Section 3.2. By using the definition of Eq. (3.7-11), we can express the relation between the field vector **E** and **D** as

$$\mathbf{E} = \frac{1}{\varepsilon_0} \eta \mathbf{D} \tag{3.7-13}$$

Substitution of Eq. (3.7-13) for **E** in the wave equation (3.2-5), leads to

$$\mathbf{s} \times [\mathbf{s} \times \eta \mathbf{D}] + \frac{1}{n^2} \mathbf{D} = 0 \tag{3.7-14}$$

where we have used $\mathbf{k} = (\omega/c)n\mathbf{s}$ and $\mathbf{s}$ is a unit vector in the direction of propagation. Since $\mathbf{D}$ is always transverse to the direction of propagation $(\mathbf{s} \cdot \mathbf{D} = 0)$, it is convenient to use a new coordinate system with one axis in the direction of propagation of the wave, and denote the two transverse axes by 1 and 2. In this coordinate system the unit vector $\mathbf{s}$ is given by

$$\mathbf{s} = \begin{pmatrix} 0 \\ 0 \\ 1 \end{pmatrix} \qquad (3.7\text{-}15)$$

and the wave equation (3.7-14) becomes

$$\begin{pmatrix} \eta_{11} & \eta_{12} & \eta_{13} \\ \eta_{21} & \eta_{22} & \eta_{23} \\ 0 & 0 & 0 \end{pmatrix} \mathbf{D} = \frac{1}{n^2} \mathbf{D} \qquad (3.7\text{-}16)$$

Since $\mathbf{s} \cdot \mathbf{D} = 0$, the third component of $\mathbf{D}$ is always zero. We can ignore η_{13}, η_{23} and define a transverse impermeability tensor η_t as

$$\eta_t = \begin{pmatrix} \eta_{11} & \eta_{12} \\ \eta_{21} & \eta_{22} \end{pmatrix} \qquad (3.7\text{-}17)$$

The wave equation then becomes

$$\left(\eta_t - \frac{1}{n^2} \right) \mathbf{D} = 0 \qquad (3.7\text{-}18)$$

where $\mathbf{D}$ is the displacement field vector.

The polarization vectors of the normal modes are eigenvectors of the transverse impermeability tensor with eigenvalues $1/n^2$. Since η_t is a symmetric 2×2 tensor $(\eta_{21} = \eta_{12})$, there are two orthogonal eigenvectors. These two eigenvectors, $\mathbf{D}_1$ and $\mathbf{D}_2$, correspond to the two normal modes of propagation with refractive indices n_1 and n_2, respectively.

To prove the equivalence, let ξ_1, ξ_2, ξ_3 be the coordinates of an arbitrary point in the new coordinate system. The index ellipsoid in this coordinate system is expressed by

$$\eta_{\alpha\beta} \xi_\alpha \xi_\beta = 1 \qquad (3.7\text{-}19)$$

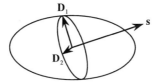

Figure 3.10. Method of the index ellipsoid. The inner ellipse is the intersection of the index ellipsoid with the plane normal to **s**.

where summation over repeated indices $\alpha,\beta(1,2,3)$ is assumed. The intersection ellipse between a plane ($\xi_3 = 0$) through the origin that is normal to the direction of propagation and the index ellipsoid is obtained by putting $\xi_3 = 0$ in Eq. (3.7-19). Thus we obtain the following equation for the intersection ellipse:

$$\eta_{11}\xi_1^2 + \eta_{22}\xi_2^2 + 2\eta_{12}\xi_1\xi_2 = 1 \qquad (3.7\text{-}20)$$

We note that the coefficients of this ellipse are exactly the transverse impermeability tensor η_t. The eigenvectors of this 2×2 tensor therefore are along the principal axes of this ellipse. In addition, the lengths of the principal axes determine the values of n according to Eq. (3.7-18). This proves the equivalence of the method of the index ellipsoid and the method described in Section 3.2.

To illustrate the method of index ellipsoid, we consider a uniaxial crystal where the index ellipsoid is an ellipsoid of revolution (see Fig. 3.10). Let the direction of propagation be given by

$$\mathbf{s} = \begin{pmatrix} \sin\theta\cos\phi \\ \sin\theta\sin\phi \\ \cos\theta \end{pmatrix} \qquad (3.1\text{-}21)$$

Following the recipe described above, the eigenvectors $\mathbf{D}_1,\mathbf{D}_2$ can be obtained by finding the direction of the major and minor axes of the intersection ellipse. The eigen refractive indices are obtained from the length of the major and minor axes. A simple inspection of the intersection ellipse yields

$$\mathbf{D}_1 = \begin{pmatrix} \sin\phi \\ -\cos\phi \\ 0 \end{pmatrix} \qquad (3.7\text{-}22)$$

$$\mathbf{D}_2 = \begin{pmatrix} \cos\theta\cos\phi \\ \cos\theta\sin\phi \\ -\sin\theta \end{pmatrix} \qquad (3.7\text{-}23)$$

$$n_1 = n_o \qquad (3.7\text{-}24)$$

$$\frac{1}{n_2^2} = \frac{\cos^2\theta}{n_o^2} + \frac{\sin^2\theta}{n_e^2} \qquad (3.7\text{-}25)$$

We note that the $\mathbf{D}$ vector of the ordinary wave is in the xy plane and is perpendicular to the c axis. The eigen refractive index of the ordinary wave is n_o. The $\mathbf{D}$ vector of the extraordinary wave is perpendicular to both $\mathbf{s}$ and $\mathbf{D}_1$, with an eigen refractive index such as that given by Eq. (3.7-25). These results are in agreement with those obtained in Section 3.3.

In what follows, we describe an alternative approach that yields the direction of polarization states (for $\mathbf{D}$ vectors) of the normal modes. We recall that the $\mathbf{D}$ vectors are always perpendicular to the direction of propagation. This approach is particularly useful for medium with a small biaxiality.

3.7.2. Perturbation Approach

In a biaxial crystal, the three principal refractive indices n_x, n_y, n_z are all different. Let n_x, n_y be the two that are closest in their refractive indices (i.e., $n_y - n_x \ll n_z - n_y$). The impermeability tensor can be written

$$\eta = \begin{pmatrix} \dfrac{1}{n_x^2} & 0 & 0 \\ 0 & \dfrac{1}{n_y^2} & 0 \\ 0 & 0 & \dfrac{1}{n_z^2} \end{pmatrix} = \begin{pmatrix} \dfrac{1}{n_o^2} & 0 & 0 \\ 0 & \dfrac{1}{n_o^2} & 0 \\ 0 & 0 & \dfrac{1}{n_e^2} \end{pmatrix} + \Delta\eta \qquad (3.7\text{-}26)$$

where

$$\frac{1}{n_o^2} = \frac{1}{2}\left(\frac{1}{n_x^2} + \frac{1}{n_y^2}\right) \qquad (3.7\text{-}27)$$

$$\frac{1}{n_e^2} = \frac{1}{n_z^2} \qquad (3.7\text{-}28)$$

$$\Delta\eta = \frac{1}{2}\left(\frac{1}{n_x^2} - \frac{1}{n_y^2}\right)\begin{pmatrix} 1 & 0 & 0 \\ 0 & -1 & 0 \\ 0 & 0 & 0 \end{pmatrix} \qquad (3.7\text{-}29)$$

where $\Delta\eta$ can be treated as a small perturbation. In other words, $\Delta\eta$ is a measure of the deviation from the uniaxial symmetry. Since the $\mathbf{D}$ vector is always perpendicular to the direction of propagation, we can write the eigenvectors of the uniaxial crystal ($\Delta\eta = 0$) $\mathbf{D}_1, \mathbf{D}_2$ as the base vectors. The eigenvector of the biaxial crystal can now be expressed in terms of $\mathbf{D}_1$ and $\mathbf{D}_2$ as

$$\mathbf{D}' = \alpha\mathbf{D}_1 + \beta\mathbf{D}_2 \qquad (3.7\text{-}30)$$

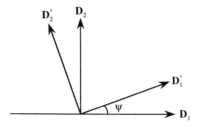

Figure 3.11. Eigenvectors of a biaxial crystal $\mathbf{D}'_1$ and $\mathbf{D}'_2$, where the eigenvectors of the uniaxial crystal $\mathbf{D}_1$, $\mathbf{D}_2$ are used as the base vectors.

where α and β are constants to be determined (see also Fig. 3.11). Thus, the problem of finding the normal modes in the biaxial crystal becomes the following eigenvalue problem:

$$(\eta + \Delta\eta)\mathbf{D}' = \frac{1}{n^2}\mathbf{D}' \tag{3.7-31}$$

Using the unperturbed $\mathbf{D}_1, \mathbf{D}_2$ as the base vectors, the preceding matrix equation becomes

$$\begin{pmatrix} \dfrac{1}{n_1^2} - \dfrac{1}{n^2} + \Delta\eta_{11} & \Delta\eta_{12} \\[2ex] \Delta\eta_{21} & \dfrac{1}{n_2^2} - \dfrac{1}{n^2} + \Delta\eta_{22} \end{pmatrix} \begin{pmatrix} \alpha \\ \beta \end{pmatrix} = 0 \tag{3.7-32}$$

The solutions are

$$\frac{1}{n^2} = \frac{1}{2}\left(\frac{1}{n_1^2} + \frac{1}{n_2^2} + \Delta\eta_{11} + \Delta\eta_{22}\right)$$

$$\pm \frac{1}{2}\sqrt{\left(\frac{1}{n_1^2} - \frac{1}{n_2^2} + \Delta\eta_{11} - \Delta\eta_{22}\right)^2 + 4\Delta\eta_{12}\Delta\eta_{21}} \tag{3.7-33}$$

$$\tan\psi = \frac{\beta}{\alpha} = \frac{\Delta\eta_{21}}{\dfrac{1}{n^2} - \dfrac{1}{n_2^2} - \Delta\eta_{22}} \tag{3.7-34}$$

where

$$\Delta\eta_{11} = \mathbf{D}_1^* \cdot \Delta\eta\mathbf{D}_1, \qquad \Delta\eta_{12} = \mathbf{D}_1^* \cdot \Delta\eta\mathbf{D}_2,$$
$$\Delta\eta_{21} = \mathbf{D}_2^* \cdot \Delta\eta\mathbf{D}_1, \qquad \Delta\eta_{22} = \mathbf{D}_2^* \cdot \Delta\eta\mathbf{D}_2$$

The angle ψ, according to Eq. (3.7-34), is the angle between the polarization vector of the normal modes measured from the unperturbed polarization vector $\mathbf{D}_1$ (see Fig. 3.11).

REFERENCES

1. S. D. Jacobs, "Liquid crystals for laser applications," in *Handbook of Laser Science and Technology*, M. J. Weber, ed., CRC Press, 1995, pp. 409–465.
2. E. U. Condon, "Theories of optical rotatory power," *Rev. Mod. Phys.* **9**, 432 (1937); *Handbook of Physics*, E. U. Condon and H. Odishaw, eds., McGraw-Hill, New York, 1958, pp. 6–12.
3. G. Szivessy and C. Munster, "Lattice optics of active crystals," *Ann. Phys.* (Leipzig) **20**, 703–736 (1934).

SUGGESTED READINGS

A. Yasiv and P. Yeh, *Optical Waves in Crystals*, Wiley, 1984.

P. Yeh, 'Optical activity,' in *Encylopedia of Materials Science and Engineering*, M. B. Bever, ed., Pegamon Press, 1986, p. 3292.

PROBLEMS

3.1. *Eigenpolarization*:
   (a) Derive the expression for the eigenpolarization of the electric field vector (3.2-8), (3.2-10).
   (b) Use the relation $\mathbf{D} = \varepsilon \mathbf{E}$ and obtain an expression for the corresponding eigenpolarization of the displacement vector $\mathbf{D}$.
   (c) Let n_1 and n_2 be the solution of the Fresnel equation (3.2-9), and let $\mathbf{E}_1, \mathbf{E}_2, \mathbf{D}_1, \mathbf{D}_2$ the corresponding eigen field vectors. Evaluate $\mathbf{E}_1 \cdot \mathbf{E}_2$, and $\mathbf{D}_1 \cdot \mathbf{D}_2$, and show that $\mathbf{D}_1$ and $\mathbf{D}_2$ are always mutually orthogonal whereas $\mathbf{E}_1$ and $\mathbf{E}_2$ are mutually orthogonal only in a uniaxial or an isotropic medium.
   (d) Show that $\mathbf{E}_1 \cdot \mathbf{D}_2 = 0$ and $\mathbf{E}_2 \cdot \mathbf{D}_1 = 0$.

3.2. *Fresnel equation*:
   (a) Derive the Fresnel equation (3.2-9) directly from Eq. (3.2-7).
   (b) Show that the Fresnel equation (3.2-9) is a quadratic equation in n^2, that is, $An^4 + Bn^2 + C = 0$, and obtain expressions for A, B, and C.
   (c) Show that $B^2 - 4AC > 0$ for the case of pure dielectrics with real $\varepsilon_x, \varepsilon_y, \varepsilon_z$.
   (d) Derive Eq. (3.2-36) from (3.2-7) for the case of uniaxial crystals.
   (e) Show that in an isotropic medium, Eq. (3.2-7) reduces to

$$k^2 - \frac{\varepsilon}{\varepsilon_0} \left(\frac{\omega}{c}\right)^2 = 0.$$

3.3. Show that the group velocity for a wavepacket propagation in an anisotropic medium also represents the transport of energy, that is, $\mathbf{v}_g = \mathbf{v}_e$. Show that the equality $\mathbf{v}_g = \mathbf{v}_e$ also holds for complex field amplitudes $\mathbf{E}$ and $\mathbf{H}$.

3.4. *Group velocity and phase velocity:*
   (a) Derive an expression for the group velocity of the extraordinary wave in a uniaxial crystal as a function of the polar angle θ of the propagation vector.
   (b) Derive an expression for the angle α between the phase velocity and the group velocity. This angle is also the angle between the field vectors $\mathbf{E}$ and $\mathbf{D}$.
   (c) Show that $\alpha = 0$ when $\theta = 0$, $\pi/2$. Find the angle θ at which α is maximized, and obtain an expression for α_{max}. Calculate this angle α_{max} for ZLI-1646 with $n_o = 1.478$, $n_e = 1.558$.
   (d) Show that for $n_o \cong n_e$ the maximum angular separation α_{max} occurs at $\theta \cong 45°$, and show that α_{max} is proportional to $|n_o - n_e|$.

3.5. (a) Show that the normal surface Eq. (3.2-7) [or Eq. (3.2-9)] can also be written

$$\frac{k_x^2}{k^2 - \omega^2 \mu \varepsilon_x} + \frac{k_y^2}{k^2 - \omega^2 \mu \varepsilon_y} + \frac{k_z^2}{k^2 - \omega^2 \mu \varepsilon_z} = 1$$

   where $k^2 = k_x^2 + k_y^2 + k_z^2$.
   (b) Find the normal vector to the normal surface by taking the gradient of the equation in (a) and show that the normal vector is perpendicular to the eigenvectors in Eqs. (3.2-8). This proves that the eigenvectors for $\mathbf{E}$ are tangent to the normal surface.

3.6. A biaxial crystal is characterized by its three principal indices $n_1 < n_2 < n_3$.
   (a) Show that the intersection of the normal surface with any coordinate plane consists of an ellipse and a circle.
   (b) Show that only the ellipse and the circle in the xz plane intersect. These points of intersection define the optic axes of the crystal.
   (c) Find the points of intersection of the ellipse and the circle in the xz plane, and derive the expression (3.7-5) for the angle between one of the optic axes and the z axis. Show that this angle approaches zero when $n_1 = n_2$. Calculate this angle for mica with $n_1 = 1.552$, $n_2 = 1.582$, and $n_3 = 1.588$.
   (d) The aperture angle of conical refraction can be estimated by the angle between the normal vectors of the ellipse and the circle at the point of intersection. Derive Eq. (3.7-9).

(e) Show that this cone angle vanishes when $n_1 = n_2$ (or $n_2 = n_3$). Calculate this cone angle for mica.

(f) Show that a rotation of an angle $\pm \chi$ in the $\xi'\zeta'$ plane will transform the cone (3.7-10) into

$$\left(\zeta'' \pm \tfrac{1}{2}\xi''\tan2\chi\right)^2 + \eta^2 = \left(\tfrac{1}{2}\xi''\tan2\chi\right)^2$$

The intersection of this cone with the plane $\xi'' = $ constant is a circle.

3.7. The phenomenon of double refraction in an anisotropic crystal may be utilized to produce polarized light. Consider a light beam incident on a plane boundary from the inside of a calcite crystal ($n_o = 1.658$, $n_e = 1.486$). Suppose that the c axis of the crystal is normal to the plane of incidence.

(a) Find the range of the internal angle of incidence such that the ordinary wave is totally reflected. The transmitted wave is thus completely polarized.

(b) Calculate values for a Glan prism. Use the basic principle described in (a) to design a calcite Glan prism shown in the following:

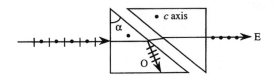

Find the range of the apex angle α.

3.8. A linearly polarized electromagnetic wave at $\lambda = 6328\,\text{Å}$ is incident normally at $x = 0$ on the yz face of a liquid crystal E7 (with $n_e = 1.75$, $n_o = 1.52$) so that it propagates along the x axis. Suppose that the wave is linearly polarized initially so that it has equal components along y and z.

(a) Determine the state of polarization at the plane x where

$$(k_z - k_y)x = \frac{\pi}{2}$$

Plot the position of the electric field vector in this plane at times $t = 0$, $\pi/6\omega$, $\pi/3\omega$, $\pi/2\omega$, $2\pi/3\omega$, $5\pi/6\omega$.

(b) A plate that satisfies the condition in (a) is known as a *quarter-wave plate* because the difference in the phase shift for the two orthogonal polarization states is a quarter of 2π. Find the thickness of a E7 quarter-wave plate at $\lambda = 6328\,\text{Å}$.

(c) Determine values for a half-wave plate. What is the state of polarization at the plane x where

$$\frac{2\pi}{\lambda}(n_e - n_o)x = \pi$$

3.9. Dichroic polarizers are made of materials whose absorption properties and reflection properties are strongly dependent on the direction of vibration of the electric field. If the two absorption coefficients are very different, a thin sheet of the material will be sufficient to transform unpolarized light into a nearly linearly polarized light.

(a) Let α_1, α_2 be the two abosrption coefficients corresponding to two independent polarizations. Derive an expression for the ratio of the two transmitted components as a function of the thickness of the medium.

(b) Show that, strictly speaking, the normal modes of propagation are no longer linearly polarized in the presence of absorption.

3.10. For a Hermitian dielectric tensor, consider the addition of a small antisymmetric term in the dielectric tensor:

$$D_i = (\varepsilon_{ij} - i\gamma_{ij})E_j \tag{A}$$

where γ_{ij} is the antisymmetric (i.e., $\gamma_{ij} = -\gamma_{ji}$) part, which is not to have an ohmic, dissipative character as in the complex dielectric constant of a metal. Rather, it is to be conservative, which means that γ_{ij} must not contribute to the electric energy density $U_e = \frac{1}{2}(\mathbf{D \cdot E})$ as computed from Maxwell's equations.

(a) Evaluate $\mathbf{D \cdot E}$ and show that $\mathbf{D \cdot E} = \varepsilon_{ij}E_iE_j$, provided γ_{ij} is an antisymmetric tensor, that is, $\gamma_{ij} = -\gamma_{ji}$.

(b) Show that Eq. (A) can be written $D_i = \varepsilon_{ij}E_j + i(\gamma \times \mathbf{E})_i$, where γ is a vector with components

$$\gamma_1 = -\gamma_{23} = \gamma_{32}, \qquad \gamma_2 = -\gamma_{31} = \gamma_{13}, \qquad \gamma_3 = -\gamma_{12} = \gamma_{21}$$

(c) Show that the inverse of a Hermitian tensor is also a Hermitian tensor, and show that

$$E_i = \left\{ \left(\frac{1}{\varepsilon}\right)_{ij} - i\left(\frac{1}{\varepsilon}\right)_{ik} \gamma_{kl} \left(\frac{1}{\varepsilon}\right)_{lj} \right\} D_j \tag{B}$$

(d) Show that Eq. (B) can be written

$$E_i = \left(\frac{1}{\varepsilon}\right)_{ij} D_j - i(\mathbf{A} \times \mathbf{D})_i \tag{C}$$

and show that the components of the vector $\mathbf{A}$ in the principal coordinate system are given by

$$A_1 = \frac{1}{\varepsilon_{11}\varepsilon_{22}\varepsilon_{33}} \gamma_1 \varepsilon_{11}, \qquad A_2 = \frac{1}{\varepsilon_{11}\varepsilon_{22}\varepsilon_{33}} \gamma_2 \varepsilon_{22},$$

$$A_3 = \frac{1}{\varepsilon_{11}\varepsilon_{22}\varepsilon_{33}} \gamma_3 \varepsilon_{33}.$$

(e) Show that the wave equation (3.6-18) can be written

$$\mathbf{s} \times \left[\mathbf{s} \times \left(\frac{1}{\varepsilon} - i\mathbf{A}\times\right)\mathbf{D}\right] = -\frac{1}{n^2}\mathbf{D} \qquad \text{(D)}$$

(f) Let the vector $\mathbf{A}$ be decomposed into $\mathbf{A} = (\mathbf{A}\cdot\mathbf{s})\mathbf{s} + \mathbf{A}_\perp$, where $\mathbf{A}_\perp$ is the component of $\mathbf{A}$ perpendicular to the direction of propagation. Show from the wave equation (D) that the optical activity is independent of $\mathbf{A}_\perp$. In other words, only $\mathbf{A}\cdot\mathbf{s}$ affects the optical rotation.

(g) In an anisotropic medium the vector $\mathbf{A}$ is given by $\mathbf{A} = \alpha\mathbf{s}$, where α is a symmetric matrix. Show that the optical activity parameter is given by $A = \alpha_{ij}s_i s_j$.

(h) By following the argument of (g) and assuming that $\mathbf{s}\cdot\mathbf{E}=0$, show that only the component of $\mathbf{G}$ along the direction of propagation will affect the optical activity. In an anisotropic medium, $\mathbf{G} = g\mathbf{s}$, where g is the gyration tensor. Show that the parameter G is Eq. (3.6-10) is $\mathbf{G}\cdot\mathbf{s}$ and derive Eq. (3.6-15).

3.11. Determine displacement eigenmodes. The wave equation (3.2-5) can be written

$$\mathbf{s} \times \left[\mathbf{s} \times \frac{\varepsilon_0}{\varepsilon}\mathbf{D}\right] = -\frac{1}{n^2}\mathbf{D}$$

where $\mathbf{s}$ is the unit vector in the direction of propagation. Let $\mathbf{D}_1, \mathbf{D}_2$ be the normalized eigenvectors with eigenvalues $1/n_1^2, 1/n_2^2$, respectively. We asume that $\varepsilon_0/\varepsilon$ is a Hermitian tensor.

(a) Show that

$$\left(\frac{\varepsilon_0}{\varepsilon}\right)_{11} \equiv \mathbf{D}_1^* \cdot \frac{\varepsilon_0}{\varepsilon}\mathbf{D}_1 \equiv \frac{1}{n_1^2}, \qquad \left(\frac{\varepsilon_0}{\varepsilon}\right)_{22} \equiv \mathbf{D}_2^* \cdot \frac{\varepsilon_0}{\varepsilon}\mathbf{D}_2 \equiv \frac{1}{n_2^2},$$

$$\left(\frac{\varepsilon_0}{\varepsilon}\right)_{12} \equiv \mathbf{D}_1^* \cdot \frac{\varepsilon_0}{\varepsilon}\mathbf{D}_2 = 0.$$

(b) Show that $\mathbf{D}_1^* \cdot \mathbf{D}_2 = 0$.

3.12. To calculate displacement eigenmodes in gyrotropic media, start from Eq. (3.6-20).

(a) Derive Eqs. (3.6-22) and (3.6-24).

(b) Show that the Jones vectors (3.6-25) satisfy the vector equation (3.6-22).

(c) Show that the Jones vectors (3.6-25) are orthogonal, that is, $\mathbf{J}_+^* \cdot \mathbf{J}_- = 0$.

(d) Let $e_\pm$ be the ellipticities of the two eigenmodes. Show that $e_+ e_- = -1$.

3.13. *Conservation of power flow*:
  (a) Show that the power flow in the direction of propagation is given by

$$\mathbf{S}\cdot\mathbf{s} = \frac{1}{2}\frac{c}{\varepsilon_0}\left[\frac{1}{n_1^3}|\mathbf{D}_1|^2 + \frac{1}{n_2^3}|\mathbf{D}_2|^2\right]$$

  where $\mathbf{D}_1,\mathbf{D}_2$ are the amplitudes of the displacement fields of the eigenmodes and n_1, n_2 are the corresponding refractive indices.
  (b) Use the relation (3.2-13) to show that

$$\mathbf{S}\cdot\mathbf{s} = \frac{1}{2\mu c}[n_1|\mathbf{A}_1|^2 + n_2|\mathbf{A}_2|^2]$$

  where $\mathbf{A}_1,\mathbf{A}_2$ are the transverse parts of the electric field vectors.
  (c) Show that the total power flow along the direction of propagation is a constant of integration:

$$\frac{d}{d\zeta}[n_1|\mathbf{A}_1|^2 + n_2|\mathbf{A}_2|^2] = 0$$

  where ζ is the distance along the direction of propagation $\mathbf{s}$ (i.e., $\zeta = \mathbf{s}\cdot\mathbf{r}$).

3.14. *Crossed polarizers*:
  (a) Show that, for normal incidence, the transmittance for unpolarized light is given by $T_x = \frac{1}{2}T_1T_2$.
  (b) Using the following identity

$$(\mathbf{A}\times\mathbf{B})\cdot(\mathbf{C}\times\mathbf{D}) = (\mathbf{A}\cdot\mathbf{C})(\mathbf{B}\cdot\mathbf{D}) - (\mathbf{A}\cdot\mathbf{D})(\mathbf{B}\cdot\mathbf{C})$$

  show that

$$\mathbf{o}_1\cdot\mathbf{o}_2 = -\frac{(\mathbf{k}\cdot\mathbf{c}_1)(\mathbf{k}\cdot\mathbf{c}_2)}{|\mathbf{k}\times\mathbf{c}_1\,\|\,\mathbf{k}\times\mathbf{c}_2|}$$

  for crossed polairzers.
  (c) Show that

$$T_x(\theta,\phi) = \frac{1}{2}\frac{\sin^4\theta\sin^2\phi\cos^2\phi}{(1-\sin^2\theta\sin^2\phi)(1-\sin^2\theta\cos^2\phi)}.$$

3.15. Show that for elliptically polarized light in an optically active medium
  (a) The polarization ellipse remains unchanged.
  (b) The specific rotatory power Eq. (3.6-4) remains valid for the principal axes of the ellipse.

3.16. Using the method discussed in Section 3.7 regarding biaxial crystals, calculate the normal modes of propagation for the following crystal configurations:

(a) $n_z = 1.65, n_x = 1.5, n_y = 1.55$, with $\theta = 45°, \phi = 45°$

(b) $n_z = 1.75, n_x = 1.5, n_y = 1.55$, with $\theta = 45°, \phi = 45°$

You need to calculate the angle of rotation ψ in the $\mathbf{D}_1 - \mathbf{D}_2$ plane, as well as the eigen-indices of refraction.

4

Jones Matrix Method

In the last chapter we discussed the propagation of plane waves in homogeneous and anisotropic media. Any plane-wave propagation can be easily described in terms of a superposition (or linear combination) of the propagation of two orthogonally polarized components. In the case of LC displays, the LC layer (e.g., TN-LC) is often inhomogeneous. In addition, the display may consist of several birefringent plates that are employed as phase retardation compensators. In a situation like this, a systematic method is needed to treat the propagation of light through the display. In this chapter, we describe such a powerful technique, known as the *Jones matrix method* [1].

4.1. JONES MATRIX FORMULATION

We have shown in the previous chapter that light propagation in a birefringent medium consists of a linear superposition of two normal modes. These normal modes have well defined phase velocities and directions of polarization. The birefringent media may be either uniaxial or biaxial. However, most commonly used materials for LC displays are uniaxial. In a uniaxial medium, these normal modes are the ordinary and the extraordinary waves. The directions of polarization for these normal modes are mutually orthogonal and are called the "slow" and "fast" axes of the medium for that direction of propagation. In traditional birefringent optics, retardation plates are usually cut in such a way that the c axis lies in the plane of the plate surfaces. Thus the propagation direction of normally incident light is perpendicular to the c axis. In LCDs, the orientation of the c axis (director) depends on the voltage applied, as well as on the boundary conditions.

Retardation plates (also called *wave plates*) and LC cells are polarization-state converters, or transformers. The polarization state of a light beam can be converted to any other polarization state by means of a suitable retardation plate. In formulating the Jones matrix method, we assume that there is no reflection of light from either surface of the plate and the light is totally transmitted through the plate surfaces. In practice, there is reflection, although most retardation plates are coated so as to reduce the surface reflection loss. The Fresnel reflections at the plate surfaces not only decrease the transmitted intensity but also affect the fine structure of the spectral transmittance because of multiple-reflection

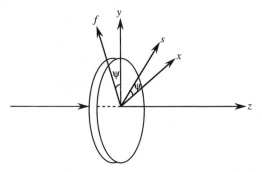

Figure 4.1. A retardation plate with azimuth angle ψ.

interference (Fabry–Perot effect). Referring to Figure 4.1, we consider an incident beam of light with a polarization state described by the Jones vector

$$\mathbf{V} = \begin{pmatrix} V_x \\ V_y \end{pmatrix} \tag{4.1-1}$$

where V_x and V_y are two complex numbers. The x and y axes are fixed laboratory axes. To determine how the light propagates in the retardation plate, we need to decompose the light into a linear combination of the "fast" and "slow" normal modes of the medium. This is done by the coordinate transformation

$$\begin{pmatrix} V_s \\ V_f \end{pmatrix} = \begin{pmatrix} \cos\psi & \sin\psi \\ -\sin\psi & \cos\psi \end{pmatrix} \begin{pmatrix} V_x \\ V_y \end{pmatrix} \equiv R(\psi) \begin{pmatrix} V_x \\ V_y \end{pmatrix} \tag{4.1-2}$$

where $R(\psi)$ is the coordinate rotation matrix, and V_s is the slow component of the polarization vector $\mathbf{V}$, whereas V_f is the fast component. The "slow" and "fast" axes are fixed in the medium. These two components are normal modes of the retardation plate and will propagate with their own phase velocities and polarizations. The azimuth ψ is defined as the angle between the sf-coordinate and the xy coordinate (see Fig. 4.1) with the z axis as the axis of the coordinate rotation. Because of the difference in phase velocity, one component is retarded relative to the other. This retardation changes the polarization state of the emerging beam. Table 4.1 lists the Jones vector of various polarization states.

Let n_s and n_f be the refractive indices associated with the propagation of the "slow" and "fast" components, respectively. The polarization state of the emerging beam in the medium sf coordinate system is given by

$$\begin{pmatrix} V_s' \\ V_f' \end{pmatrix} = \begin{pmatrix} \exp\left(-in_s \dfrac{2\pi}{\lambda} d\right) & 0 \\ 0 & \exp\left(-in_f \dfrac{2\pi}{\lambda} d\right) \end{pmatrix} \begin{pmatrix} V_s \\ V_f \end{pmatrix} \tag{4.1-3}$$

Table 4.1. Jones Vectors

Polarization State	Jones Vector
	$\begin{pmatrix} \cos\phi \\ \sin\phi \end{pmatrix}$
	$\dfrac{1}{\sqrt{2}}\begin{pmatrix} 1 \\ -i \end{pmatrix}$
	$\dfrac{1}{\sqrt{2}}\begin{pmatrix} 1 \\ i \end{pmatrix}$
	$\begin{pmatrix} a\cos\phi + ib\sin\phi \\ a\sin\phi - ib\cos\phi \end{pmatrix}$
	$\begin{pmatrix} a\cos\phi - ib\sin\phi \\ a\sin\phi + ib\cos\phi \end{pmatrix}$

Note: a and b are the principal axes of the ellipse.

where d is the thickness of the plate and λ is the wavelength of the light beam. The phase retardation is the difference of the exponents in Eq. (4.1-3) and is defined as

$$\Gamma = \frac{2\pi}{\lambda}(n_s - n_f)d \qquad (4.1\text{-}4)$$

Note that the phase retardation Γ is a measure of the relative change in phase as a result of the propagation, not the absolute change. The birefringence of a typical retardation plate is small, that is, $|n_s - n_f| \ll n_s, n_f$. Consequently, the absolute change in phase caused by the plate may be hundreds of times greater than the phase retardation. Let ϕ be the mean absolute phase change:

$$\phi = \frac{1}{2}(n_s + n_f)\frac{2\pi}{\lambda}d \qquad (4.1\text{-}5)$$

Then Eq. (4.1-3) can be written in terms of ϕ and Γ as

$$\begin{pmatrix} V'_s \\ V'_f \end{pmatrix} = e^{-i\phi}\begin{pmatrix} e^{-i\Gamma/2} & 0 \\ 0 & e^{i\Gamma/2} \end{pmatrix}\begin{pmatrix} V_s \\ V_f \end{pmatrix} \qquad (4.1\text{-}6)$$

The Jones vector of the polarization state of the emerging beam in the xy coordinate is given by transforming back from the sf coordinate system:

$$\begin{pmatrix} V'_x \\ V'_y \end{pmatrix} = \begin{pmatrix} \cos\psi & -\sin\psi \\ \sin\psi & \cos\psi \end{pmatrix}\begin{pmatrix} V'_s \\ V'_f \end{pmatrix} \qquad (4.1\text{-}7)$$

By combining Eqs. (4.1-1), (4.1-6), and (4.1-7), we can write the transformation due to the retardation plate as

$$\begin{pmatrix} V'_x \\ V'_y \end{pmatrix} = R(-\psi)W_0R(\psi)\begin{pmatrix} V_x \\ V_y \end{pmatrix} \qquad (4.1\text{-}8)$$

where $R(\psi)$ is the coordinate rotation matrix and W_0 is the Jones matrix for the retardation plate. These are given, respectively, by

$$R(\psi) = \begin{pmatrix} \cos\psi & \sin\psi \\ -\sin\psi & \cos\psi \end{pmatrix} \qquad (4.1\text{-}9a)$$

and

$$W_0 = e^{-i\phi}\begin{pmatrix} e^{-i\Gamma/2} & 0 \\ 0 & e^{i\Gamma/2} \end{pmatrix} \qquad (4.1\text{-}9b)$$

The phase factor $e^{-i\phi}$ can be neglected if interference effects due to multiple reflections are not important, or not observable. A retardation plate is characterized by its phase retardation Γ and its azimuth angle ψ, and is represented by the product of three matrices (4.1-8):

$$W = R(-\psi)W_0R(\psi) \qquad (4.1\text{-}10)$$

Note that the Jones matrix of a wave plate is a unitary matrix, that is, $W^\dagger W = 1$, where the dagger† denotes Hermitian conjugate. The passage of a beam of polarized light through a wave plate is mathematically described as a unitary transformation. Many physical properties are invariant under unitary transformations; these include the orthogonal relationship between the Jones vectors and the magnitude of the Jones vectors. Thus, if the polarization states of two beams are mutually orthogonal, they will remain orthogonal after passing through an arbitrary wave plate. The general properties of the Jones matrices will be further discussed later in this section.

The Jones matrix of an ideal homogeneous linear sheet polarizer oriented with its transmission axis parallel to the laboratory x axis is

$$P_0 = e^{-i\phi'} \begin{pmatrix} 1 & 0 \\ 0 & 0 \end{pmatrix} \qquad (4.1\text{-}11)$$

where ϕ' is the absolute phase accumulated as a result of the finite optical thickness of the polarizer. The Jones matrix of a polarizer rotated by an angle ψ about the z axis is given by

$$P = R(-\psi)P_0 R(\psi) \qquad (4.1\text{-}12)$$

Thus, if we neglect the absolute phase ϕ' the Jones matrix representations of the polarizers transmitting light with electric field vectors parallel to the x and y axes, respectively, are given by

$$P_x = \begin{pmatrix} 1 & 0 \\ 0 & 0 \end{pmatrix} \quad \text{and} \quad P_y = \begin{pmatrix} 0 & 0 \\ 0 & 1 \end{pmatrix}$$

To find the effect of a train of retardation plates and polarizers on the polarization state of a polarized light, we write down the Jones vector of the incident beam, and then write down the Jones matrices of the various elements. The Jones vector of the emerging beam is obtained by carrying out the matrix multiplication in sequence. Table 4.2 lists the Jones matrices of wave plates and polarizers at various azimuth angles.

4.1.1. Half-Wave Retardation Plate and Quarter-Wave Plate

Example 4.1: A Half-Wave Retardation Plate. A half-wave plate has a phase retardation of $\Gamma = \pi$. According to Eq. (4.1-4), an a-cut* uniaxial plate will act as

---

*A crystal plate is considered a-cut if its faces are perpendicular to the principal a axis.

Table 4.2. Jones Matrices

Optical Element	Jones Matrices

Wave Plates

$$\Gamma \equiv \frac{2\pi}{\lambda}(n_e - n_o)d$$

$$\begin{pmatrix} e^{-i\Gamma/2} & 0 \\ 0 & e^{i\Gamma/2} \end{pmatrix}$$

$$\begin{pmatrix} e^{i\Gamma/2} & 0 \\ 0 & e^{-i\Gamma/2} \end{pmatrix}$$

$$\begin{pmatrix} \cos\frac{\Gamma}{2} & -i\sin\frac{\Gamma}{2} \\ -i\sin\frac{\Gamma}{2} & \cos\frac{\Gamma}{2} \end{pmatrix}$$

$$R(-\psi)\begin{pmatrix} e^{-i\Gamma/2} & 0 \\ 0 & e^{i\Gamma/2} \end{pmatrix}R(\psi)$$

$$= \begin{pmatrix} \cos\psi & -\sin\psi \\ \sin\psi & \cos\psi \end{pmatrix}\begin{pmatrix} e^{-i\Gamma/2} & 0 \\ 0 & e^{i\Gamma/2} \end{pmatrix}\begin{pmatrix} \cos\psi & \sin\psi \\ -\sin\psi & \cos\psi \end{pmatrix}$$

Polarizers

$$\begin{pmatrix} 1 & 0 \\ 0 & 0 \end{pmatrix}$$

$$\begin{pmatrix} 0 & 0 \\ 0 & 1 \end{pmatrix}$$

$$\begin{pmatrix} 1/2 & 1/2 \\ 1/2 & 1/2 \end{pmatrix}$$

$$R(-\psi)\begin{pmatrix} 1 & 0 \\ 0 & 0 \end{pmatrix}R(\psi)$$

$$= \begin{pmatrix} \cos\psi & -\sin\psi \\ \sin\psi & \cos\psi \end{pmatrix}\begin{pmatrix} 1 & 0 \\ 0 & 0 \end{pmatrix}\begin{pmatrix} \cos\psi & \sin\psi \\ -\sin\psi & \cos\psi \end{pmatrix}$$

$$= \begin{pmatrix} \cos\psi & -\sin\psi \\ \sin\psi & \cos\psi \end{pmatrix}\begin{pmatrix} \cos\psi & \sin\psi \\ 0 & 0 \end{pmatrix}\begin{pmatrix} \cos^2\psi & \cos\psi\sin\psi \\ \sin\psi\cos\psi & \sin^2\psi \end{pmatrix}$$

* Here we assume that the c axis is perpendicular to the z axis (i.e., $\theta = 90°$). In general, Eq. (3.3-4) must be used for n_e if $\theta \neq 90°$.

a half-wave plate provided the thickness is $t = \lambda/2|n_e - n_o|$ (or old multiples thereof). We will determine the effect of a half-wave plate on the polarization state of a transmitted light beam. The azimuth angle of the wave plate is taken as 45°, and the incident beam as vertically polarized. The Jones vector for the incident beam can be written as

$$V = \begin{pmatrix} 0 \\ 1 \end{pmatrix} \tag{4.1-13}$$

and the Jones matrix for the half-wave plate is obtained by using Eqs. (4.1-9) and (4.1-10):

$$W = \frac{1}{\sqrt{2}} \begin{pmatrix} 1 & -1 \\ 1 & 1 \end{pmatrix} \begin{pmatrix} -i & 0 \\ 0 & i \end{pmatrix} \frac{1}{\sqrt{2}} \begin{pmatrix} 1 & 1 \\ -1 & 1 \end{pmatrix} = \begin{pmatrix} 0 & -i \\ -i & 0 \end{pmatrix}. \tag{4.1-14}$$

The Jones vector for the emerging beam is obtained by multiplying Eqs. (4.1-14) and (4.1-13); the result is

$$V' = \begin{pmatrix} -i \\ 0 \end{pmatrix} = -i \begin{pmatrix} 1 \\ 0 \end{pmatrix} \tag{4.1-15}$$

This represents a beam of horizontally polarized light. The effect of the half-wave plate is to rotate the polarization by 90°. It can be shown that for a general azimuth angle ψ, the half-wave plate will rotate the polarization by an angle 2ψ (see Problem 4.1). In other words, linearly polarized light remains linearly polarized, except that the plane of polarization is rotated by an angle of 2ψ.

When the beam of incident light is circularly polarized, a half-wave plate will covert a beam of right-hand circularly polarized light into a beam of left-hand circularly polarized light and vice versa, regardless of the azimuth angle. The proof is left as an exercise (see Problem 4.1). Figure 4.2 illustrates the effect of half-wave plate. ∎

Example 4.2: A Quarter-Wave Plate. A quarter-wave plate has a phase retardation of $\Gamma = \pi/2$. If the plate is made of an a-cut (or b-cut) uniaxially anisotropic medium, the thickness is $t = \lambda/4|n_e - n_o|$ (or odd multiples thereof). Suppose again that the azimuth angle of the plate is $\psi = 45°$ and the incident beam is vertically polarized. The Jones vector for the incident beam is given again by Eq. (4.1-13). The Jones matrix for this quarter-wave plate, according to Eq. (4.1-10), is

$$W = \frac{1}{\sqrt{2}} \begin{pmatrix} 1 & -1 \\ 1 & 1 \end{pmatrix} \begin{pmatrix} e^{-i\pi/4} & 0 \\ 0 & e^{i\pi/4} \end{pmatrix} \frac{1}{\sqrt{2}} \begin{pmatrix} 1 & 1 \\ -1 & 1 \end{pmatrix} = \frac{1}{\sqrt{2}} \begin{pmatrix} 1 & -i \\ -i & 1 \end{pmatrix} \tag{4.1-16}$$

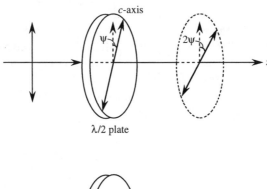

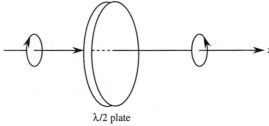

$\lambda/2$ plate

Figure 4.2. Effect of a half-wave plate on the polarization state of a beam.

The Jones vector for the emerging beam is obtained by multiplying Eqs. (4.1-16) and (4.1-13) and is given by

$$\mathbf{V}' = \frac{1}{\sqrt{2}}\begin{pmatrix} -i \\ 1 \end{pmatrix} = \frac{-i}{\sqrt{2}}\begin{pmatrix} 1 \\ i \end{pmatrix} \tag{4.1-17}$$

This represents a beam of left-hand circularly polarized light. The effect of a 45°-oriented quarter-wave plate is to convert a beam of vartically polorized light into a beam of left-hand circularly polarized light. If the incident beam is horizontally polarized, the emerging beam will be right-hand circularly polarized. In general, a quarter-wave plate can convert a linearly polarized light into an elliptically polarized light and vice versa. The effect of the quarter-wave plate is illustrated in Figure 4.3.   ∎

4.1.2. General Properties of Jones Matrix

Here, we consider some general properties of the Jones matrix. Specifically, we consider the Jones matrix under the transformation of retroreflection, mirror reflection, and time reversal and then discuss the principle of reciprocity. We also consider the unimodular and unitary nature of the Jones matrix.

We consider a birefringent system (between $z = 0$ and $z = L$), which consists of a series of anisotropic wave plates (e.g., LC cells) with arbitrary orientations of the optical axes. For the purpose of discussion, we define the following two Jones matrices.

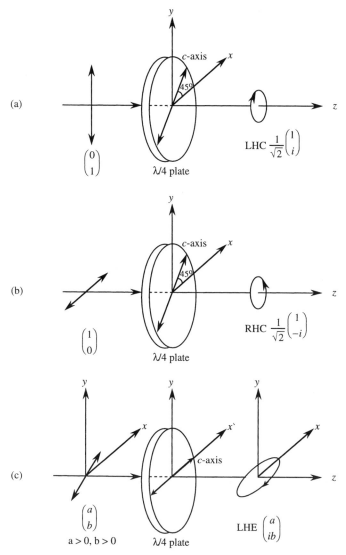

Figure 4.3. (a)–(f) Effect of a quarter-wave plate (assuming $\Gamma = k(n_e - n_o)d = \pi/2$) on the polarization state of a linearly polarized beam and an elliptically polarized beam.

1. Incidence from left side $(z = 0)$—the relationship between the input Jones vector $\mathbf{V}^{in}$ at $(z = 0)$ and the output Jones vector $\mathbf{V}^{out}$ at $(z = L)$ is written

$$\begin{pmatrix} V_x^{out} \\ V_y^{out} \end{pmatrix} = \begin{pmatrix} M_{11} & M_{12} \\ M_{21} & M_{22} \end{pmatrix} \begin{pmatrix} V_x^{in} \\ V_y^{in} \end{pmatrix} \qquad (4.1\text{-}18)$$

where M is the Jones matrix for incidence from the left side.

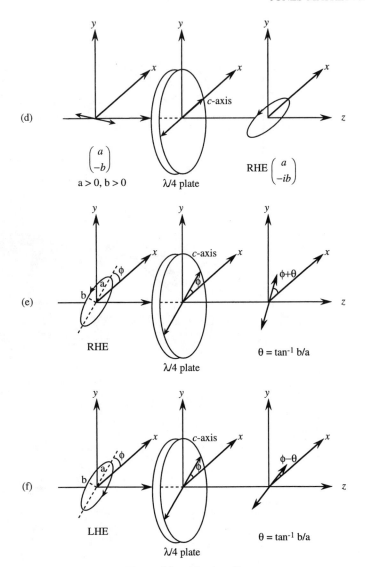

Figure 4.3. (*Continued*).

2. Incidence from right side ($z = L$)—the relationship between the input Jones vector $\mathbf{V}^{\text{in}}$ at ($z = L$) and the output Jones vector $\mathbf{V}^{\text{out}}$ at ($z = 0$) is written

$$\begin{pmatrix} V_x^{\text{out}} \\ V_y^{\text{out}} \end{pmatrix} = \begin{pmatrix} N_{11} & N_{12} \\ N_{21} & N_{22} \end{pmatrix} \begin{pmatrix} V_x^{\text{in}} \\ V_y^{\text{in}} \end{pmatrix} \qquad (4.1\text{-}19)$$

where N is the Jones matrix for incidence from the right side. Here we assume that the beam of light retrace the path defined by the incidence from $z = 0$ in (1). In other words, case 2 can be obtained by a retroreflection of the beam of light in case 1.

The following fundamental properties of these Jones matrices are based on the preceding definitions:

Time Reversal Symmetry. Consider the situation when the time is reversed. The output beam will retrace the beam path, propagate through the birefringent system, and become the phase conjugate of the input beam. As a result of the time reversal symmetry, it can be shown that

$$NM^* = 1 \qquad (4.1\text{-}20)$$

Principle of Reciprocity. From the fundamental principle of reciprocity in physics and the definition of the Jones matrices in Eqs. (4.1-18) and (4.1-19), we have

$$N_{11} = M_{11}, \qquad N_{22} = M_{22}, \qquad N_{21} = M_{12}, \qquad N_{12} = M_{21}$$
$$(4.1\text{-}21a)$$

In other words, N is the transpose of M:

$$N = \tilde{M} \qquad (4.1\text{-}21b)$$

where the notation $\sim$ indicates a transpose operation. This property is particularly useful in LCDs that involve the use of a reflector (mirror). Using Eqs. (4.1-20, 21), it can be shown that both N and M are unitary matrices:

$$M^\dagger M = 1, \qquad N^\dagger N = 1 \qquad (4.1\text{-}22)$$

where $N^\dagger, M^\dagger$ are Hermitian conjugate of N and M, respectively. This proves the unitary nature of the Jones matrix. Thus, if the Jones matrix M is written

$$M = \begin{pmatrix} A & B \\ C & D \end{pmatrix} \qquad (4.1\text{-}23)$$

where A, B, C, and D are the matrix elements, then the inverse of M can be written

$$M^{-1} = M^\dagger = \begin{pmatrix} A^* & C^* \\ B^* & D^* \end{pmatrix} \qquad (4.1\text{-}24)$$

Furthermore, the Jones matrix is unimodular based on the definition of Eq. (4.1-10), provided all the birefringent plates are lossless. In other words

$$\det(M) = AD - BC = 1 \qquad (4.1\text{-}25)$$

Thus, the inverse of the Jones matrix can also be written

$$M^{-1} = \begin{pmatrix} D & -B \\ -C & A \end{pmatrix} \tag{4.1-26}$$

Using Eqs. (4.1-24) and (4.1-26), we obtain the following relationships for the matrix elements:

$$C = -B^*, \qquad D = A^* \tag{4.1-27}$$

These relationships are very useful in simplifying the calculations involving the use of Jones matrix method. Using the relationships in Eq. (4.1-27), the Jones matrix can be written

$$M = \begin{pmatrix} A & B \\ -B^* & A^* \end{pmatrix} \tag{4.1-28}$$

We note that the collection of all Jones matrices forms a mathematical group. In other words, the relationships remain true after the multiplication of any two Jones matrices.

Mirror Reflection. By invoking the principle of reciprocity, we now consider the situation when the beam of light undergoes retroreflection. In retroreflection, the wavevector of the reflected beam of light is an inverse of that of the incident beam of light. In other words

$$\mathbf{k}_r = -\mathbf{k} \qquad \text{(retroreflection)} \tag{4.1-29}$$

where $\mathbf{k}_r$ is the wavevector of the retroreflected beam of light and $\mathbf{k}$ is the wavevector of the original incident beam of light. If we denote (θ, ϕ) as the angle of incidence in spherical coordinate with the z axis as the polar axis, we can express the wavevectors as

$$\mathbf{k} = k(\sin\theta\cos\phi, \sin\theta\sin\phi, \cos\theta)$$
$$\mathbf{k}_r = k(-\sin\theta\cos\phi, -\sin\theta\sin\phi, -\cos\theta) \tag{4.1-30}$$

where k is the wavenumber, θ is the polar angle, and ϕ is the azimuth angle. A mirror reflection is different from the retroreflection. Let the mirror be oriented with its normal parallel to the z axis. The wavevector of the mirror-reflected beam of light is given by

$$\mathbf{k}_m = k(\sin\theta\cos\phi, \sin\theta\sin\phi, -\cos\theta) \tag{4.1-31}$$

We note that only the z component of the wavevector is reversed in sign, as the mirror is normal to the z axis. These three wavevectors are illustrated in Figure 4.4. It is important to note that, in general, these three wavevectors are coplanar, but not collinear.

As we know, the normal surface of a general birefringent medium is invariant under the inversion operation. Thus the phase retardation of an arbitrarily

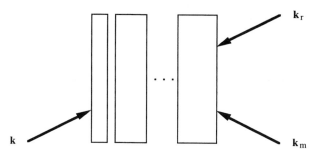

Figure 4.4. Three incident wavevectors in a general birefringent system that consists of a series of arbitrarily oriented birefringent plates, where **k** is the wavevector of the original incident beam of light, $\mathbf{k}_r$ represents the wavevector of the retroreflected beam of light, and $\mathbf{k}_m$ represents the wavevector of the mirror-reflected beam of light.

oriented birefringent plate is also invariant under the inversion operation. This is consistent with the principle of reciprocity. Let the Jones matrices be written M, N, M_m for the three incident wavevectors $\mathbf{k}, \mathbf{k}_r, \mathbf{k}_m$, respectively. According to Eqs. (4.1-30) and (4.1-31), we find that $\mathbf{k}_m$ can be obtained from $\mathbf{k}_r$ by replacing ϕ with $\phi + \pi$. Thus, these three Jones matrices are related by the following relationships:

$$N(\theta, \phi) = \tilde{M}(\theta, \phi) \tag{4.1-32}$$

$$M_m = N(\theta, \phi + \pi) \tag{4.1-33}$$

Consequently, the Jones matrix for the mirror-reflected beam M_m can be written

$$M_m = \tilde{M}(\theta, \phi + \pi) \tag{4.1-34}$$

This equation is particularly useful for investigating the reflection properties of reflective LCDs that involve a mirror after the LC cell.

4.2. INTENSITY TRANSMISSION SPECTRUM

Our development of the Jones matrix method has thus far been concerned with the polarization state of the light beam. In many cases, we need to determine the transmitted intensity. A narrowband filter, for example, transmits radiation only in a small spectral regime and rejects (or absorbs) radiation at other wavelengths. To change the intensity of the transmitted beam, an analyzer is usually required. An analyzer is basically a polarizer. It is called an "analyzer" simply because of its location in the optical system. In most birefringent optical systems, a polarizer is placed in front of the system in order to "prepare" a beam of polarized light. A second polarizer (analyzer) is placed at the output to analyze the polarization state of the emerging beam. Because the phase retardation of each wave plate is wavelength dependent, the polarization state of the emerging beam depends on the wavelength of the light. A polarizer at the rear will cause the overall transmitted intensity to be wavelength dependent.

The Jones vector representation of a beam of light contains information about not only the polarization state but also the intensity of light. Let us now consider the light beam after it passes through the polarizer. Its electric vector can be written as a Jones vector

$$\mathbf{E} = \begin{pmatrix} E_x \\ E_y \end{pmatrix} \qquad (4.2\text{-}1)$$

where E_x, E_y are the components in the xy coordinate. The intensity is calculated as follows:

$$I = \mathbf{E}^\dagger \cdot \mathbf{E} = |E_x|^2 + |E_y|^2 \qquad (4.2\text{-}2)$$

where the dagger indicates the Hermitian conjugate. If the Jones vector of the emerging beam after it passes through the analyzer is written as

$$\mathbf{E}' = \begin{pmatrix} E'_x \\ E'_y \end{pmatrix}, \qquad (4.2\text{-}3)$$

the transmittance of the birefringent optical system is calculated as follows:

$$T = \frac{|E'_x|^2 + |E'_y|^2}{|E_x|^2 + |E_y|^2} \qquad (4.2\text{-}4)$$

Example 4.3: A Birefringent Plate Sandwiched between Parallel Polarizers (a Plate between Parallel Polarizers). Referring to Figure 4.5, we consider a birefringent plate sandwiched between a pair of parallel polarizers. The plate is oriented so that the "slow" and "fast" axes are at $45°$ with respect to the transmission axes of polarizers. Let the birefringence be $n_e - n_o$ and the plate thickness be d. The phase retardation is then given by

$$\Gamma = \frac{2\pi}{\lambda}(n_e - n_o)d \qquad (4.2\text{-}5)$$

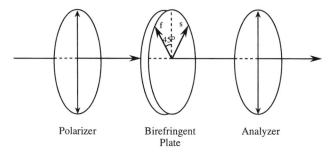

Figure 4.5. A birefringent plate sandwiched between a pair of parallel polarizers.

and the corresponding Jones matrix is, according to Eq. (4.1-10) (or Table 4.2)

$$W = \begin{pmatrix} \cos \dfrac{\Gamma}{2} & -i \sin \dfrac{\Gamma}{2} \\[2mm] -i \sin \dfrac{\Gamma}{2} & \cos \dfrac{\Gamma}{2} \end{pmatrix}. \tag{4.2-6}$$

Let the incident beam be unpolarized, so that after it passes through the front polarizer, the electric field vector can be represented by the following Jones vector:

$$\frac{1}{\sqrt{2}} \begin{pmatrix} 0 \\ 1 \end{pmatrix}, \tag{4.2-7}$$

where we assume that the intensity of the incident beam is unity and only half of the intensity passes through the polarizer where transmission axis is parallel to the y axis. The Jones vector representation of the electric field vector of the transmitted beam is obtained as follows:

$$\mathbf{E'} = \begin{pmatrix} 0 & 0 \\ 0 & 1 \end{pmatrix} \begin{pmatrix} \cos \dfrac{\Gamma}{2} & -i \sin \dfrac{\Gamma}{2} \\[2mm] -i \sin \dfrac{\Gamma}{2} & \cos \dfrac{\Gamma}{2} \end{pmatrix} \frac{1}{\sqrt{2}} \begin{pmatrix} 0 \\ 1 \end{pmatrix}$$

$$= \frac{1}{\sqrt{2}} \begin{pmatrix} 0 \\ \cos \dfrac{\Gamma}{2} \end{pmatrix}. \tag{4.2-8}$$

The transmitted beam is vertically (y) polarized with an intensity given by

$$I = \frac{1}{2} \cos^2 \frac{\Gamma}{2} = \frac{1}{2} \cos^2 \left[\frac{\pi(n_e - n_o)d}{\lambda} \right] \tag{4.2-9}$$

It can be seen from Eq. (4.2-9) that the transmitted intensity is a sinusoidal function of the wavenumber and peaks at $\lambda = (n_e - n_o)d$, $(n_e - n_o)d/2$,

$(n_e - n_o)d/3, \dots$. These wavelengths correspond to $\Gamma = 2\pi, 4\pi, 6\pi, \dots$. In other words, maximum transmission occurs when the plate is an integral number of full waves. The wavenumber separation between transmission maxima increases with decreasing plate thickness. ■

Example 4.4: A Birefringent Plate Sandwiched between a Pair of Crossed Polarizers (a Plate between Crossed Polarizers). If we rotate the analyzer shown in Figure 4.5 by 90°, then the input and output polarizers are crossed. The transmitted beam for this case is obtained as follows:

$$\mathbf{E}' = \begin{pmatrix} 1 & 0 \\ 0 & 0 \end{pmatrix} \begin{pmatrix} \cos\dfrac{\Gamma}{2} & -i\sin\dfrac{\Gamma}{2} \\ -i\sin\dfrac{\Gamma}{2} & \cos\dfrac{\Gamma}{2} \end{pmatrix} \frac{1}{\sqrt{2}} \begin{pmatrix} 0 \\ 1 \end{pmatrix} = \frac{-i}{\sqrt{2}} \begin{pmatrix} \sin\dfrac{\Gamma}{2} \\ 0 \end{pmatrix} \qquad (4.2\text{-}10)$$

The transmitted beam is horizontally (x) polarized with an intensity given by

$$I = \frac{1}{2}\sin^2\frac{\Gamma}{2} = \frac{1}{2}\sin^2\left[\frac{\pi(n_e - n_o)d}{\lambda}\right]. \qquad (4.2\text{-}11)$$

This is again a sinusoidal function of the wavenumber. The transmission spectrum consists of a series of maxima at $\lambda = 2(n_e - n_o)d, 2(n_e - n_o)d/3, \dots$. These wavelengths correspond to phase retardations of $\pi, 3\pi, 5\pi, \dots$, that is, when the wave plate becomes a half-wave plate or an odd integral multiple of a half-wave plate. ■

Example 4.5: A Birefringent Plate Sandwiched between a Pair of Polarizers (a Plate between Two Polarizers). If we rotate the analyzer shown in Figure 4.5 so that the transmission axis of the analyzer forms an angle ψ with respect to the x axis, then the input and output polarizers are neither parallel nor crossed. The transmitted beam for this case is obtained as follows (see Table 4.2):

$$\mathbf{E}' = \begin{pmatrix} \cos^2\psi & \cos\psi\sin\psi \\ \sin\psi\cos\psi & \sin^2\psi \end{pmatrix} \begin{pmatrix} \cos\dfrac{\Gamma}{2} & -i\sin\dfrac{\Gamma}{2} \\ -i\sin\dfrac{\Gamma}{2} & \cos\dfrac{\Gamma}{2} \end{pmatrix} \frac{1}{\sqrt{2}} \begin{pmatrix} 0 \\ 1 \end{pmatrix}$$

$$= \frac{-i\cos\psi\sin\dfrac{\Gamma}{2} + \sin\psi\cos\dfrac{\Gamma}{2}}{\sqrt{2}} \begin{pmatrix} \cos\psi \\ \sin\psi \end{pmatrix} \qquad (4.2\text{-}12)$$

The transmitted beam is polarized in the same direction as the transmission axis of the analyzer with an intensity

$$I = \frac{1}{2}\cos^2\psi\sin^2\frac{\Gamma}{2} + \frac{1}{2}\sin^2\psi\cos^2\frac{\Gamma}{2} \qquad (4.2\text{-}13)$$

where

$$\Gamma = \frac{2\pi(n_e - n_o)d}{\lambda}$$

We note that Eq. (4.2-13) reduces to Eq. (4.2-9) when $\psi = \pi/2$, (i.e., when the analyzer is parallel to the polarizer) and Eq. (4.2-13) reduces to Eq. (4.2-11) when $\psi = 0$, (i.e., when the analyzer is perpendicular to the polarizer).   ■

Example 4.6: A Birefringent Plate Sandwiched between a Pair of Polarizers (c Plate between Two Polarizers). If we replace the birefringent plate in Figure 4.5 by a c plate (with c axis along the z direction) and rotate the analyzer so that the transmission axes of the polarizer and the analyzer form an angle θ with respect to each other, since the indices of refraction of the two polarization eigenstates of a wave propagating along the c axis are the same, the phase retardation is zero. Thus, the c plate does not affect the polarization state of the wave. Therefore, the system is equivalent to two polarizers with their transmission directions rotated by θ with respect to each other. As discussed before, the transmitted beam has an intensity

$$I = \tfrac{1}{2}\cos^2\theta \qquad (4.2\text{-}14)$$

and the polarization state of the emerging beam is along the transmission direction of the analyzer.   ■

Example 4.7: A Birefringent Plate Sandwiched between a Pair of Crossed Polarizers (c Plate between Crossed Polarizers). In Example 4.6, if $\theta = \pi/2$, the c plate is sandwiched between a pair of crossed polarizers. According to Eq. (4.2-14), the intensity of the transmitted beam is $I = 0$. It is important to remember that this is valid only for incident beam along the c axis, where the phase retardation is zero. This is no longer true for oblique incidence.   ■

Example 4.8: A Birefringent Plate Sandwiched between a Pair of Parallel Polarizers (c Plate between Parallel Polarizers). In Example 4.6, if $\theta = 0$, the c plate is sandwiched between a pair of parallel polarizers. According to Eq. (4.2-14), the intensity of the transmitted beam is $I = \tfrac{1}{2}$. Again, we note that this is valid for incident beam along the c axis, where the phase retardation is zero.

■

4.3. OPTICAL PROPERTIES OF A TWISTED NEMATIC LIQUID CRYSTAL (TN-LC)

In this section the propagation of electromagnetic radiation through a slowly twisting anisotropic medium is described by the Jones matrix method. The transmission of light through a twisted nematic liquid crystal (TN-LC) is a typical example. In such a medium, the orientation of the local c axis (or director) is a function of position in the medium. As a result of the twisting of the c axis, the medium is not homogeneous. The Jones matrix method described earlier can still be used provided we subdivide the medium into a large number of thin plates ($N \geq 20$) so that each of the thin plates can be approximated by a homogeneous medium. In fact, we are going to subdivide the twisted anisotropic medium into N plates of equal thickness and assume that each plate is a wave plate with a phase retardation and an azimuth angle. The overall Jones matrix can then be obtained by multiplying together all the matrices associated with these plates in sequence.

We will limit ourselves to the case when the twisting is linear and the azimuth angle of the axes is

$$\psi(z) = \alpha z \qquad (4.3\text{-}1)$$

where z is the distance in the direction of propagation and α is a constant.

Let Γ be the phase retardation of the plate when it is untwisted. In particular, for the case of nematic LC with c axis parallel to the plate surfaces, Γ is given by

$$\Gamma = \frac{2\pi}{\lambda}(n_e - n_o)d \qquad (4.3\text{-}2)$$

where d is the thickness of the plate. In most twisted nematic LCs, the c axes (or directors) are tilted at a small angle θ ($\sim 1°$) relative to the plate surface. If this angle is a constant other than zero, then n_e in Eq. (4.3-2) must be replaced with the eigen refractive index given by Eq. (3.3-4). The total twist angle is

$$\phi \equiv \psi(d) = \alpha d \qquad (4.3\text{-}3)$$

To derive the Jones matrix for such a structure, we now divide this TN-LC plate into N equally thin plates. Each plate has a phase retardation of Γ/N. The plates are oriented at azimuth angles $\rho, 2\rho, 3\rho, \ldots, (N-1)\rho, N\rho$ with $\rho = \phi/N$. The overall Jones matrix for these N plates is given by

$$M = W_N W_{N-1} \cdots W_3 W_2 W_1 = \prod_{m=1}^{N} W_m = \prod_{m=1}^{N} R(-m\rho) W_0 R(m\rho), \qquad (4.3\text{-}4)$$

where R is the coordinate rotation matrix [defined by Eq. (4.1-9a)] and W_m is the Jones matrix for the mth plate; W_0 can be written

$$W_0 = \begin{pmatrix} e^{-i\Gamma/2N} & 0 \\ 0 & e^{i\Gamma/2N} \end{pmatrix} \qquad (4.3\text{-}5)$$

Using $\rho = \phi/N$ and $R(\psi_1)R(\psi_2) = R(\psi_1 + \psi_2)$, the overall Jones matrix Eq. (4.3-4) can be written

$$M = R(-\phi)\left[W_0 R\left(\frac{\phi}{N}\right)\right]^N \tag{4.3-6}$$

where ϕ is the total twist angle ($\phi = \alpha d$). Using Eqs. (4.1-9a) and (4.3-6), we obtain

$$M = R(-\phi)\begin{pmatrix} \cos\dfrac{\phi}{N}e^{-i\Gamma/2N} & \sin\dfrac{\phi}{N}e^{-i\Gamma/2N} \\[2mm] -\sin\dfrac{\phi}{N}e^{i\Gamma/2N} & \cos\dfrac{\phi}{N}e^{i\Gamma/2N} \end{pmatrix}^N \tag{4.3-7}$$

Equation (4.3-7) can be further simplified by using Chebyshev's identity for unimodular matrices [2]:

$$\begin{pmatrix} A & B \\ C & D \end{pmatrix}^m = \begin{pmatrix} \dfrac{A\sin mZ - \sin(m-1)Z}{\sin Z} & B\dfrac{\sin mZ}{\sin Z} \\[3mm] C\dfrac{\sin mZ}{\sin Z} & \dfrac{D\sin mZ - \sin(m-1)Z}{\sin Z} \end{pmatrix} \tag{4.3-8}$$

with

$$Z = \cos^{-1}\left[\frac{1}{2}(A+D)\right] \tag{4.3-9}$$

In the limit when N tends to infinite ($N \to \infty$), the result is given by (see Problem 4.9)

$$M = \begin{pmatrix} \cos\phi & -\sin\phi \\ \sin\phi & \cos\phi \end{pmatrix}\begin{pmatrix} \cos X - i\dfrac{\Gamma}{2}\dfrac{\sin X}{X} & \phi\dfrac{\sin X}{X} \\[3mm] -\phi\dfrac{\sin X}{X} & \cos X + i\dfrac{\Gamma}{2}\dfrac{\sin X}{X} \end{pmatrix} \tag{4.3-10}$$

where

$$X = \sqrt{\phi^2 + \left(\frac{\Gamma}{2}\right)^2} \tag{4.3-11}$$

Here we have an exact expression for the Jones matrix of a linearly twisted nematic LC plate.

Let $\mathbf{V}$ be the initial polarization state; then the polarization state $\mathbf{V}'$ after exiting the plate can be written

$$\mathbf{V}' = M\mathbf{V} \tag{4.3-12}$$

where M is as given by Eq. (4.3-10).

In the preceding results Eqs. (4.3-10)–(4.3-12), the polarization states are expressed in terms of the x and y components. It is often useful and desirable to examine the polarization states in the local principal coordinate system where the e component is along the direction of the director and the o component is perpendicular to the director. Thus, the result can also be written

$$\begin{pmatrix} V'_e \\ V'_o \end{pmatrix} = \begin{pmatrix} \cos X - i\dfrac{\Gamma \sin X}{2X} & \phi\dfrac{\sin X}{X} \\ -\phi\dfrac{\sin X}{X} & \cos X + i\dfrac{\Gamma \sin X}{2X} \end{pmatrix} \begin{pmatrix} V_e \\ V_o \end{pmatrix} \tag{4.3-13}$$

where $\begin{pmatrix} V'_e \\ V'_o \end{pmatrix}$ and $\begin{pmatrix} V_e \\ V_o \end{pmatrix}$ are polarization states in the local principal coordinate system (*eo* coordinate system). For most LCs with positive birefringence $(n_o < n_e)$, the e axis is the "slow" axis whereas the o axis is the "fast" axis. Equations (4.3-10), (4.3-12), and (4.3-13) can be considered general analytical results. Many of the results on twisted nematic liquid crystal displays obtained earlier can be shown to be special cases of these two equations. In what follows, we discuss some important properties of the transmission of polarized light in TN-LC. First, we discuss a phenomenon known as the *adiabatic following* [2] (also known as the *waveguiding*) of the state of polarization of the beam inside the twisted nematic LC medium.

4.3.1. Adiabatic Following (Waveguiding in TN-LC)

Consider an input beam of light with its electric field vector polarized parallel to the director of the LC (c axis) at the entrance plane. In TN-LCDs, this is known as the *E-mode operation*. In the principal coordinate system, the input polarization state can be written as the following Jones vector:

$$\begin{pmatrix} V_e \\ V_o \end{pmatrix} = \begin{pmatrix} 1 \\ 0 \end{pmatrix} \qquad (E \text{ mode}) \tag{4.3-14}$$

The output polarization state in terms of the Jones vector in the local principal coordinate at the exit plane can be written, according to Eq. (4.3-13), as follows:

$$\begin{pmatrix} V'_e \\ V'_o \end{pmatrix} = \begin{pmatrix} \cos X - i\dfrac{\Gamma \sin X}{2X} \\ -\phi\dfrac{\sin X}{X} \end{pmatrix} \qquad (4.3\text{-}15)$$

It often happens in TN-LC that the twist angle ϕ is much smaller than the phase retardation angle Γ. For example, consider a LC layer of E7 with $20\,\mu m$ thickness and a twist angle of $\pi/2$. With a birefringence of $\Delta n = 0.23$, we have $\phi/\Gamma = 37$ at $\lambda = 500\,nm$. Note that for TN liquid crystal cells with $\phi \ll \Gamma$, the second term (o component) of the Jones vector in Eq. (4.3-15) is near zero. Thus, the output Jones vector can be written approximately as

$$\begin{pmatrix} V'_e \\ V'_o \end{pmatrix} = \begin{pmatrix} e^{-i\Gamma/2} \\ 0 \end{pmatrix} \qquad (4.3\text{-}16)$$

In other words, the electric field vector of the beam remains parallel to the local director as the beam propagates in the twisted nematic LC medium. If the electric field vector of the incident beam is polarized perpendicular to the director of the LC at the entrance plane. (This is known as the *O-mode operation* in TN-LCDs). We will obtain an output polarization state perpendicular to the local director at the output plane. This explains the phenomenon of adiabatic following (or waveguiding). It appears that the polarization vector of the incident beam follows the twist of the liquid crystal directors. Figure 4.6a,b shows the waveguiding in TN-LC. This phenomenon has an important application in LCDs and light valves (spatial light modulators). The principle of operation of the light valves and LCD will be discussed further in this book. It is important to remember that adiabatic following occurs provided the incident beam is polarized either parallel or perpendicular to the director at the entrance plane, and the twist rate is small. Strictly speaking, the polarization state in the medium is elliptical as illustrated in Figure 4.6c.

4.3.2. 90° Twisted Nematic Liquid Crystal

Equation (4.3-13) can be employed to obtain the transmission formula for a 90° twisted nematic liquid crystal cell sandwiched between a pair of parallel polarizers. This is the so-called normally black (NB) configuration. We consider the *E*-mode operation. The front polarizer is oriented with its transmission axis parallel to the local director (*e* axis), so that the input Jones vector is written

$$\begin{pmatrix} V_e \\ V_o \end{pmatrix} = \begin{pmatrix} 1 \\ 0 \end{pmatrix} \qquad (E\,\text{mode}) \qquad (4.3\text{-}17)$$

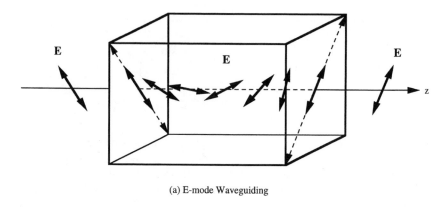

(a) E-mode Waveguiding

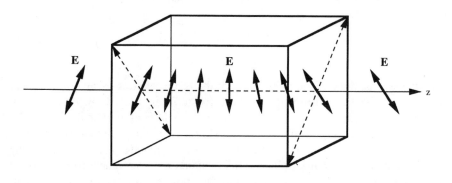

(b) O-mode Waveguiding

Figure 4.6. Wave guiding and polarization states in a 90° TN-LC cell with a right-handed twist. (*a*) *E*-mode waveguiding; (*b*) *O*-mode waveguiding; (*c*) actual polarization states in the medium.

The second polarizer is oriented with the transmission axis perpendicular to the local director at the exit plane (*o* axis). According to Eq. (4.3-15), this leads to an output intensity of [2,3]

$$T = \frac{\sin^2[\phi\sqrt{1+u^2}]}{1+u^2} \tag{4.3-18}$$

where ϕ is the twist angle of the cell, u is the so-called Mauguin parameter given by

$$u = \frac{\Gamma}{2\phi} = \frac{2}{\lambda}(n_e - n_o)d \qquad \left(\phi = \frac{\pi}{2}\right) \tag{4.3-19}$$

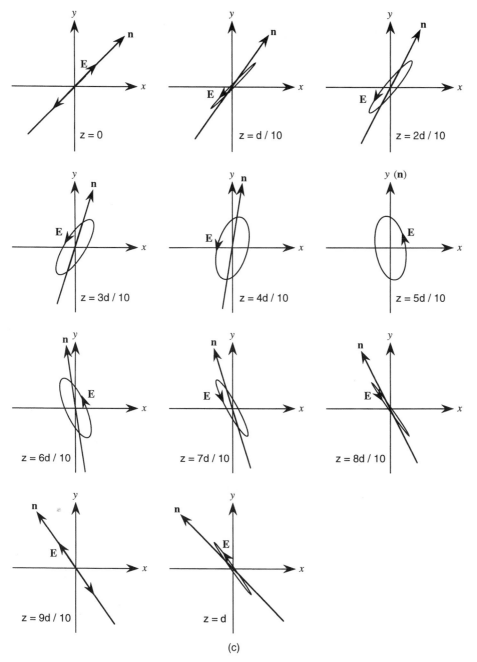

Figure 4.6. (*Continued*)

We note that the transmission is near 0 for LCs with $\phi \ll \Gamma$. This result is in agreement with the phenomenon of adiabatic following. The polarization vector remains parallel to the local director (e axis) as the beam propagates inside the TN-LC medium. Thus, the output polarization vector is parallel to the director at the output plane. In other words, the output polarization state has been rotated by 90° following the twist of the liquid crystal. Since this direction is perpendicular to the transmission axis of the analyzer, the transmission is zero. A residual transmission is due to the small ellipticity of the polarization state of the output beam. The formula (4.3-18) will be used later in Chapter 5 for the design of TN-LCDs. We now investigate the polarization state of the output beam by examining the polarization ellipse.

Polarization Ellipse

The Jones vector in Eq. (4.3-15) corresponds to a polarization ellipse with an ellipticity and azimuth orientation given by

$$e = \tan\left(\frac{1}{2}\sin^{-1}\left[\frac{\Gamma\phi}{X^2}\sin^2 X\right]\right) \qquad (4.3\text{-}20)$$

$$\tan 2\psi = \frac{2\phi X \tan X}{\left(\phi^2 - \dfrac{\Gamma^2}{4}\right)\tan^2 X - X^2} \qquad (4.3\text{-}21)$$

where ψ is the angle of the major axis of the polarization ellipse measured from the local director axis (e axis) at the exit plane. We also note that for liquid crystals with $\phi \ll \Gamma$, the ellipticity is approximately 0, and the angle ψ is approximately 0, reflecting the adiabatic following (or waveguiding phenomenon). In Eq. (4.3-20) we adhere to the convention that e is positive for RHC polarized light.

Now, consider an input beam of light with its electric field vector polarized perpendicular to the director of the liquid crystal at the entrance plane. In the principal coordinate system, the input polarization state can be written

$$\begin{pmatrix} V_e \\ V_o \end{pmatrix} = \begin{pmatrix} 0 \\ 1 \end{pmatrix} \qquad (O\text{ mode}) \qquad (4.3\text{-}22)$$

The output polarization state in terms of the Jones vector in the local principal coordinate can be written, according to Eq. (4.3-13)

$$\begin{pmatrix} V'_e \\ V'_o \end{pmatrix} = \begin{pmatrix} \phi\dfrac{\sin X}{X} \\ \cos X + i\dfrac{\Gamma \sin X}{2X} \end{pmatrix} \qquad (4.3\text{-}23)$$

Note again that for liquid crystals with $\phi \ll \Gamma$, this Jones vector is approximately polarized perpendicular to the local director. This is in agreement with the phenomenon of waveguiding. The Jones vector in Eq. (4.3-23) corresponds to a polarization ellipse with an ellipticity and azimuth orientation given by

$$e = \tan\left(\frac{1}{2}\sin^{-1}\left[-\frac{\Gamma\phi}{X^2}\sin^2 X\right]\right) \qquad (4.3\text{-}24)$$

$$\tan 2\psi = \frac{2\phi X \tan X}{X^2 - \left(\phi^2 - \frac{\Gamma^2}{4}\right)\tan^2 X} \qquad (4.3\text{-}25)$$

On the basis of the unitary property of the Jones matrix, the two polarization ellipses described in Eqs. (4.3-20) and (4.3-21) and Eqs. (4.3-24) and (4.3-25) are mutually orthogonal. In other words, the principal axes of the polarization ellipses are mutually perpendicular, the senses of revolution of the ellipse are opposite, and the magnitudes of the ellipticity are identical (with ellipticity defined as the ratio of short axis to long axis). In what follows, we examine the transmission properties of a general TN-LCD.

4.3.3.   Transmission Properties of a General TN-LCD

The general expression for the Jones matrix derived earlier can now be applied to obtain the transmission properties of a general TN–LCD. The transmission of polarized light through a general TN-LCD system has been investigated by several previous workers. An essential part of the previous analysis involves the derivation of the output polarization state for an arbitrary input polarization state. This can be done by using a differential form of the Jones calculus, to obtain a Riccati differential equation that describes the evolution of the polarization state in a general TN-LC (including cholesteric LCs) [4,5]. In this section, we investigate the same problem by using the Jones matrix method. Using Eqs. (4.3-12) and (4.3-13), we can obtain an analytic expression for the transmission of a general TN-LCD. This analytic expression is extremely useful for explaining the working mechanism of the conventional TN-LCDs and various STN-LCDs (super twisted nematic), OMI-LCDs (optical mode interference), and LTN-LCDs (lower twisted nematic). In addition they are also useful for the experimental measurement of the cell gap and twist angle of a general TN-LCD [6].

Referring to Figure 4.7, we investigate the transmission of a beam of polarized light through a general TN-LCD. According to Figure 4.7, the input and output polarization states, as determined by the orientation angles of the polarizer transmission axes, are given by

$$\begin{pmatrix} V_x \\ V_y \end{pmatrix} = \begin{pmatrix} \cos \Phi_{\text{ent}} \\ \sin \Phi_{\text{ent}} \end{pmatrix}, \qquad \begin{pmatrix} V_x' \\ V_y' \end{pmatrix} = \begin{pmatrix} \cos \Phi_{\text{exit}} \\ \sin \Phi_{\text{exit}} \end{pmatrix} \qquad (4.3\text{-}26)$$

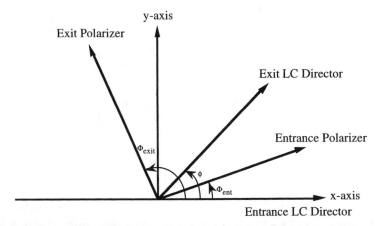

Figure 4.7. Schematic drawing showing the orientation of the polarizer axes, LC directors of a general TN-LCD in the xy plane. The x axis is chosen as the entrance LC director orientation. The entrance polarizer has a transmission axis along the direction ϕ_{ent}. The exit LC director is along the direction ϕ. The exit polarizer has a transmission axis along the direction ϕ_{exit}. All angles are measured from the x axis.

where Φ_{ent}, Φ_{exit} are the angles of the plane of polarization of the input and output beams, respectively. The intensity transmission of the system is given by, according to Jones matrix method

$$T = |V'^* \cdot MV|^2 \qquad (4.3\text{-}27)$$

where V and V' are the input and output Jones vectors, respectively. Here, we assume an incident beam of polarized light. A factor of $\frac{1}{2}$ must be included for unpolarized light. Using Eq. (4.3-10) for M, and Eq. (4.3-26) for V and V', we obtain the transmission T, after few steps of matrix multiplication

$$T = |A + iB|^2$$
$$A = \cos X \cos (\phi - \Phi_{exit} + \Phi_{ent}) + \phi \frac{\sin X}{X} \sin (\phi - \Phi_{exit} + \Phi_{ent})$$
$$B = -\frac{\Gamma \sin X}{2X} \cos (\phi - \Phi_{exit} - \Phi_{ent}) \qquad (4.3\text{-}28)$$

where X is as given by Eq. (4.3-11). After few more steps of algebra, the transmitted intensity can be written

$$
\begin{aligned}
T = {} & \cos^2(\phi - \Phi_{exit} + \Phi_{ent}) \\
& + \sin^2 X \sin 2(\phi - \Phi_{exit}) \sin 2\Phi_{ent} \\
& + \frac{\phi}{2X} \sin 2X \sin 2(\phi - \Phi_{exit} + \Phi_{ent}) \\
& - \phi^2 \frac{\sin^2 X}{X^2} \cos 2(\phi - \Phi_{exit}) \cos 2\Phi_{ent}
\end{aligned}
\tag{4.3-29}
$$

This equation describes the dependence of the transmission of a general TN-LCD on the orientation angles of the polarizers relative to the director at the input surface, as well as the total twist angle ϕ of the LC. The analytic expression is extremely useful for explaining the working mechanism of the conventional TN-LCDs and various STN, OMI, and LTN-LCDs. In addition it is also useful for the experimental measurement of the cell gap and twist angle of a general TN-LCD.

The transmission formula Eq. (4.3-29) can also be written in various forms. If we define the azimuth angle of the transmission axes of the polarizers (α, β) as measured from the local director (see Figure 4.7)

$$
\alpha = \Phi_{ent}
$$
$$
\beta = \Phi_{exit} - \phi
$$

then the transmission formula can be written

$$
\begin{aligned}
T = {} & \cos^2(\alpha - \beta) - \sin^2 X \sin 2\beta \sin 2\alpha \\
& + \frac{\phi}{2X} \sin 2X \sin 2(\alpha - \beta) - \phi^2 \frac{\sin^2 X}{X^2} \cos 2\alpha \cos 2\beta
\end{aligned}
\tag{4.3-29a}
$$

We now expand the term $\sin 2X$ and perform few steps of algebra, the transmission formula can also be written

$$
T = \cos^2(\alpha + \beta) - \cos^2 X \cos 2\alpha \cos 2\beta \left[\frac{\phi}{X} \tan X - \tan 2\alpha \right] \left[\frac{\phi}{X} \tan X + \tan 2\beta \right]
\tag{4.3-29b}
$$

The last expression is particularly useful in the design of TN-LCDs or STN-LCDs. For example, a visual examination of the transmission formula (4.3-29b) yields the conditions for a maximum transmission of $T = 1$. According to the formula, maximum transmission of $T = 1$ occurs when the polarizer's transmission angles (α, β) are given by

$$\tan 2\alpha = \frac{\phi}{X} \tan X$$

$$\beta = -\alpha$$

It will be shown later in this section that an input beam of light linearly polarized at an azimuth angle α will be transformed by the TN-LC cell into an output beam of light linearly polarized at an azimuth angle of β.

Example 4.9. Consider a normally black (NB) 90° TN-LCD ($\phi = \pi/2$), with $\Phi_{ent} = 0$ and $\Phi_{exit} = 0$. Using Eq. (4.3-29), we obtain

$$T = \phi^2 \frac{\sin^2 X}{X^2}$$

which yields $T = 0$ when $X = \pi$, 2π, 3π, ... Note that the results is exactly the same as Eq. (4.3-18). ∎

Example 4.10. Consider a normally white (NW) 90° TN-LCD ($\phi = \pi/2$), with $\Phi_{ent} = 0$ and $\Phi_{exit} = \pi/0$. Using Eq. (4.3-29), we obtain

$$T = 1 - \phi^2 \frac{\sin^2 X}{X^2}$$

which yields $T = 1$ when $X = \pi$, 2π, 3π, ... ∎

90° Rotation Symmetry

We consider the situation when both polarizers are rotated by 90° relative to the TN-LC cell. The new azimuth angles for the polarizers become

$$\Phi'_{ent} = \Phi_{ent} + \frac{\pi}{2}, \qquad \Phi'_{exit} = \Phi_{exit} + \frac{\pi}{2}$$

Substituting the above equations into the general transmission formula, Eq. (4.3-29), we find that the transmission remains unchanged. In other words, the transmission is invariant under the rotation of both the input and the output polarizers by 90°. This is, in fact, a general property of lossless birefringent systems, and is a result of the unitary nature of the Jones matrix.

4.3.4. Normal Modes of Propagation in a General TN-LC

Generally speaking, an input beam of linearly polarized light is transformed into an output beam of elliptically polarized light by the TN-LC cell. Here we are looking for the normal modes of propagation of the structure whose polarization ellipses remain the same as they propagate in the medium. Using Eq. (4.3-13), the Jones vector of the normal modes must obey the following equation

$$
\begin{pmatrix} \cos X - i\dfrac{\Gamma \sin X}{2X} & \phi\dfrac{\sin X}{X} \\[3mm] -\phi\dfrac{\sin X}{X} & \cos X + i\dfrac{\Gamma \sin X}{2X} \end{pmatrix} \begin{pmatrix} V_e \\ V_o \end{pmatrix} = \gamma \begin{pmatrix} V_e \\ V_o \end{pmatrix}
\tag{4.3-30}
$$

where γ is the eigenvalue and $\begin{pmatrix} V_e \\ V_o \end{pmatrix}$ is the polarization state of the normal modes in the local principal coordinate system (*eo* coordinate system). We recall that the local *e* axis is parallel to the local director, and the local *o* axis is perpendicular to the local director. An input polarization state satisfying Eq. (4.3-30) will be transformed into an output polarization state identical to that of the input one in the local principal coordinate system (*eo* coordinate system). By solving Eq. (4.3-30), we obtain the following eigenvalues and eigenvectors for the normal modes of propagation:

te mode : $\begin{pmatrix} V_e \\ V_o \end{pmatrix} = \begin{pmatrix} 1 \\ -i(\sqrt{1+u^2} - u) \end{pmatrix}$ $\gamma = e^{-i\phi\sqrt{1+u^2}}$ (4.3-31)

to mode : $\begin{pmatrix} V_e \\ V_o \end{pmatrix} = \begin{pmatrix} -i(\sqrt{1+u^2} - u) \\ 1 \end{pmatrix}$ $\gamma = e^{+i\phi\sqrt{1+u^2}}$ (4.3-32)

with the Mauguin parameter u given by

$$
u = \frac{\Gamma}{2\phi} = \frac{p}{2\lambda}(n_e - n_o)
\tag{4.3-33}
$$

where Γ = phase retardation
ϕ = cell twist angle
p = pitch of the twist
$\Delta n = n_e - n_o$ is the birefringence.

These two normal modes of propagation are designated as twisted extraordinary (*te* mode) and twisted ordinary (*to* mode). In the limit of slow twist $(1 \ll u)$, these modes reduce to the extraordinary and ordinary modes of the uniaxial medium. In a 90° twist cell, the cell thickness is one quarter of a pitch $(d = p/4)$; thus u is given by

$$u = \frac{2d}{\lambda} \Delta n \qquad (90° \text{ twist}) \qquad (4.3\text{-}34)$$

where $\Delta n = (n_e - n_o)$. We note that the normal modes are, in general, elliptically polarized except when $u = 0$ or ∞. The major and minor axes of the ellipses are parallel to the local principal axes of the LC medium. We also note that the twisted extraordinary (*te*) mode has its major axis parallel to the local director (*e* axis), whereas the twisted ordinary (*to*) mode has its major axis perpendicular to the local director. These polarization ellipses are mutually orthogonal with opposite senses of rotation. Figure 4.8 shows the polarization ellipses of the normal modes.

With the normal modes of propagation available, any polarized beam of light in TN-LC can now be decomposed into a linear combination of these two modes. As a result of the propagation, a phase retardation between theses two modes exists. According to Eqs. (4.3-31) and (4.3-32), the phase retardation is given by

$$\Gamma_N = 2\phi\sqrt{1 + u^2} = \sqrt{\Gamma^2 + 4\phi^2} = 2X \qquad (4.3\text{-}35)$$

where a subscript N is added to distinguish the difference between Γ_N and Γ.

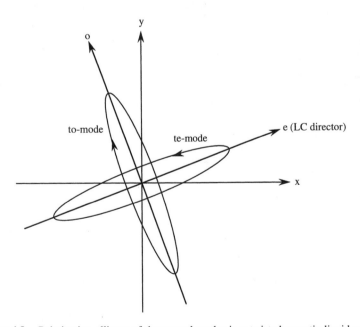

Figure 4.8. Polarization ellipses of the normal modes in a twisted nematic liquid crystal.

The ratio of the principal axes (minor axis/major axis) of the polarization ellipse is given by

$$e = \sqrt{1 + u^2} - u$$

We also note that the Jones vectors of the normal modes are independent of the cell thickness for a given birefringence and twist rate. Thus, the polarization ellipses of the normal modes will follow the twist (rotation) of the molecular axes as the beam propagates in the cell. In the regime of slow twist rate when $1 \ll u$, the polarization ellipses are almost linear. This is the Mauguin regime when linearly polarized light will rotate following the twist of the local molecular axes. Generally speaking, the waveguiding phenomenon always occurs for the normal modes, even outside the Mauguin regime, provided the Jones matrix method is valid. Strictly speaking, the Jones matrix method is valid only when the reflected waves can be neglected. Exact solutions of the wave equation, including the reflected waves, will be discussed later in this book.

Example 4.11. Consider a liquid crystal layer of E7 with $20\,\mu m$ thickness and a twist angle of $\pi/2$. Using a birefringence of $\Delta n = 0.23$, we have $u = \Gamma/2\phi = 18.4$ at $\lambda = 500\,nm$. The Jones vectors of the normal modes are

te mode: $\begin{pmatrix} V_e \\ V_o \end{pmatrix} = \begin{pmatrix} 1 \\ -0.027i \end{pmatrix}$ $\gamma = e^{-9.21i\pi}$

to mode: $\begin{pmatrix} V_e \\ V_o \end{pmatrix} = \begin{pmatrix} -0.027i \\ 1 \end{pmatrix}$ $\gamma = e^{+9.21i\pi}$

In the limit when u tends to infinity (twist rate approaches zero), the normal modes of propagation becomes, according to Eqs. (4.3-31) and (4.3-32)

te mode: $\begin{pmatrix} V_e \\ V_o \end{pmatrix} = \begin{pmatrix} 1 \\ 0 \end{pmatrix}$ $\gamma = e^{-iX} = e^{-i\Gamma/2}$

to mode: $\begin{pmatrix} V_e \\ V_o \end{pmatrix} = \begin{pmatrix} 1 \\ 0 \end{pmatrix}$ $\gamma = e^{iX} = e^{i\Gamma/2}$

where we note that X approaches $\Gamma/2$ when $u = \infty$. This is the case of a homogeneous crystal.   ■

Linearly Polarized Beam

In most LCDs involving the use of sheet polarizers, the input beam arriving at the front end of the TN cell (or STN cell) is often linearly polarized. It is thus important to know the output polarization state of the output beam at the rear end of the TN cell (or STN cell) for a given linearly polarized input beam. As we mentioned before, the output beam is often elliptically polarized. It is known that an elliptically polarized beam cannot be eliminated completely by using a linear sheet polarizer. An incomplete extinction at the rear polarizer may lead to a poor contrast ratio in the display. Thus a good understanding of the polarization state of the output beam is important in the design of high contrast LCDs. We now investigate the polarization state of the output beam. Specifically, we are interested in the possibility of obtaining a linearly polarized output beam given a linearly polarized input beam. For this purpose, we introduce a parameter

$$\chi = \frac{V_o}{V_e} \qquad (4.3\text{-}36)$$

which can be employed to describe the polarization state. Linearly polarized beams are characterized by a real χ. Using Eqs. (4.3-13) and (4.3-36), the output polarization state can be written

$$\chi' = \frac{-\phi \dfrac{\sin X}{X} + \left(\cos X + i \dfrac{\Gamma \sin X}{2X} \right) \chi}{\cos X - i \dfrac{\Gamma \sin X}{2X} + \phi \dfrac{\sin X}{X} \chi} \qquad (4.3\text{-}37)$$

Given a real χ for the input beam, this equation yields the output polarization state χ'. Solutions for real χ and χ' exist, provided

$$\chi = -v \pm \sqrt{1 + v^2} \qquad (4.3\text{-}38)$$

$$\chi' = -\chi \qquad (4.3\text{-}39)$$

where

$$v = \frac{X}{\phi \tan X} \qquad \text{with} \qquad X = \sqrt{\phi^2 + \frac{\Gamma^2}{4}} \qquad (4.3\text{-}40)$$

In other words, if the input beam is linearly polarized with a polarization state χ given by Eq. (4.3-38), then the output beam will be linearly polarized with a polarization state χ' given by Eq. (4.3-39). From Eq. (4.3-38), the two linearly polarized input beams are mutually orthogonal. These two linearly polarized input states are transformed into linearly polarized output states by the TN (or STN) cell. Let ψ_1 and ψ_2 be the azimuth angles of the input polarization states measured from the local director at the input end. These azimuth angles can be written

$$\tan \psi_1 = \sqrt{1 + v^2} - v \qquad (4.3\text{-}41)$$

$$\psi_2 = \frac{\pi}{2} + \psi_1 \qquad (4.3\text{-}42)$$

The azimuth angles of the corresponding output polarization states are

$$\psi'_1 = -\psi_1 \tag{4.3-43}$$

$$\psi'_2 = -\psi_2 \tag{4.3-44}$$

Again, these angles (ψ'_1, ψ'_2) are measured from the local director (e axis) at the output end. On the basis of the preceding discussion, high contrast operations can be achieved by orienting the polarizers along the proper directions relative to the local directors. The following is a list of these operations:

Normally white (NW) operations:

Case 1. Input linear polarization: $\psi_1 = \tan^{-1}(\sqrt{1+v^2} - v)$
    Output linear polarization: $\psi'_1 = -\tan^{-1}(\sqrt{1+v^2} - v)$
Case 2. Input linear polarization: $\psi_2 = \pi/2 + \tan^{-1}(\sqrt{1+v^2} - v)$
    Output linear polarization: $\psi'_2 = \pi/2 - \tan^{-1}(\sqrt{1+v^2} - v)$

Normally black (NB) operations:

Case 3. Input linear polarization: $\psi_1 = \tan^{-1}(\sqrt{1+v^2} - v)$
    Output linear polarization: $\psi'_1 = \pi/2 - \tan^{-1}(\sqrt{1+v^2} - v)$

Case 4. Input linear polarization: $\psi_2 = \pi/2 + \tan^{-1}(\sqrt{1+v^2} - v)$
    Output linear polarization: $\psi'_2 = -\tan^{-1}(\sqrt{1+v^2} - v)$

It is important to note that these azimuth angles are measured from the local director axis adjacent to the polarizers.

Example 4.12. Consider a LC layer of ZLI-1646 with 14 μm thickness and a twist angle of $3\pi/2$. Using a birefringence of $\Delta n = 0.08$, we have $v = 0.404$ at $\lambda = 589$ nm. The azimuth angles for the transmission axes of the input and output polarizers should be

Case 1. NW operation: $\psi_1 = 34°$, $\psi'_1 = -34°$
Case 2. NW operation: $\psi_2 = 124°$, $\psi'_2 = 56°$
Case 3. NW operation: $\psi_3 = 34°$, $\psi'_3 = 56°$
Case 4. NW operation: $\psi_4 = 124°$, $\psi'_4 = -34°$

Figure 4.9 illustrates the azimuth angles of the input linear polarization states and the output linear polarization states.

Example 4.13. Consider the special case of a TN-LC cell with $X = \pi, 2\pi, 3\pi, \ldots$. This is an interesting case, as the phase retardation Γ_N according to Eq. (4.3-35) is $2\pi, 4\pi, 6\pi, \ldots$. Thus, the output polarization state in the local *eo* coordinate is exactly the same as the input polarization state in local *eo* coordinate. Specifically, the output polarization will be linear if the input polarization is linear. In fact, the output polarization state is, according to Eq. (4.3-37), $\chi' = \chi$, which is valid for all χ. ∎

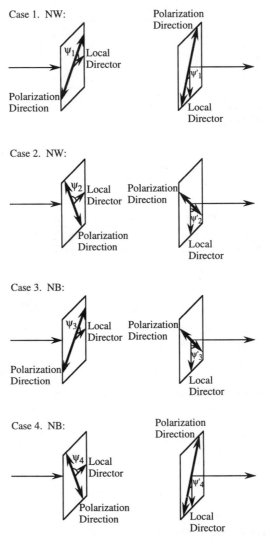

Figure 4.9. Azimuth angles of linear polarization states.

4.4. PHASE RETARDATION AT OBLIQUE INCIDENCE

Generally speaking, there are always two independent modes of propagation given a direction of propagation in an anisotropic medium. In uniaxially birefringent media, these two modes are the ordinary mode and the extraordinary mode, which are mutually orthogonal. The eigen indices of refraction of these two modes depends on the direction of propagation. In uniaxially birefringent media, the refractive index of the ordinary mode is independent of the direction of propagation, whereas the refractive index for the extraordinary mode depends

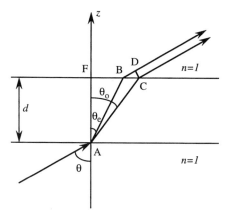

Figure 4.10. Phase retardation at oblique incidence. Note that CD is perpendicular to the direction of propagation in air.

on the direction of propagation. Thus the phase retardation may depend on the direction of propagation. In this section, we investigate the dependence of the phase retardation on the direction of incidence.

First, we consider a plate of homogeneous and uniaxially birefringent medium with its c axis parallel to the plate surfaces. Such a plate is known as an *a plate*. A cell of LC with a homogeneously parallel alignment is a good example. To describe the phase retardation at oblique incidence, we choose the z axis perpendicular to the plate surfaces (see Fig. 4.10). For a normally incident beam, the phase retardation is given by

$$\Gamma = \frac{2\pi}{\lambda}(n_e - n_o)d \qquad (4.4\text{-}1)$$

A general expression for the phase retardation is given by

$$\Gamma = (k_{ez} - k_{oz})d \qquad (4.4\text{-}2)$$

which is valid for both normal incidence and oblique incidence. Such a general expression can be derived as follows:

1. *Wave Approach.* Given an arbitrary incident plane wave, both ordinary and extraordinary waves are generated in the medium. The electric field amplitude in the medium can be written

$$\mathbf{E} = \mathbf{E}_e \exp[-i(\alpha x + \beta y + k_{ez}z)] + \mathbf{E}_o \exp[-i(\alpha x + \beta y + k_{oz}z)] \qquad (4.4\text{-}3)$$

where $\mathbf{E}_e, \mathbf{E}_o$ are constants representing the amplitudes of the modal components and α, β and k_{oz}, k_{ez} are components of the wavevectors. We note that the tangential components (α, β) of the wavevectors are the same because of the continuity condition at the boundary. As the waves propagate in the medium from $z = 0$ to $z = d$, the phase difference between the modes is exactly

$$\Gamma = (k_{ez} - k_{oz})d \tag{4.4-4}$$

2. *Ray Approach.* Referring to Figure 4.10, we consider the incidence of a beam of light at the surface of the medium. As a result of the birefringence, double refraction occurs. Let θ_e, θ_o be the ray angles and n'_e, n_o be the eigen indices of refraction for the rays. We note that n_o is independent of the angle of incidence, whereas n'_e depends on the angle of incidence, according to Eq. (3.3-4). The phase retardation can be written (see Fig. 4.10)

$$\Gamma = kn'_e AB + kBD - kn_o AC \tag{4.4-5}$$

where k is the wavenumber in vacuum, and we assume, without loss of generality, that the beam is incident from air with a refractive index of $n = 1$. Using the continuity condition (Snell's law)

$$\sin \theta = n'_e \sin \theta_e = n_o \sin \theta_o \tag{4.4-6}$$

and simple trigonometry, we obtain

$$\Gamma = k[n'_e \cos \theta_e - n_o \cos \theta_o]d \tag{4.4-7}$$

which is exactly $\Gamma = (k_{ez} - k_{oz})d$.

4.5. CONOSCOPY

A conoscope is a polarizing microscope designed to examine the transmission property of a sample plate at various directions of incidence simultaneously. For uniaxial media, a conoscope can be employed to determine the orientation of the c axis. The basic principle can be described as follows.

Referring to Figure 4.11, we consider the apparatus that employs a convergent (or divergent) beam of light as the illuminating source. As we know, a beam of convergent (or divergent) light consists of a linear superposition of many plane waves over a cone of solid angles. A lens is employed to convert each component of the plane waves into a point at the focal plane. By examining the intensity distribution at the focal plane, we know the transmission of the sample–polarizer combination at various angles simultaneously. By virtue of the anisotropic nature, the transmission of an anisotropic sample sandwiched between polarizers depends on the sample orientation as well as the angle of incidence. The intensity distribution at the focal plane can be employed to determine the sample orientation and the phase retardation as a function of angle of incidence. To illustrate this, we consider the following examples.

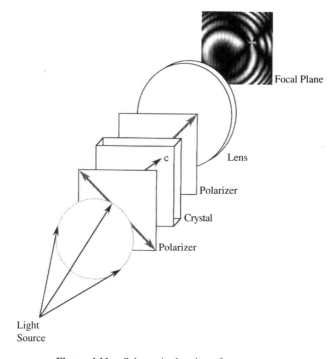

Figure 4.11. Schematic drawing of a conoscope.

4.5.1. *a* Plate of Uniaxial Crystals

Referring to Figure 4.12, we consider an *a* plate of uniaxial medium sandwiched between a pair of crossed polarizers. The *c* axis of the medium is parallel to the surface of the plate. Furthermore, the *c* axis is oriented at 45° relative to the transmission axes of the polarizers. According to the discussion earlier, the phase retardation can be written $\Gamma = (k_{ez} - k_{oz})d$, where k_{ez}, k_{oz} are *z* components of the wavevectors inside the medium and *d* is the thickness of the plate. The *z* axis is perpendicular to the surface of the plate.

For small angles of incidence $\theta \ll 1$, the phase retardation can be written (see Problem 4.6)

$$\Gamma = \Gamma_0 \left[1 + \frac{\sin^2 \theta}{2n_o^2} \left(\frac{n_o}{n_e} - \frac{n_e + n_o}{n_e} \cos^2 \phi \right) \right] \qquad (4.5\text{-}1)$$

where Γ_0 is the phase retardation at normal incidence $[\Gamma_0 = 2\pi(n_e - n_o)d/\lambda]$, θ is the angle of incidence measured relative to the normal to the plate, and ϕ is the angle between the *c*-axis and the projection of the incident wavevector on the surface of the medium. For materials with a small birefringence, the phase

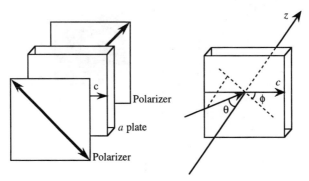

Figure 4.12. *a* plate between crossed polarizers.

retardation can be written

$$\Gamma = \Gamma_0 \left[1 + \frac{\sin^2\theta}{2n_o^2} \left(1 - 2\cos^2\phi \right) \right] \tag{4.5-2}$$

or

$$\Gamma = \Gamma_0 \left[1 + \frac{1}{2n_o^2} \left(\sin^2\theta \sin^2\phi - \sin^2\theta \cos^2\phi \right) \right] \tag{4.5-3}$$

The transmission at the focal plane can be written

$$T = \frac{1}{2} \sin^2 \frac{\Gamma}{2} \tag{4.5-4}$$

If the phase retardation is a multiple of 2π (full wave), the emergent beam at the exit face of the crystal will remain at the same polarization state as that at the entrance face. This beam will be blocked by the second polarizer, leading to a dark fringe of hyperbola at the focal plane. So the intensity pattern at the focal plane consists of a family of dark hyperbolic fringes. Between these dark fringes, there will be fringes of maximum intensity. The scale of the hyperbola depends on the wavelength. For white light illumination, colored hyperbolic fringes will be observed. The color is constant along each of the hyperbolas. Each of the lines is called an isochromatic line. Figure 4.13 shows the conoscopic intensity pattern of the a-plate discussed above using a monochromatic source. If a pair of parallel polarizers are used, a complementary conoscopic intensity pattern will be obtained at the focal plane.

4.5.2. *c* Plate of Uniaxial Crystals

Referring to Figure 4.14, we consider a *c* plate of a uniaxial medium sandwiched between a pair of crossed polarizers. At normal incidence, the wave is

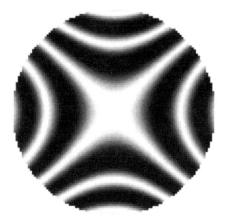

Figure 4.13. Conoscopic intensity pattern of an *a* plate between crossed polarizers.

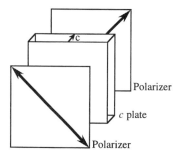

Figure 4.14. *c* plate between crossed polarizers.

propagating along the *c* axis of the crystal. Both components of the wave are propagating at exactly the same phase velocity. Thus, the polarization state remains unchanged exiting the medium. So, the transmission is zero at normal incidence. At a general angle of incidence, a phase retardation is introduced between the two components of the wave in the plate. Because of the symmetry of the structure, the phase retardation is a function of the angle of incidence θ only. Zero transmission occurs only when the phase retardation is an integral multiple of 2π. Thus, a dark ring appears at an angle when the phase retardation is an integral multiple of 2π. In the solid cone of angles, there will be a series of concentric dark rings at the focal plane. Between these dark rings, there will be rings of intensity maxima. The radii of the bright and dark rings depend on the wavelength and the thickness of the plate. From the discussion on the phase retardation at oblique incidence, we can write

$$\Gamma = \Gamma_0 \frac{n_o + n_e}{n_o n_e^2} \sin^2\theta \qquad (4.5\text{-}5)$$

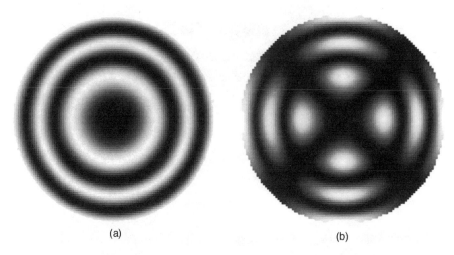

(a)   (b)

Figure 4.15. Conoscopic intensity pattern of c plate between crossed polarizers. (*a*) Theoretical calculation using ordinary Jones matrix method [Eqs. (4.5-4) and (4.5-5)]. (*b*) Theoretical calculation using extended Jones matrix method.

provided $\theta \ll 1$. This equation can be used to determine the angular radius of the dark rings (or the bright rings). Colored rings will be observed when white light is used. Again, color is constant around any circle centered on the axis. Such a line (circle in this case) is again, called an *isochromatic line*. Figure 4.15*a* shows the intensity pattern of the dark and bright rings of a c-cut crystal plate using a monochromatic source. This picture is obtained by using Eqs. (4.5-4) and (4.5-5).

Figure 4.15*b* shows the conoscopic intensity pattern calculated using a more accurate method, the extended Jones matrix method (which will be discussed in Chapter 8). Notice that, two perpendicular diameters of zero intensity exist in the intensity pattern. These dark diameters are due to the excitation of a single mode (either ordinary or extraordinary) of propagation in the medium at certain special angle of incidence. Since only a single mode is excited, there will be no change of polarization state, regardless of the phase retardation. Thus, the emergent beam is blocked by the rear polarizer. Single wave excitation occurs when the plane of incidence (plane formed by the incident wavevector and the normal to the surface) is parallel to the transmission axis of the front or rear polarizer. In the case when the transmission axis is parallel to the plane of incidence, only extraordinary wave is excited in the medium. On the other hand, when the transmission axis is perpendicular to the plane of incidence, only the ordinary wave is excited in the medium. In both cases, the polarization state remains unchanged and the emergent beam at the exit face of the medium is blocked by the rear polarizer. Since these dark lines will be the same for all wavelengths, they are called the *achromatic lines*. The isochromatic rings of intensity maxima are, of course, interrupted by the achromatic dark lines. The isochromatic lines are often called "rings", and the achromatic lines are called the "brushes" as they are not as sharp. When parallel polarizers are used, a complementary conoscopic intensity pattern is obtained at the focal plane.

4.5.3. *c* Cut Biaxial Plate

We now consider a plate of biaxial material cut perpendicular to the direction bisecting the optic axes. The plate is sandwiched between a pair of crossed polarizers. Generally speaking, isochromatic rings are visible for directions near

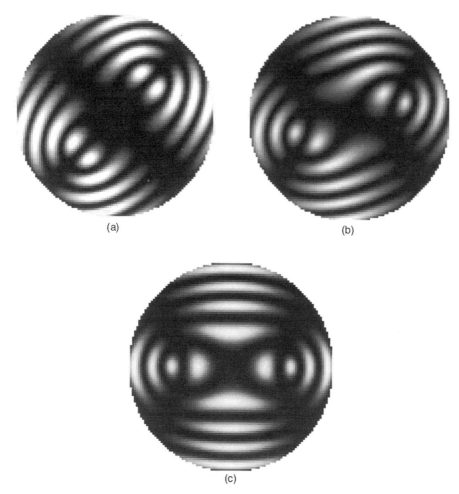

(a)   (b)

(c)

Figure 4.16. Conoscopic intensity pattern of a *c*-cut biaxial plate between crossed polarizers. (*a*) The transmission axis of the front polarizer is parallel to the plane of the optic axes (*xz* plane). (*b*) The plate is rotated around the *c* axis (note that the *c* axis in a biaxial crystal is not an optic axis) by a small angle. (*c*) The plane of the optic axes is 45° relative to the transmission axis of the front polarizer.

the optic axes. As there are two optic axes in this arrangement, there are two sets of concentric rings. This is a unique aspect of biaxial media. It is important to note that the observation of two sets of rings requires a plate of substantial thickness. The shape of the achromatic lines (brushes) depends on the orientation of the polarizers relative to the plane of the optic axes. Figure 4.16a shows the conoscopic intensity pattern for the case when the transmission axis of the front polarizer is parallel to the plane of the optic axes (xz plane). In this case, the achromatic lines are two perpendicular straight lines. When the crystal is rotated around the c axis (note that the c axis in a biaxial medium is not an optic axis) by a small angle, the conoscopic intensity pattern is illustrated in Figure 4.16b. Note that the isochromatic rings are rotated accordingly, whereas the achromatic lines (brushes) are somewhat distorted. Figure 4.16c shows the conoscopic intensity pattern for the case when the plane of the optic axes is 45° relative to the transmission axis of the front polarizer. We note that the isochromatic rings are rotated accordingly, and the intensity pattern is symmetric with respective to the plane of the optic axes.

4.6. REFLECTION PROPERTY OF A GENERAL TN-LCD WITH A REAR MIRROR

So far we have discussed the transmissive modes of LCDs that require two polarizers to obtain the display of intensity modulation at each pixel. The input polarizer prepares a linear polarization state before entering the LC cell. The analyzer is employed to achieve the polarization interference to produce the black or dark state. If the backlight is replaced with a mirror and the lighting is illuminated from the front, only one polarizer is needed as the light passes through the LC cell twice. For direct-view applications, the lighting is provided by the ambient light. Reflective mode LCDs can also be employed for projection displays (see Section 5.7 for more discussions). Here we discuss the reflection property of a general TN-LCD with a back mirror. Referring to Figure 4.17, we consider a reflective display that employs a TN-LC or STN-LC sandwiched between a polarizing beamsplitter (PBS) and a mirror. This particular configuration is useful for projection displays. For direct view applications, the PBS is often replaced with a sheet polarizer. The use of a polarizing beamsplitter allows us to investigate both the reflection and transmission properties.

We now consider the reflection and transmission properties of the reflective LCD in the field-OFF state at normal incidence. According to Figure 4.17b, the incident and reflected polarization states, as determined by the orientation angles of the polarizer reflection axis, are given by

$$\begin{pmatrix} V_x \\ V_y \end{pmatrix} = \begin{pmatrix} \cos\theta \\ \sin\theta \end{pmatrix}, \quad \begin{pmatrix} V_x' \\ V_y' \end{pmatrix} = \begin{pmatrix} \cos\theta \\ \sin\theta \end{pmatrix} \qquad (4.6\text{-}1)$$

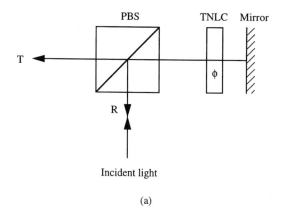

Incident light

(a)

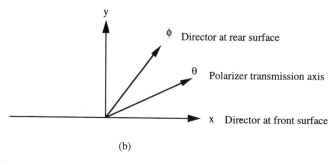

(b)

Figure 4.17. (*a*) Schematic drawing of a reflective mode LCD. (*b*) Azimuth angle of various components in the reflective mode LCD. The *x* axis is chosen to be parallel to the director at the input surface. ϕ is the total twist angle of the LC cell, θ is the angle of the reflection axis of the polarizing beamsplitter (PBS). For a sheet polarizer, θ is the transmission axis of the polarizer.

where θ is the angle of the plane of polarization of the incident and reflected beam. Here we assume an incident beam of polarized light. The intensity reflectivity of the LCD shown in Figure 4.17 is given by, according to Jones matrix method and Eq. (4.1-32)

$$R = |V' \cdot \tilde{M}MV|^2 \tag{4.6-2}$$

where M is the Jones matrix of the LC cell, the tilde $\sim$ indicates a transpose operation, and V and V' are the incident and reflected Jones vectors, respectively. A factor of $\frac{1}{2}$ must be included in Eq. (4.6-2) if the incident beam is unpolarized. Using Eq. (4.3-13) for M, and Eq. (4.6-1) for V and V', we obtain, after a few steps of matrix multiplication

$$R = |A + iB|^2$$

$$A = \cos^2 X + \phi^2 \frac{\sin^2 X}{X^2} - \frac{\Gamma^2}{4} \frac{\sin^2 X}{X^2} \qquad (4.6\text{-}3)$$

$$B = \frac{\Gamma \sin X}{X} \left[\cos 2\theta \cos X + \sin 2\theta \frac{\phi \sin X}{X} \right]$$

where we recall that θ is the angle between the reflection axis of the polarizing beamsplitter and the director at input surface, ϕ is the total twist angle of the LC cell, and X is given by

$$X = \sqrt{\phi^2 + \left(\frac{\Gamma}{2}\right)^2} \qquad (4.6\text{-}4)$$

The intensity reflectivity can also be written

$$R = \left[\cos^2 X + \phi^2 \frac{\sin^2 X}{X^2} - \frac{\Gamma^2}{4} \frac{\sin^2 X}{X^2} \right]^2$$
$$+ \left\{ \frac{\Gamma \sin X}{X} \left[\cos 2\theta \cos X + \sin 2\theta \frac{\phi \sin X}{X} \right] \right\}^2 \qquad (4.6\text{-}5)$$

By using the Mauguin parameter u such that

$$X = \phi\sqrt{1 + u^2} \qquad \text{with} \qquad u = \frac{\Gamma}{2\phi} = \frac{\pi \Delta n d}{\phi \lambda} \qquad (4.6\text{-}6)$$

the intensity reflectivity can be written

$$R = \left\{ \cos^2 X + \frac{1 - u^2}{1 + u^2} \sin^2 X \right\}^2 + 4u^2 \left\{ \frac{\cos 2\theta \cos X \sin X}{\sqrt{1 + u^2}} + \frac{\sin 2\theta \sin^2 X}{1 + u^2} \right\}^2 \qquad (4.6\text{-}7)$$

The intensity transmission is given by

$$T = 1 - R \qquad (4.6\text{-}8)$$

In the case when a sheet polarizer is used, Eq. (4.6-7) is the reflectivity of the display system, where θ is the angle of the transmission axis of the polarizer measured from the local director at the input surface.

By examining Eq. (4.6-7) for the reflectivity, we find that for a given TN-LC or STN-LC cell with a twist angle ϕ and a Mauguin parameter u, the reflectivity R is a periodic function of the angle θ. By differentiating Eq. (4.6-7) with respect to the angle θ, we obtain the following maximum and minimum reflectivity:

$$R_{\min} = \left\{ \cos^2 X + \frac{1-u^2}{1+u^2} \sin^2 X \right\}^2, \qquad \text{when} \qquad \tan 2\theta_{\min} = -\frac{\sqrt{1+u^2}}{\tan X}$$

$$(4.6\text{-}9)$$

and

$$R_{\max} = 1 \qquad \text{when} \qquad \tan 2\theta_{\max} = \frac{\tan X}{\sqrt{1+u^2}} \qquad (4.6\text{-}10)$$

Furthermore, we obtain, according to Eq. (4.6-9)

$$R_{\min} = 0 \qquad \text{when} \qquad \tan X = \pm\sqrt{\frac{u^2+1}{u^2-1}} \qquad (4.6\text{-}11)$$

which is possible only when the Mauguin parameter is greater than one $(1 < u)$. In this case, the polarizer azimuth angle required is given by

$$\tan 2\theta = -\frac{\sqrt{u^2+1}}{\tan X} = \mp\sqrt{u^2-1} \qquad (4.6\text{-}12)$$

In summary, $R_{\max} = 1$ is always possible provided the polarizer orientation angle θ satisfies the condition (4.6-10). On the other hand, zero reflectivity $(R_{\min} = 0)$ is possible provided the Mauguin parameter satisfies the condition (4.6-11) and the polarizer is aligned along the azimuth given by Eq. (4.6-12). By examining the polarizer angles in Eqs. (4.6-9) and (4.6-10), we also find that

$$\theta_{\max} - \theta_{\min} = 45° \qquad (4.6\text{-}13)$$

In the other words, a rotation of the polarizer by 45° in either direction can change the reflectivity from minimum to maximum, or from maximum to minimum.

4.6.1. Normally Black Reflection (NBR) Operation

Since both reflection and transmission (see Fig. 4.17a) can be employed for display applications. Normally black reflectivity ($R = R_{min}$ or 0) corresponds to normally white transmission (NWT) ($T = T_{max}$). We now examine the possibility of zero intensity reflectivity ($R_{min} = 0$, or $T_{max} = 1$). This corresponds to an optimum normally black reflection (NBR) configuration in the reflection mode, or an optimum normally white transmission (NWT) configuration in the transmission mode. According to Eqs. (4.6-11) and (4.6-12), zero reflectivity occurs when the Mauguin parameter satisfies the following condition

$$\tan X = \pm \sqrt{\frac{u^2 + 1}{u^2 - 1}} \qquad (4.6\text{-}14)$$

and the polarizer is aligned along the azimuth angle given by

$$\tan 2\theta = \mp \sqrt{u^2 - 1} \qquad (4.6\text{-}15)$$

We note that zero reflectivity in NBR mode (or 100% transmission in NWT) is possible only when $1 < u$. Given an LC cell with a total twist angle ϕ, Eq. (4.6-14) can be solved for the Mauguin parameter u, which is proportional to the cell thickness. Once u is obtained, Eq. (4.6-15) can then be employed to obtain the angle θ required for zero reflectivity. It can be shown that the conditions in Eqs. (4.6-14) and (4.6-15) correspond to the situation when the output polarization state after a single pass through the cell is circularly polarized. In other words, the LC cell is equivalent to a quarter-wave plate [Problem 4.10]. Table 4.3 lists the Mauguin parameters u and the required polarizer reflection axis θ needed to obtain a zero reflectivity (normally black reflection) for an LC cell with total twist angles of $\phi = 45°$ and $\phi = 90°$. For example, let us consider a 45° TN cell. $R = 0$ is possible when $\Delta nd / \lambda$ is 0.29 and with the polarizer aligned at an azimuth angle of $-15.26°$, or 0.68 and with the polarizer aligned at an azimuth angle of 34.29°. For $\lambda = 0.55 \, \mu m$, these correspond to a TN cell of $\Delta nd = 0.16$ or $0.37 \, \mu m$.

Figure 4.18 plots the twist angle ϕ as a function of $\Delta nd / \lambda$ for NBR operation. The straight line shown is when the Mauguin parameter is unity ($u = 1$). Given a total twist angle, we can draw a horizontal line for the twist angle. The intersection with the curves in the figure are the solutions for normally black state. We note that in the case of zero twist ($\phi = 0$), the Mauguin parameter u is infinite. Zero reflectivity (normally black state) occurs when, according to Eqs. (4.6-14) and (4.6-15)

$$\frac{\Gamma}{2} = \frac{\pi}{4}, \quad 3\frac{\pi}{4}, \quad 5\frac{\pi}{4}, \quad 7\frac{\pi}{4}, \quad 9\frac{\pi}{4} \qquad (4.6\text{-}16)$$

$$\theta = 45° \qquad (4.6\text{-}17)$$

Table 4.3.   LC Parameters for Normally Black Reflection ($R = 0$) Operation

Twist Angle	u	$\Delta nd/\lambda$	θ
45°	1.16092	.29023	− 15.2637
45°	2.73792	.68448	34.2888
45°	4.92575	1.23144	− 39.1433
45°	6.91475	1.72869	40.8424
45°	8.95226	2.23806	− 41.7932
45°	10.9491	2.73728	42.3799
45°	12.9653	3.24132	− 42.7882
45°	14.9638	3.74095	43.0841
45°	16.9728	4.24319	− 43.3111
45°	18.9719	4.74297	43.4893
90°	2.35416	1.17708	− 32.4317
90°	3.32399	1.66200	36.2459
90°	4.60431	2.20216	− 38.4383
90°	5.39721	2.69861	39.6612
90°	6.43040	3.21520	− 40.5268
90°	7.42721	3.71361	41.1311
90°	8.44545	4.22273	− 41.5999
90°	9.44363	4.72182	41.9607

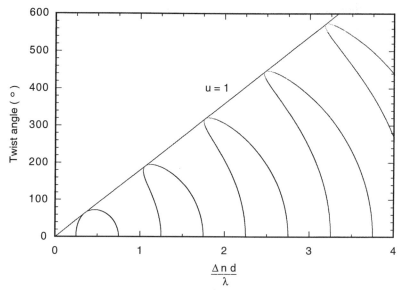

Figure 4.18.   Twist angle ϕ as a function of $\Delta nd/\lambda$ for normally black reflection (NBR) operation.

These parameters correspond to the cases when the LC cell becomes a quarter-wave plate. A linearly polarized beam of light with a polarization angle of 45° will be converted into a circularly polarized beam of light. A double pass due to the rear mirror leads to a linearly polarized beam of light with a rotation of polarization direction by 90°. This then leads to a zero reflectivity.

When the LC cell is thin so that the Mauguin parameter is less than unity ($u < 1$), the polarizer must be aligned along the azimuth given by, according to Eq. (4.6-9)

$$\tan 2\theta = -\frac{\sqrt{u^2 + 1}}{\tan X} \tag{4.6-18}$$

to ensure a minimum reflectivity in the field-OFF state.

4.6.2. Normally White Reflection (NWR) Operation

Most reflective mode LCDs for direct viewing are operated in the normally black (NBR) configuration. When a strong field is applied to the LC cell, the phase retardation becomes zero as a result of homeotropic alignment of the LC director. Thus, maximum reflectivity ($R = 1$) occurs in the white state (field-ON). In fact, maximum reflectivity always occurs when the LC director is homeotropically aligned, since there is only one polarizer. In the transmissive mode LCDs, there is a choice of using either crossed or parallel polarizers. This allows the possibility of either white ($T = 1$) or black ($T = 0$) state when the LC director is homeotropically aligned. This is the main difference between the transmissive and reflective modes. For projection displays, polarizing beamsplitters can be employed to tap the transmission output ($T = 1 - R$). The polarizing beamsplitter shown in Figure 4.17a is functioning as a pair of crossed polarizers for transmission output. Thus, zero transmission ($T = 0, R = 1$) is obtained when the LC molecules are homeotropically aligned.

To operate in the normally white reflection (NWR) mode with $R = 1$ in the field-OFF state for direct viewing, the LC cell must be thick enough to function like a half-wave plate in the field-OFF case, leading to a 100% reflectivity. A half-wave plate can rotate the polarization state of a beam of linearly polarized light by 90°. A double pass through the cell would restore the incident polarization state, leading to a 100% reflectivity. When an electric field is applied to the cell, the reorientation (tilt) of the LC director leads to a decrease of the phase retardation. If the retardation is such that the LC cell is functioning like a quarter-wave plate, then zero reflectivity is achieved.

We study the condition of unity reflectivity ($R = 1$) in the field-OFF state. According to Eq. (4.6-7), it is obvious that 100% reflectivity occurs when

$$X = \pi, 2\pi, 3\pi, 4\pi, \ldots \tag{4.6-19}$$

with an arbitrary angle θ for the polarizer. We note that for a 90° TN-LC cell, the condition (4.6-19) corresponds to Mauguin parameters of $u = \sqrt{3}, \sqrt{15}, \sqrt{35}, \ldots$, which yield a normally black operation in a parallel-polarizer configuration after a single pass through the cell [see Eq. (4.3-18)]. The condition (4.6-19) also corresponds to the case when the phase retardation between the normal modes (2X) is exactly an integral number of full waves, according to Eq. (4.3-35). Thus, the polarization state remains the same after a single pass through the LC cell, regardless of the incident polarization state. A mirror reflection simply repeats the process and results in a 100% reflectivity.

One hundred percent reflectivity ($R_{max} = 1$) also occurs in situations when X is not an integral number of π. This will occur at specific orientations θ of the polarizer given by, according to Eq. (4.6-10)

$$\tan 2\theta_{max} = \frac{\tan X}{\sqrt{1 + u^2}} \qquad (4.6\text{-}20)$$

The preceding discussion is limited to the field-OFF state. When a voltage is applied to the LC cell, the LC director is reoriented by the applied field. This leads to a decrease in the phase retardation. The reflection properties of the reflective LCDs in the field-ON states will be discuss in the next chapter.

REFERENCES

1. R. C. Jones, *J. Opt. Soc. A.* **31**, 488 (1941).
2. See, for example, A. Yariv and P. Yeh, *Optical Waves in Crystals*, Wiley, 1984.
3. C. H. Gooch and H. A. Tarry, *J. Phys. D: Appl. Phys.* **8**, 1575–1584 (1975)
4. R. M. A. Azzam and N. M. Bashara, *J. Opt. Soc. Am.* **62**, 1252–1257 (1972).
5. H. L. Ong, *J. Appl. Phys.* **64**(2), 614–628 (1988).
6. H. L. Ong, *SID'94 Digest* 787–789 (1994).
7. B. Lyot, "Optical apparatus with wide field using interference of polarized light," *C. R. Acad. Sci.* (Paris) **197**, 1593 (1933).
8. B. Lyot, "Filter monochromatique polarisant et ses applications en physique solaire," *Ann. Astrophys.* **7**, 31 (1944).
9. Y. Öhman, "A new monochromator," *Nature* **41**, 157, 291 (1938).
10. Y. Öhman, "On some new birefringent filter for solar research," *Ark. Astron.* **2**, 165 (1958).

SUGGESTED READINGS

See, for example P. Yeh, *Optical Waves in Layered Media*, Wiley, 1988, pp. 289–291.
  [Note that there is a sign correction for ϕ in Eq. (10.6-8), (ϕ is replaced with $-\phi$).]
S.-T. Wu and C.-S Wu, *Appl. Phys. Lett.* **68**, 1455–1457 (1996).

E. Beynon, K. Saynor, M. Tillin, and M. Towler, *IDRC'97 Digest* L-34 (1997).

P. Yeh, "Optical properties of general twisted-nematic LCDs," *J. Soc. Info. Displays* **5**, 289–292 (1997).

PROBLEMS

4.1. A *half-wave plate* has a phase retardation of $\Gamma = \pi$. Assume that the plate is oriented so that the azimuth angle (i.e., the angle between the x axis and the slow axis of the plate) is ψ.

   (a) Find the polarization state of the transmitted beam, assuming that the incident beam is linearly polarized in the y direction.

   (b) Show that a half-wave plate will convert right-hand circularly polarized light into left-hand circularly polarized light, and vice versa, regardless of the azimuth angle of the plate.

   (c) E7 is a nematic liquid crystal with $n_o = 1.52$ and $n_e = 1.75$ at $\lambda = 577$ nm. Find the half-wave-plate thickness at this wavelength, assuming that the plate is made in such a way that the surfaces are parallel to the directors (i.e., a plate).

4.2. A *quarter-wave plate* has a phase retardation of $\Gamma = \pi/2$. Assume that the plate is oriented in a direction with azimuth angle ψ

   (a) Find the polarization state of the transmitted beam, assuming that the incident beam is linearly polarized in the y direction.

   (b) Assuming that the polarization state resulting from (a) is represented by a complex number on the complex plane, show that the locus of these points as ψ varies from 0 to $\pi/2$ is a branch of a hyperbola. Obtain the equation of the hyperbola.

   (c) ZLI-1646 is a nematic liquid crystal with $n_o = 1.478$ and $n_e = 1.558$ at $\lambda = 589$ nm. Find the thickness of an a plate at this wavelength.

4.3. *Polarization transformation by a wave plate*:

   A wave plate is characterized by its phase retardation Γ and azimuth angle ψ.

   (a) Find the polarization state of the emerging beam, assuming that the incident beam is polarized in the x direction.

   (b) Use a complex number to represent the resulting polarization state obtained in (a).

   (c) The polarization state of the incident x-polarized beam is represented by a point at the origin of the complex plane. Show that the transformed polarization state can be anywhere on the complex plate, provided Γ can be varied from 0 to 2π and ψ can be varied from 0 to $\pi/2$. Physically, this means that any polarization state can be produced from linearly polarized light, provided a proper wave plate is available.

(d) Show that the locus of these points in the complex plane obtained by rotating a wave plate from $\psi = 0$ to $\psi = \pi/2$ is a hyperbola. Derive the equation of this hyperbola.

(e) Show that the Jones matrix W of a wave plate is unitary, that is, $W^\dagger W = 1$, where the dagger indicates Hermitian conjugation.

(f) Let $\mathbf{V}_1'$ and $\mathbf{V}_2'$ be the transformed Jones vectors from $\mathbf{V}_1$ and $\mathbf{V}_2$, respectively. Show that if $\mathbf{V}_1$ and $\mathbf{V}_2$ are orthogonal, so are $\mathbf{V}_1'$ and $\mathbf{V}_2'$.

4.4. *Polarizers and projection operators*:

An ideal polarizer can be considered as a projection operator that acts on the incident polarization state and projects the polarization vector along the transmission axis of the polarizer.

(a) Neglecting the absolute phase factor in Eq. (4.1-11), show that $P_o^2 = P_0$ and $P^2 = P$. Operators satisfying these conditions are called *projection operators* in linear algebra.

(b) Show that if $\mathbf{E}_1$ is the amplitude of the electric field, the amplitude of the beam after it passes through the polarizer is given by $\mathbf{p}(\mathbf{p} \cdot \mathbf{E}_1)$, where $\mathbf{p}$ is the unit vector along the transmission axis of the polarizer.

(c) If the incident beam is vertically polarized (i.e., $\mathbf{E}_1 = \hat{y} E_0$), the polarizer transmission axis is in the x direction (i.e., $\mathbf{p} = \hat{x}$). The transmitted beam has zero amplitude, since $\hat{x} \cdot \hat{y} = 0$. However, if a second polarizer is placed in front of the first polarizer and is oriented at 45° with respect to it, the transmitted amplitude is not zero. Find this amplitude.

(d) Consider a series of N polarizers with the first one oriented at $\psi_1 = \pi/2N$, the second one at $\psi_2 = 2(\pi/2N)$, the third one at $\psi_3 = 3(\pi/2N), \ldots$, and the Nth one at $N(\pi/2N)$. Let the incident beam be horizontally polarized. Show that the transmitted beam is vertically polarized with an amplitude of $[\cos(\pi/2N)]^N$. Evaluate the amplitude for $N = 1, 2, 3, \ldots, 10$. Show that in the limit of $N \to \infty$, the amplitude becomes one. In other words, a series of polarizers oriented like a fan can rotate the polarization of the light without attenuation.

4.5. *Lyot–Öhman filter*:

In solar physics, the distribution of hydrogen in the solar corona is measured by photographing at the wavelength of H_α line ($\lambda = 6563\text{Å}$). To enhance the signal-to-noise ratio, a filter of extremely narrow bandwidth ($\sim 1\text{Å}$) is required. The polarization filter devised by Lyot and Öhman consists of a set of birefringent plates separated by parallel polarizers [7–10]. The plate thicknesses are in geometric progression, that is $d, 2d, 4d, 8d, \ldots$. All the plates are oriented at an azimuth angle of 45°.

(a) Show that if n_o and n_e are the refractive indices of the plates, the transmittance of the whole stack of N plates is given by

$$T = \tfrac{1}{2}\cos^2 x \cos^2 2x \cos^2 4x \cdots \cos^2 2^{N-1}x$$

with

$$x = \frac{\pi d(n_e - n_o)}{\lambda} = \frac{\pi d(n_e - n_o)\nu}{c}$$

(b) Show that the transmission can be written

$$T = \frac{1}{2}\left(\frac{\sin 2^N x}{2^N \sin x}\right)^2$$

(c) Show that the transmission bandwidth [full width at half maximum (FWHM)] of the whole system is governed by that of the bands of the thickest plate:

$$\Delta\nu_{1/2} \sim \frac{c}{2^N d(n_e - n_o)}$$

and the free spectral range $\Delta\nu$ is governed by that of the bands of the thinnest plate:

$$\Delta\nu \sim \frac{c}{d(n_e - n_o)}$$

The finesse F of the system, defined as $\Delta\nu/\Delta\nu_{1/2}$ is then $F \sim 2^N$.

(d) Design a filter with a bandwidth of $1\,\text{Å}$ at the H_α line, using quartz as the birefringent material. Assume that $n_o = 1.5416$ and $n_e = 1.5506$ at $\lambda = 6563\,\text{Å}$. Find the required thickness of the thickest plate.

(e) Show that according to (b) the bandwidth (FWHM) is given by

$$\Delta\nu_{1/2} = 0.886 \frac{c}{2^N d(n_e - n_o)}$$

4.6. *Off-axis effect:*

A wave plate made of uniaxial crystal with its c axis parallel to the plate surfaces has a phase retardation of $2\pi(n_e - n_o)d/\lambda$ for a normal incident beam. For an off-axis beam, the light will "see" different birefringence because the refractive index for the extraordinary wave is dependent on the direction of the beam. In addition, the pathlength in the crystal plate is no longer d (see Fig. 4.19).

(a) Let θ_o, θ_e be the refractive angles for the ordinary and extraordinary waves, respectively. Show that the phase retardation is

$$\Gamma = \frac{2\pi}{\lambda}(n'_e \cos\theta_e - n_o \cos\theta_o)d$$

where n'_e is the eigen refractive index of the E-wave.

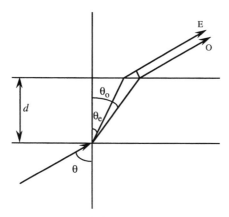

Figure 4.19.   Off-axis rays.

(b) Let θ be the angle of incidence, and ϕ be the angle between the c axis and the tangential component of the wave vector. Show that the phase retardation can be expressed in terms of θ and ϕ as

$$\Gamma = \frac{2\pi}{\lambda} d \left[n_e \sqrt{1 - \frac{\sin^2\theta \sin^2\phi}{n_e^2} - \frac{\sin^2\theta \cos^2\phi}{n_o^2}} - n_o \sqrt{1 - \frac{\sin^2\theta}{n_o^2}} \right]$$

(c) Show that in the case when $\sin^2\theta$ is much smaller than n_o^2 and n_e^2, the phase retardation can be written

$$\Gamma = \frac{2\pi}{\lambda} (n_e - n_o) d \left\{ 1 + \sin^2\theta \left[\frac{\sin^2\phi}{2n_o n_e} - \frac{\cos^2\phi}{2n_o^2} \right] \right\}$$

(d) According to the result in (c), a Lyot–Öhman filter with a passband of λ_0 at normal incidence will have a passband of $\lambda_0 \pm \Delta\lambda$ for off-axis light. Show that with $n_{\text{eff}}^2 = n_o^2 \approx n_e^2$, $\Delta\lambda$ is given by

$$\Delta\lambda = \frac{\lambda \sin^2\theta}{2n_{\text{eff}}^2}$$

(e) Show that a narrowband Lyot–Öhman filter has a limited field-of-view of

$$\theta \approx \pm n_o \left(\frac{2\Delta\lambda_{1/2}}{\lambda} \right)^{1/2}$$

where we assume $n_e - n_o \ll n_o$.

4.7. *Equivalent circuit of a general birefringent network*:

It can be shown that a general birefringent network (e.g., a liquid crystal cell) is equivalent to a wave plate and a polarization rotator. A polarization rotator will rotate the polarization ellipse of an input beam by an angle without changing the shape of the ellipse. The Jones matrix of a polarization rotator can be written

$$A(\rho) = \begin{pmatrix} \cos\rho & -\sin\rho \\ \sin\rho & \cos\rho \end{pmatrix}$$

where ρ is the angle of rotation. We note that the polarization rotation of ρ is equivalent to a coordinate rotation of $-\rho$.

(a) Find the Jones matrix of a birefringent wave plate with phase retardation Γ and azimuth angle ψ followed by a polarization rotator. Show that the Jones matrix is unitary.

(b) The Jones of a general birefringent network is unitary. Based on the result in (a), an arbitrary birefringent network is equivalent to a phase retardation wave plate followed by a polarization rotator. Let the Jones matrix of the general birefringent network be written

$$M = \begin{pmatrix} a+ib & c+id \\ -c+id & a-ib \end{pmatrix}$$

where a, b, c and d are real. Show that the phase retardation Γ, azimuth angle ψ, and the rotation angle ρ of the equivalent network (a wave plate and a rotator) can be written,

$$\cos^2\frac{\Gamma}{2} = a^2 + c^2 \qquad \sin^2\frac{\Gamma}{2} = b^2 + d^2$$

$$\tan(\rho + 2\psi) = \frac{d}{b} \qquad \tan\rho = \frac{c}{a}$$

(c) Show that a uniformly TN-LC cell is equivalent to a retardation plate followed by a polarization rotator. Show that the phase retardation Γ_e, azimuth angle ψ, and the rotation angle ρ of the equivalent network can be written,

$$\sin\frac{\Gamma_e}{2} = -\frac{\Gamma}{2}\frac{\sin X}{X} \qquad \tan(2\psi) = \phi\frac{\tan X}{X} \qquad \rho = \phi - 2\psi$$

Based on the above results in (a)–(c), an input beam of linearly polarized light with electric field vector oriented along the principal axes of the equivalent wave plate (ψ or $\psi + \pi/2$) will be transformed into an output beam of linearly polarized light.

4.8. *Existence of linearly polarized input and output states*:

Given a general birefringent network, there exist two linearly polarized input states such that the corresponding output polarization states are linear. These two input polarization states are mutually orthogonal. So are the two output polarization states.

(a) A beam of linearly polarized light is transformed into a beam of elliptically polarized light by using a wave plate (uniaxial) of phase retardation Γ. Let θ be the angle between the polarization vector and the c axis of the plate. Show that as θ varies from 0 to π, the output polarization state traces out a great circle on the Poincaré sphere. This great circle can be obtained by rotating the equator around the S_1 axis by an angle Γ.

(b) If a polarization rotator (ρ) is placed after the wave plate in Problem 4.8 (a), show that as θ varies from 0 to π, the output polarization state also traces out a great circle on the Poincaré sphere. This great circle is obtained by rotating the great circle in 4.8(a) around the polar axis by an angle 2ρ. Based on the result, any input polarization state of a beam of monochromatic light can be transformed into any output polarization state by using the combination of a single wave plate and a polarization rotator.

(c) Using the equivalent circuit in Problem 4.7, show that given a general birefringent network, there exist two linearly polarized input states such that the output polarization states are linear. These two polarization states are mutually orthogonal. Find the polarization states of the input and output beams.

The existence of these linearly polarized states can be proven geometrically as follows. Given an input beam of linearly polarized light, an output beam of elliptically polarized light is obtained. By varying the angle of polarization of the input linearly polarized state from 0 to π, the output polarization state will trace out a great circle on the Poincaré sphere. This great circle is either the equator itself, or will intersect with the equator at two points. These two points correspond to output states of linear polarization.

4.9. *Twisted nematic liquid crystals*:

(a) Use Chebyshev's identity (4.3-8) to expand Eq. (4.3-7).

(b) Derive Eq. (4.3-10).

(c) Derive Eqs. (4.3-20) and (4.3-21).

(d) Derive Eq. (4.3-29).

(e) Derive Eqs. (4.3-31) and (4.3.32).

(f) Derive Eqs. (4.3-38) and (4.3-39).

4.10. Using Eq. (4.3-37) for the output polarization state, show that an output circular polarization state is obtained when the conditions Eqs. (4.6-11) and (4.6-12) are satisfied. [Hint: Set $\chi' = i$ and then find χ.]

4.11. In Example 4.2.2, we assumed that the c axis of the a plate is oriented at $45°$ from the transmission axis of the polarizer. Show that if the orientation is at a general angle θ, the transmission can be written

$$T = \frac{1}{2} \sin^2 2\theta \sin^2 \frac{\Gamma}{2}$$

which reduces to Eq. (4.2-11) when $\theta = 45°$.

4.12. *Polarization ellipse in TN-LC*:

The polarization ellipse inside a uniformly TN-LC cell can be obtained by using Eq. (4.3-20) and (4.3-21), for an input wave polarized along the input director.

(a) Using the Mauguin parameter $u = \Gamma/2\phi$, show that the ellipticity and the orientation of the ellipse can be written

$$e = \tan\left\{\frac{1}{2}\sin^{-1}\left[\frac{2u}{1+u^2}\sin^2\alpha z\sqrt{1+u^2}\right]\right\}$$

$$\tan 2\psi = \frac{2\sqrt{1+u^2}\tan\alpha z\sqrt{1+u^2}}{(1-u^2)\tan^2\alpha z\sqrt{1+u^2} - 1 - u^2}$$

where α is the twist rate. Show that the ellipticity and the orientation are periodic functions of z.

(b) Let the ellipticity vary between 0 and e_{max}. Show that, in the Mauguin regime $(1 \ll u)$, e_{max} is given by

$$e_{max} = \frac{u}{1+u^2}$$

Show that maximum ellipticity occurs when $X = \alpha z\sqrt{1+u^2} = \pi/2, 3\pi/2, \ldots$. Minimum ellipticity occurs when $X = \alpha z\sqrt{1+u^2} = \pi, 2\pi, \ldots$

Show that maximum and minimum ellipticity occur when the principal axes of the polarization ellipse are the eo-axes (i.e. $\psi = 0$).

(c) The azimuth angle ψ of the ellipse relative to the director is a periodic functions of z, and varies between $(-\psi_{max}, \psi_{max})$. Show that, in the Mauguin regime $(1 \ll u)$, ψ_{max} can be written

$$\psi_{max} = \frac{\sqrt{1+u^2}}{2u^2}$$

4.13. *Mixed mode operation of TN-LC*:

In the *E*-mode or *O*-mode operations, the input polarization states are either parallel or perpendicular to the director at the input surface. In a mixed mode operation, the polarization state is oriented at a general angle relative to the director at the surface. Consider a special case when the angle between the director and the input polarization state is 45°. In the *eo*-coordinate, the input Jones vector can be written

$$V = \frac{1}{\sqrt{2}}\begin{pmatrix} 1 \\ 1 \end{pmatrix}$$

(a) Show that, in this case, the *te*-mode and *to*-mode are excited equally. In other words, this input wave can be decomposed into a linear combination of *te*- and *to*-modes with equal weight.

(b) Using an analyzer oriented in the direction of

$$V' = \frac{1}{\sqrt{2}} \begin{pmatrix} 1 \\ 1 \end{pmatrix}$$

in the *eo*-coordinate at the output surface, show that the transmission can be written

$$T = \cos^2 X$$

according to Eq. (4.3-13). This formula resembles the transmission of a wave plate between a pair of parallel polarizers. If the analyzer is rotated by an angle of $90°$, the transmission formula becomes $\sin^2 X$. We note that in the mixed mode operation, the phenomenon of waveguiding disappears. The transmission in this case is a result of the interference between the two normal modes.

(c) Derive the same result of $T = \cos^2 X$ by using the general transmission formula Eq. (4.3-29). What are the values of Φ_{ent}, Φ_{exit}?

4.14. *Excitation of* te- *or* to-*modes*:

The *te* or *to* modes in a TN-LC cell can be excited, provided the input polarization state has a polorization ellipse which is identical to that of these normal modes. A quarter wave plate can be employed to convert a beam of linearly polarized light into a beam of elliptically polarized light. Find the appropriate azimuth angle of the input polarizer and the required azimuth angle of the quarter wave plate, so that a pure *te*-mode can be excited in the TN-LC cell.

[Answer: The *c* axis of the wave plate is either parallel or perpendicular to the local director (*e* axis) at the input surface. The azimuth of the input polarizer must be oriented at the ellipticity angle $\psi = \tan^{-1} e = \tan^{-1}(\sqrt{1 + u^2} - u)$ relative to the local director (*e* axis).]

4.15. According to the transmission formula Eq. (4.3-29b), maximum transmission of $T = 1$ occurs when the azimuth angles of the polarizer's transmission angles are given by

$$\tan 2\alpha = \frac{\phi}{X} \tan X \qquad \alpha = -\beta$$

Show that these angles are equivalent to the polarization states in Eq. (4.3-38) and (4.3-39).

4.16. Figure 4.15*b* shows the conoscopic intensity pattern of a *c* plate sandwiched between a pair of crosses polarizers. We note that a pair of brushes exists in the conoscopic intensity pattern. Let c_1, c_2 be the unit vectors representing the c axes of the polarizers (assuming O-type). The c axis of the wave plate is parallel to the z axis. Let $\mathbf{k}$ be the wave vector of the incident beam inside the media (assuming the refractive indices of the polarizers and the wave plate are nearly the same.) The polarization states $\mathbf{o}_1, \mathbf{o}_2$ of the transmitted O-waves in the polarizers are given by Eq. (3.5-6) and (3.5-7).

(a) Show that only the O-wave is excited in the wave plate when $\mathbf{k}, \mathbf{z}$ are coplanar with $\mathbf{c}_1$.

(b) Let the polarization state of the O-wave in the wave plate be written $\mathbf{E}_o$. Show that

$$\mathbf{E}_o \cdot \mathbf{o}_2 = 0$$

In other words, the transmission is zero when $\mathbf{k}, \mathbf{z}$ are coplanar with either $\mathbf{c}_1$ or $\mathbf{c}_2$.

5

Liquid Crystal Displays

Liquid crystals can be employed for display applications using various configurations. Most of the displays produced recently involve the use of either twisted nematic (TN) or supertwisted nematic (STN) liquid crystals. In TN liquid crystal, the director of the LC is twisted by an angle of about 90°, whereas in STN the LC director is twisted by an angle larger than 90° (e.g., 180°, 240°, or 270°). As we mentioned before, each display system consists of a two-dimensional array of pixels ($M \times N$) that can be turned ON or OFF electrically. Each of pixels consists of a LC cell sandwiched between transparent electrodes. In the early days of TN displays, the liquid crystal cells are driven by two sets of electrodes ($M + N$) using multiplexing technique. The advantage of multiplexing is that $M \times N$ pixels can be addressed by using only $M + N$ electrical contacts. The simplicity of multiplexing leads to a degradation of the device performance, especially in the viewing contrast and the limited pixel resolution. The performance of TN displays was later improved by using a different driving technique that involves the use of thin film transistor (TFT) arrays. In TFT arrays, there are transistors associated with each of the pixels. This significantly increases the performance as well as the production cost. The subjects of multiplexing and active matrix (AM) addressing will be discussed in detail in the next chapter.

The STN displays was introduced recently to improve the performance of LCD without the use of TFT. It will be shown later that a larger twist angle can lead to a significantly larger electrooptical distortion. This leads to a substantial improvement in the contrast and viewing angles over TN displays.

In this chapter, we discuss the principles of operation of various displays including TN-LCD, STN-LCD, PD-LCD, reflective LCD and so on. In addition, we also discuss important properties of these displays, including the contrast and viewing angles.

5.1. TWISTED NEMATIC (TN) DISPLAYS

5.1.1. Principle of Operation

The operation of a twisted nematic display is based on the waveguiding property discussed in Chapter 4. As we recall, a beam of linearly polarized light can propagate in a TN liquid crystal cell along the direction of the twist axis. If the

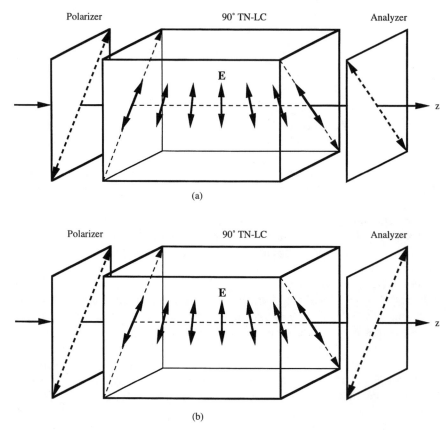

Figure 5.1. Principle of operation of TN displays (E mode): (*a*) NW operation; (b) NB operation. The dashed arrows on the surface of the TN-LC cell indicate the rubbing directions.

direction of polarization is parallel (or perpendicular) to the director at incidence, the polarization state of the light will follow the twist of the local director as the light propagates in the LC. This leads to a rotation of the output polarization state. In a 90° TN-LC cell, the direction of polarization of the beam is rotated by 90°. Figure 5.1 illustrates the principle of operation of a 90° TN-LCD.

In a typical TN display with a 90° twist, the LC is sandwiched between a pair of crossed polarizers. The polarizer is aligned so that the transmission axis is parallel to the director at the input end of the LC (E-mode operation). This ensures that the input beam of light is polarized along the direction of the director. The analyzer is also aligned so that the transmission axis is parallel to the director at the exit end of the liquid crystal. This ensures a total transmission of a beam of linearly polarized light by virtue of the waveguiding property. Thus, a beam of unpolarized light can transmit through the display suffering only a 3-dB loss at the front polarizer. This is known as the normally white (NW) configuration (see Fig. 5.1*a*).

In the TN display, each of the cells consists of a LC material sandwiched between two glass plates separated by a gap of $5-10\,\mu m$. The inner surfaces of the plates are deposited with transparent electrodes made of conductive coatings of indium tin oxide (ITO). These transparent electrodes are coated with a thin layer of polyimide with a thickness of several hundred angstroms. The polyimide films are unidirectionally rubbed to ensure that the local directors are parallel to the rubbing directions at the surfaces. In a 90° TN, the rubbing direction of the lower substrate is perpendicular to the rubbing direction of the upper surface. Thus, in the inactivated state (voltage OFF), the local director undergoes a continuous (uniform) twist of 90° in the region between the plates. Sheet polarizers are laminated on the outer surfaces of the plates. The transmission axes of the polarizers are aligned parallel to the rubbing directions of the adjacent polyimide films.

In the activated state (voltage ON) when a small voltage (3–5 V) is applied to the electrodes, a strong electric field is built up in the liquid crystal. As a result of the dielectric anisotropy, the liquid crystals are aligned parallel to the direction of the applied electric field. This leads to a homeotropically aligned configuration of the LC molecules. In other words, the directors of the liquid crystal are now perpendicular to the plates. Using the terminology of Chapter 4, this is effectively a c plate of liquid crystal. As we know, the plane of polarization remains unchanged when a beam of light propagates in the normal direction of a c plate. This leads to a zero transmission when the c plate is sandwiched between a pair of crossed polarizers (NW operation). Thus, by controlling the voltage ON or OFF, we are able to control the transmission of light through the liquid crystal cell. This is the basic principle of operation of a TN display cell.

5.1.2. Transmission Properties of Field-OFF State

The transmission property of a TN liquid crystal cell can be investigated by using the Jones matrix method developed in Chapter 4. To discuss the transmission property, we consider a 90° TN cell with a front local director parallel to the transmission axis of the polarizer. A beam of unpolarized light is converted into a beam of linearly polarized light by the front polarizer. We mentioned earlier that waveguiding occurs as the beam propagates in the LC cell. And the transmitted beam is linearly polarized with a rotation of the plane of polarization by 90°. Strictly speaking, the waveguiding is valid only in the limit.

$$\phi \ll \frac{2\pi}{\lambda}\Delta nd \qquad (5.1\text{-}1)$$

where ϕ = total twist angle
d = thickness of LC layer
Δn = birefringence of LC material
λ = wavelength of light

This is known as the *limit of slow twist*. The inequality (5.1-1) is also known as the *Mauguin condition*. For a 90° TN cell, this condition reduces to

$$\frac{\lambda}{2} \ll \Delta nd \tag{5.1-2}$$

In general cases when the condition is not satisfied, the output beam is elliptically polarized with an ellipticity given by Eq. (4.3-20).

To discuss the polarization state, we use the *eo* coordinate system in which the *e* axis is parallel to the local director and the *o* axis is perpendicular to the local director. Consider an input beam of light with its electric field vector polarized parallel to the director of the liquid crystal at the input end of the LC layer. In the *eo* coordinate system, the input polarization state can be written as the following Jones vector:

$$\begin{pmatrix} V_e \\ V_o \end{pmatrix} = \begin{pmatrix} 1 \\ 0 \end{pmatrix} \tag{5.1-3}$$

The output polarization state in terms of the Jones vector in the local *eo* coordinate at the rear end of the LC layer can be written, according to Eqs. (4.3-13) or (4.3-15), as

$$\begin{pmatrix} V_e' \\ V_o' \end{pmatrix} = \begin{pmatrix} \cos X - i\,\dfrac{\Gamma \sin X}{2X} \\ -\phi\,\dfrac{\sin X}{X} \end{pmatrix} \tag{5.1-4}$$

where

$$\Gamma = \frac{2\pi}{\lambda}\Delta nd \tag{5.1-5}$$

$$X^2 = \phi^2 + \frac{\Gamma^2}{4} \tag{5.1-6}$$

We note that the Mauguin condition, Eq. (5.1-1), is equivalent to $\phi \ll \Gamma$. We also note that for liquid crystals satisfying the Mauguin condition (with $\phi \ll \Gamma$), the second element of the column vector is near zero. Thus, Eq. (5.1-4) corresponds to a Jones vector that is approximately linearly polarized along the local director (*e* axis). This explains the phenomenon of waveguiding.

In most TN displays, the LC layers actually used may not satisfy the Mauguin condition. This leads to a series of performance degradation, including a reduction in brightness and in contrast, an undesirable coloration caused by the dependence of transmission on wavelength. Using Eq. (5.1-4), we can now discuss the transmission property of the 90° TN cell. The discussion is divided into two categories:

1. *Normally Black (NB) Mode.* In this situation the 90° TN cell is sandwiched between a pair of parallel polarizers (see Fig. 5.1*b*). The director at the input end is parallel to the transmission axis of the polarizer, whereas the director at the exit end is perpendicular to the transmission axis of the analyzer. According to Eq. (5.1-4), the transmission for unpolarized monochromatic light is given by

$$T = \frac{1}{2}\phi^2 \frac{\sin^2 X}{X^2} = \frac{1}{2} \frac{\sin^2\left[\frac{\pi}{2}\sqrt{1+u^2}\right]}{1+u^2} \tag{5.1-7}$$

where we used $\phi = \pi/2$ as the twist angle, u is the Mauguin parameter given by

$$u = \frac{\Gamma}{2\phi} = 2d\frac{\Delta n}{\lambda} \tag{5.1-8}$$

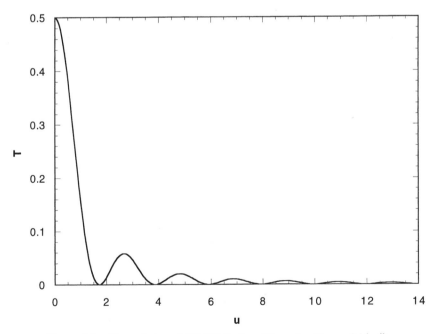

Figure 5.2. Transmission of NB TN displays (T vs. u), with $u = 2d\,\Delta n/\lambda$.

This equation was first derived by Gooch and Tarry [1]. From Eq. (5.1-7), we note that the transmission is near 0 for liquid crystals with $1 \ll u$, which is equivalent to the Mauguin condition. To illustrate the transmission property, we plot the transmission formula Eq. (5.1-7) as a function of the parameter u (see Fig. 5.2). The transmission is described by the square of a sinc function that consists of a main peak at $u = 0$, and a series of side lobes. We note that the oscillation is caused by the $\sin^2$ term in the numerator [Eq. (5.1-7)]. The envelope of the curve decreases with increasing value of u caused by the term $(1 + u^2)$ in the denominator. We also note that the transmission is very low when u is large (i.e., when the Mauguin condition is fulfilled). Zero transmission occurs when the argument of sin is an integral number of π (i.e., $X = \pi$, 2π, $3\pi,\ldots$). This corresponds to $u = \sqrt{3}, \sqrt{15}, \sqrt{35}, \ldots$. These values are known as the first, second, and third minimum conditions. In a normally black (NB) mode, it is desirable to have a zero transmission in the inactivated state (voltage OFF). According to Figure 5.2 [or Eq. (5.1-7)], zero transmission occurs at only one wavelength for a given TN cell. There is a residual transmission for other wavelengths. This is the cause of a undesirable coloration. Maximum transmission of the side lobes occurs approximately at $u = \sqrt{8}, \sqrt{24}, \sqrt{48}, \ldots$. We note that the peak of the first side lobe is $\frac{1}{9}$th of the main peak, the peak of the second side lobe is $\frac{1}{25}$th of the main peak.

2. *Normally White (NW) Mode.* In this scenario the 90° TN cell is sandwiched between a pair of crossed polarizers. The director at the input end is parallel to the transmission axis of the polarizer and the director at the exit end is parallel to the transmission axis of the analyzer. According to Eq. (5.1-4), the transmission for unpolarized monochromatic light is given by

$$T = \frac{1}{2} - \frac{1}{2} \frac{\sin^2\left[\frac{\pi}{2}\sqrt{1 + u^2}\right]}{1 + u^2} \tag{5.1-9}$$

We note that the transmission is complementary to that of the NB mode. For liquid crystals with large u, the second term in Eq. (5.1-9) is very small. This leads to a transmission of approximately $\frac{1}{2}$ for unpolarized light. Figure 5.3 shows a plot T versus u. We note that maximum transmission of $\frac{1}{2}$ occurs when the argument of sin is an integral multiple of π. This corresponds to $u = \sqrt{3}, \sqrt{15}, \sqrt{35}, \ldots$.

Transmitted Luminance

In the discussion above, we note that the transmission of a TN display cell depends on the wavelength of light. For black-and-white displays in the NB mode, it is desirable to select a LC cell with the proper parameter u such that the transmitted luminance is minimized in the black state. The integrated transmitted luminance is obtained by integrating the transmission function $T(\lambda)$ with the photopic response of the human eyes $P(\lambda)$ (see Fig. 5.4) and the

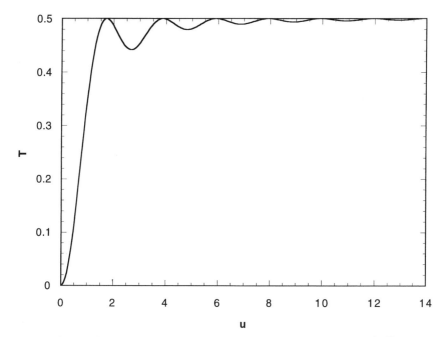

Figure 5.3.   Transmission of NW TN displays (T vs. u), with $u = 2d\,\Delta n/\lambda$.

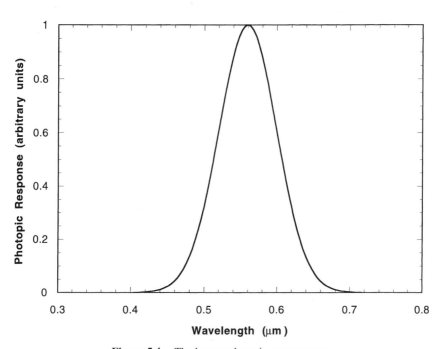

Figure 5.4.   The human photopic eye response.

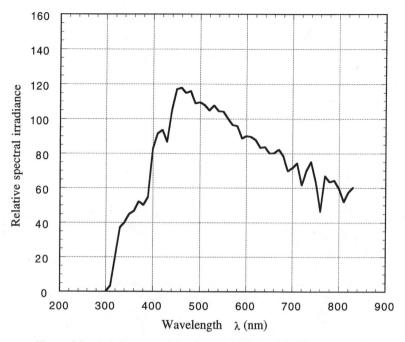

Figure 5.5. Relative spectral irradiance of CIE daylight illuminant D65.

illuminant spectral distribution $D(\lambda)$ as follows:

$$L = \frac{\displaystyle\int_{380}^{780} T(\lambda)D(\lambda)P(\lambda)d\lambda}{\displaystyle\int_{380}^{780} D(\lambda)P(\lambda)d\lambda} \tag{5.1-10}$$

where the limits of integration are from 380 nm (in the ultraviolet) and 780 nm (in the infrared). Using the D65 standard illuminating spectrum (see Fig. 5.5), a standard that approximates the emission characteristics of a blackbody held at 6500 K, we obtain a luminance that exhibits minima at $\Delta nd = 0.48$, 1.09, and 1.68 µm. These values agrees very well with the minima at $u = \sqrt{3}, \sqrt{15}$, and $\sqrt{35}$ in Figure 5.2 by selecting $\lambda = 0.55$ µm, which is close to the peak of the photopic response.

5.1.3. Transmission Properties of Field-ON State

When a small voltage is applied to the electrodes, a strong electric field is established in the LC cell. Recall that the typical cell thickness is in the range of

5–10 μm. So a small voltage of 1.5 V can lead to an electric field of 1.5–3 kV/cm. Under the influence of the electric field, the LC molecules are forced to align parallel to the electric field. On the other hand, the LC molecules at the boundaries must be parallel to the rubbing directions because of a strong molecular interaction with the surfaces. In addition, there is elastic inertia (restoring torque) in liquid crystals to resist a distortion of the director distribution. So the application of an electric voltage can lead to a redistribution of the director in the LC cell. In what follows, we will briefly describe the distribution of the director in a TN cell under the influence of an applied voltage. A detailed discussion can be found in Appendix D.

Field-Induced Director Reorientation

In the field-OFF state, the LC molecules are all aligned parallel to the surfaces of the glass plates. A uniform twist of the directors exists in the cell as a result of the boundary conditions. Recall that the rubbing directions of the inner surfaces of the two glass plates are mutually perpendicular in a 90° TN cell. When a sufficiently high voltage (few volts) is applied, the liquid crystal molecules are tilted toward the direction of the electric field. So in a field-ON state, both tilt angle and twist angle are needed to described the distribution of the director.

To describe the director distribution, we choose the z axis perpendicular to the cell surfaces. According to Eq. (D-8) in Appendix D, the elastic energy density in a distorted twist cell can be written

$$U_{EL} = \frac{1}{2}(k_1 \cos^2\theta + k_3 \sin^2\theta)\left(\frac{d\theta}{dz}\right)^2$$
$$+ \frac{1}{2}(k_2 \cos^2\theta + k_3 \sin^2\theta)\cos^2\theta\left(\frac{d\phi}{dz}\right)^2 \tag{5.1-11}$$

where k_1, k_2, k_3 are the elastic constants of the LC material, ϕ is the twist angle, and θ is the tilt angle of the director. The electrostatic energy density can be written

$$U_{EM} = \frac{1}{2}\mathbf{E}\cdot\mathbf{D} = \frac{1}{2}\frac{D^2}{\varepsilon_\parallel \sin^2\theta + \varepsilon_\perp \cos^2\theta} \tag{5.1-12}$$

where $\mathbf{D}$ is the displacement field and $\varepsilon_\parallel, \varepsilon_\perp$ are the dielectric constants of the LC material.

The final distribution of the director $[\theta(z), \phi(z)]$ as functions of z can be obtained by minimizing the total energy integrated over the LC cell (see Appendix D). This requires the technique of variational calculus.

According to Appendix D, the redistribution of the director in a TN cell requires a *threshold voltage* given by

$$V_T = V_{c1}\left[1 + \frac{\Phi^2}{\pi^2}\left(\frac{k_3 - 2k_2}{k_1}\right)\right]^{1/2} \qquad (5.1\text{-}13)$$

where Φ is the total twist angle in the field-OFF state and V_{c1} is given by

$$V_{c1} = \pi\sqrt{\frac{k_1}{\Delta\varepsilon}} \qquad (5.1\text{-}14)$$

where $\Delta\varepsilon$ is the *dielectric anisotropy* $\Delta\varepsilon = \varepsilon_\parallel - \varepsilon_\perp$. Physically, V_{c1} is a threshold voltage needed to tilt the LC molecules in a nematic cell with a planar orientation ($\Phi = 0$). The threshold voltage to tilt LC molecules is slightly higher in twisted nematic cells, according to Eq. (5.1-13) provided $0 < k_3 - 2k_2$. For a 90° TN cell, the threshold voltage is given by

$$V_T = V_{c1}\left[1 + \left(\frac{k_3 - 2k_2}{4k_1}\right)\right]^{1/2} \qquad (5.1\text{-}15)$$

For most nematic mixtures formulated for TN displays, V_T is typically around 1 V.

Example 5.1. Using Table 1.1, we obtain $k_1 = 7.7 \times 10^{-12}\,N$, $k_2 = 4 \times 10^{-12}\,N$, $k_3 = 12.2 \times 10^{-12}\,N$, $\varepsilon_\parallel = 10.6\varepsilon_0$, and $\varepsilon_\perp = 4.6\varepsilon_0$, for ZLI-1646. According to Eqs. (5.1-14,15), we obtain $V_{c1} = 1.196\,V$ and $V_T = 1.275\,V$. ■

Using the results described in Appendix D, we can obtain the distribution of the orientation of the director in a TN cell. This also requires numerical procedures. Figure 5.6 shows the tilt $\theta(z)$ and twist $\phi(z)$ angles of the director (optic axis) inside a TN cell filled with a LC mixture of ZLI-1646 for several applied voltages. For this LC mixture, the threshold voltage is around 1.275 V. The director distribution exhibits the following symmetry,

$$\theta(z) = \theta(d - z)$$
$$\phi(z) + \phi(d - z) = \frac{\pi}{2}.$$

We note that the tilt angle is zero at both ends as a result of the boundary conditions. In practice, the inner surfaces of the glass plates are treated so that

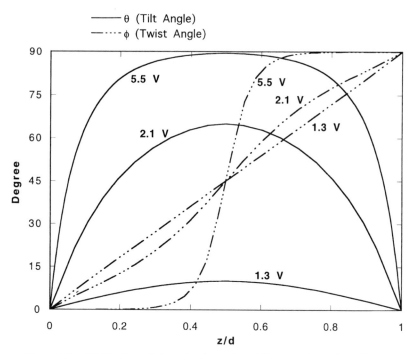

Figure 5.6.   Distribution of tilt and twist angles of director, where d = cell gap.

the LC director exhibits a small pretilt angle (about 1°) at the boundaries. As we mentioned before, the pretilt is very important to maintain the distribution stability of the LC cell. According to Eq. (5.1-12), the electrostatic energy is an even function of the tilt angle. When a voltage is applied, the tilt angle for the LC molecules can be either positive or negative. In actual operations without a pretilt, this can lead to a distribution instability in the LC cell. Thus, a small pretilt can facilitate the tilt of the LC molecules when a small voltage is applied. The tilt angle distribution resembles a sinusoidal function at small voltages and becomes squaring off at higher voltages. Because the tilt angle is symmetric about the center of the cell, it is maximum at the center. The midlayer tilt angle increase monotonically as a function of the applied voltage and asymptotically approaches 90° at higher applied voltages. Figure 5.7 shows the midlayer tilt angle as a function of the applied voltage.

The twist angle remains relatively uniform at low voltages (e.g., 1.3 V). Significant changes occur at voltages that cause a midlayer tilt of greater than 45° (e.g., 2.1 V). In the case when the midlayer tilt approaches 90°, the twist angle undergoes a sharp transition at midlayer from 0° to 90°. The twist angle remains parallel to the rubbing directions of the surfaces at both ends. A very fast twisting of 90° occurs within a thin layer at the center of the cell. We also note that the twist angle remains 45° at the center of the cell regardless of the applied

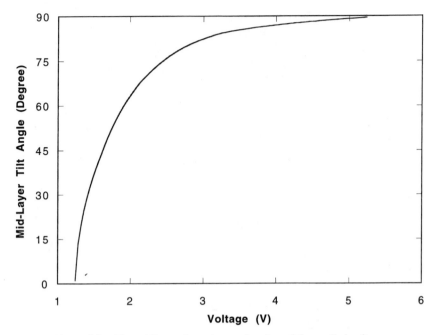

Figure 5.7. The midlayer tilt angle as a function of the applied voltage.

voltages. Table 5.1 lists the tilt and twist angles at 20 equally spaced locations in the LC layer and at the boundaries for three different applied voltages.

Transmission at Normal Incidence

Given the distribution of the tilt and twist angles, we can then use the standard Jones matrix method to calculate the transmission of the TN cell. In the numerical approach, the TN cell is divided into a large number of plates ($N = 20–100$) with an identical thickness. Each of the plates is treated as a homogeneous and uniaxially anisotropic crystal plate. In the matrix method, each of the uniaxial crystal plates can be represented by a Jones matrix that transforms an input polarization state into an output polarization state. The overall Jones matrix is then obtained by multiplying all the individual Jones matrices in sequence. In what follows, we discuss the transmission at normal incidence of a 90° TN cell as a function of the applied voltage.

Normally Black Mode. Referring to Figure 5.8, we consider the transmission of a TN cell with $\Delta nd = 0.48\,\mu m$ as a function of the applied voltage. From the discussion above, this thickness corresponds to the first minimum of a normally black TN display for green light at $\lambda = 550\,nm$ ($u_{green} \approx \sqrt{3}$). Thus, in the field-OFF state ($V = 0$), the transmission for green light at 550 nm is zero. Residual transmission exists for both red and blue (about 0.03). This is a result of an

Table 5.1. Tilt and Twist Angles versus z where d is the LC Cell Gap

	$V = 1.29$ V		$V = 2.10$ V		$V = 5.49$ V	
z/d	$\theta(°)$	$\phi(°)$	$\theta(°)$	$\phi(°)$	$\theta(°)$	$\phi(°)$
0.000	0.0	0.0	0.0	0.0	0.0	0.0
0.025	0.8	2.3	7.5	1.6	24.6	0.0
0.075	2.3	6.8	22.3	4.8	54.7	0.0
0.125	3.8	11.4	33.7	7.8	69.4	0.0
0.175	5.2	15.9	42.5	10.9	77.9	0.1
0.225	6.5	20.5	49.4	14.4	82.8	0.2
0.275	7.6	25.0	54.7	18.3	85.4	0.6
0.325	8.5	29.4	58.8	23.0	87.2	2.3
0.375	9.2	33.9	61.7	28.5	88.3	4.9
0.425	9.7	38.3	63.7	34.6	89.0	16.9
0.475	9.9	42.8	64.7	41.4	89.3	35.6
0.525	9.9	47.2	64.7	48.6	89.3	54.4
0.575	9.7	51.7	63.7	55.4	89.0	73.1
0.625	9.2	56.1	61.7	61.5	88.3	85.1
0.675	8.5	60.6	58.8	67.0	87.2	87.7
0.725	7.6	65.0	54.7	71.7	85.4	89.4
0.775	6.5	69.5	49.4	75.6	82.8	89.8
0.825	5.2	74.1	42.5	79.1	77.9	89.9
0.875	3.8	78.6	33.7	82.2	69.4	90.0
0.925	2.3	83.2	22.3	85.2	54.7	90.0
0.975	0.8	87.7	7.5	88.4	24.6	90.0
1.000	0.0	90.0	0.0	90.0	0.0	90.0

overshoot for blue light ($u_{blue} > \sqrt{3}$) and a undershoot for red light ($u_{red} < \sqrt{3}$) due to the $1/\lambda$ dependence. When the field is turned ON, the transmission remains the same for applied voltages between 0 and the threshold voltage V_T. As we know, the LC director remains unchanged for voltages under the threshold voltage. As the applied voltage increases beyond the threshold voltage V_T, the LC molecules are tilted toward the direction of the electric field. This effectively decreases the phase retardation leading to an increase in the transmission. For red light, the tilt of the directors leads to a further decrease of u_{red} below $\sqrt{3}$, leading to an increase of the transmission. For blue light, the parameter u_{blue} passes through $\sqrt{3}$ as the voltage increases, leading to a zero transmission at a voltage of around 1.5 V. Beyond this voltage, the tilt angle continues to increase as the applied voltage increases, leading to a further decrease of the effective phase retardation and the value of u_{blue}. This leads to an increase of transmission as a function of the voltage. At a very large applied voltage, all the LC molecules are practically aligned parallel to the electric field. This resembles the situation of a c plate sandwiched between parallel polarizers. Thus, all three colors of light are transmitted completely.

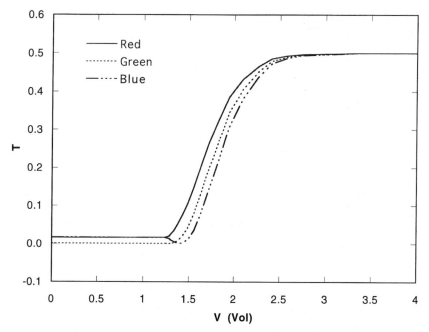

Figure 5.8.   Transmission of a TN cell in the NB mode as a function of the applied voltage.

If we were to pick a TN cell with $\Delta nd = 1.09\,\mu m$, the electrooptical transmission curves would be slightly different. A TN cell with $\Delta nd = 1.09\,\mu m$ corresponds to the second minimum of a normally black TN display for green light at $\lambda = 550\,nm$ ($u_{green} \approx \sqrt{15}$). The transmission for the green light is again zero at 0 voltage. As the applied voltage increases, the transmission of the green light will first increase when the voltage is greater than the threshold voltage as the parameter u_{green} decreases. The transmission will then decrease and become zero when the parameter reaches $u_{green} = \sqrt{3}$. Beyond this point, the transmission will increase as the applied voltage continues to increase. For blue light, the transmission curves will dip to zero 2 times, as the parameter passes through $u_{blue} = \sqrt{3}$ and $u_{blue} = \sqrt{15}$. For red light, the transmission curves will dip to zero only once, as the parameter passes through $u_{red} = \sqrt{3}$.

Normally White (NW) Mode. In this mode, the TN cell is sandwiched between a pair of crossed polarizers. Assuming ideal polarizers, the electrooptical transmission curves are complementary to those of the NB mode. Figure 5.9 shows the electrooptical transmission curves for the NW modes of operation.

In designing TN-LC cells for display application, it is desirable to select cells with $\Delta n = 0.48\,\mu m$ ($u = \sqrt{3}$) to minimize the dependence of the transmission on λ. The minimum cell gap also ensures a larger viewing angle.

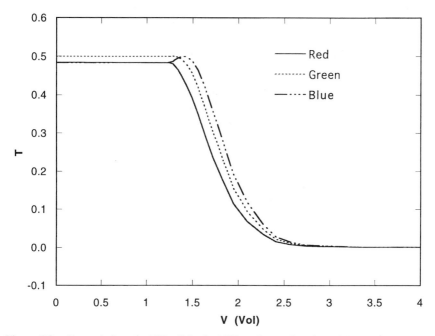

Figure 5.9. Transmission of a TN cell in the NW mode as a function of the applied voltage.

Contrast at Normal Incidence

We now discuss the contrast of the TN displays. The contrast of any display system is defined as the ratio of the transmission of the bright state (high transmission) to that of the dark state (low or zero transmission). Since the transmission depends on the applied voltage, the contrast is also a function of the applied voltage. For normally black mode, the contrast is defined as

$$C_{NB} = \frac{T_{NB}(V)}{T_{NB}(0)} \qquad (5.1\text{-}16)$$

where T_{NB} is the transmission of the TN cell. Similarly, the contrast of a normally white mode is defined as

$$C_{NW} = \frac{T_{NW}(0)}{T_{NW}(V)} \qquad (5.1\text{-}17)$$

where T_{NW} is the transmission of the TN cell. On the basis of the complementary electrooptical transmission of these two modes, C_{NW} can also be written as

$$C_{NW} = \frac{T_{NW}(0)}{T_{NW}(V)} = \frac{0.5 - T_{NB}(0)}{0.5 - T_{NB}(V)} \qquad (5.1\text{-}18)$$

Let the residual transmission leakage be written

$$T_{NB}(0) = T_0 \tag{5.1-19}$$

where T_0 is a small number in the range of 0.0–0.05, depending on whether T_{NB} is the integrated transmission over all wavelengths, or the transmission for an individual wavelength. Also, for the purpose of discussion, let

$$T_{NB}(V) = T_1 \tag{5.1-20}$$

where T_1 can vary from T_0 to 0.5 as the applied voltage increases. Using T_1 and T_0, the contrasts can be written

$$C_{NB} = \frac{T_{NB}(V)}{T_{NB}(0)} = \frac{T_1}{T_0} \tag{5.1-21}$$

$$C_{NW} = \frac{T_{NW}(0)}{T_{NW}(V)} = \frac{0.5 - T_0}{0.5 - T_1} \tag{5.1-22}$$

We note that both contrasts are monotonically increasing functions of the applied voltage. As $T_{NB}(V)$ reaches the maximum value of 0.5 at high applied voltages, we find that the contrast of the normally black mode is primarily limited by the leakage $T_{NB}(0)$ (residual transmission) because of the slight ellipticity of the output polarization state, according to Eq. (5.1-21). A contrast ratio of 100 : 1 can be easily achieved in this mode. In the normally white mode of operation, the contrast ratio can be much higher at high applied voltages as the denominator approaches zero, according to Eq. (5.1-22). As we know, at high applied voltages, all LC molecules are aligned parallel to the electric field, the TN cell becomes a homeotropically aligned cell. In practice, the leakage of light $T_{NW}(V)$ at this state is limited by the extinction ratio of the crossed polarizers. Contrast ratio of 1000 : 1 can be achieved with the use of sheet polarizers.

Assuming $T_0 = 0.0005$, we find that the contrasts are 1 at zero voltage. As the applied voltage increases, T_1 increases from 0.0005 to 0.5. At high applied voltages, $T_1 = 0.5$, the contrast C_{NB} reaches 1000, whereas C_{NW} reaches infinity as the denominator approaches zero asymptotically. Equal contrast of these modes occurs at $T_1 = 0.5 - T_0$, which corresponds to transmission of $T_{NB}(V) = 0.4995$. Based on the electrooptical transmission curves discussed above, this requires an applied voltage of about 3.3 V. So for operations below 3.3 V in this example, the normally black mode exhibits a higher contrast ratio. Figure 5.10 shows the calculated contrast ratios for these two modes of operation for green light. Since the cell is designed for minimum leakage at green light, the contrast is very good. The contrast ratio for the integrated luminance follows a similar pattern, with a lower value of the maximum contrast.

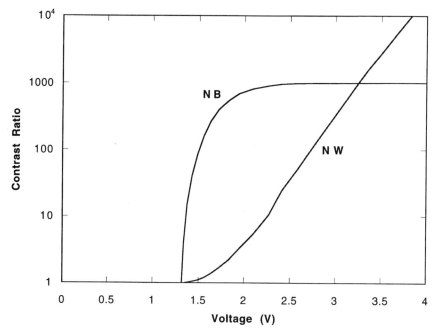

Figure 5.10. Contrast ratio as a function of the applied voltage.

Viewing Angles

In the preceding discussion, we considered only the transmission properties of TN-LCD at normal incidence. The Jones matrix method can also be employed to investigate the transmission properties at a general angle of incidence. As we know, the phase retardation depends on the angle of incidence. The transmission is now a function of several variables, including the angles of incidence θ, ϕ, wavelength λ, and the applied voltage V:

$$T = T(\theta, \phi, \lambda, V) \tag{5.1-23}$$

The angles of incidence are defined such that θ is the angle between the incidence direction and the normal vector to the surface, ϕ is the azimuth angle of the incidence direction. For the azimuthal angles we use the convention that the director at the entrance face is oriented at $\phi = 45°$ and the director at the exit face is oriented at $\phi = 135°$. We further assume that the transmission axes of the polarizers for the normally white mode are perpendicular to the LC alignment directions on the surface of the adjacent substrates (O mode; see Fig. 4.6b). In calculating the transmission for off-axis light using the Jones matrix method, it is important to use the phase retardation $\Gamma(\theta,\phi,\lambda)$ for off-axis incidence (Eq. (4.4-2)), according to the discussion in Chapter 4. More accurate results can be calculated using the extended Jones matrix method (Chapter 8). For a given

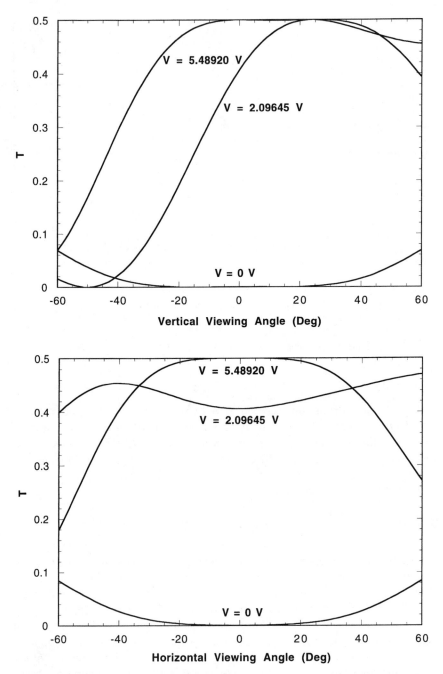

Figure 5.11. (*a*) Horizontal and vertical viewing characteristics of a NB TN-LCD at three different applied voltages ($V = 0$, 2.10, 5.49 V).

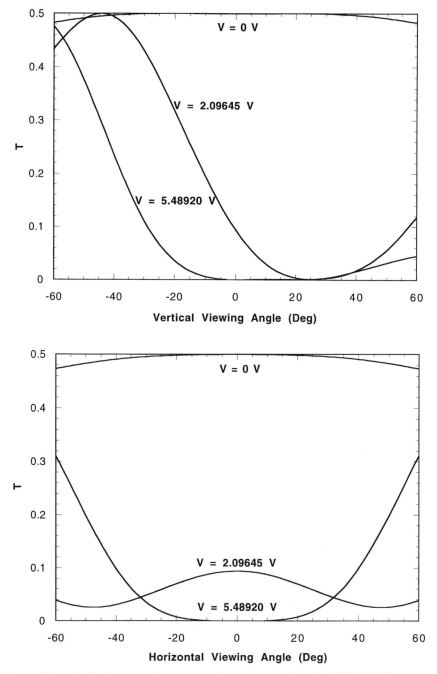

Figure 5.11. (*b*) Horizontal and vertical viewing characteristics of a NW TN-LCD at three different applied voltages ($V = 0$, 2.10, 5.49 V).

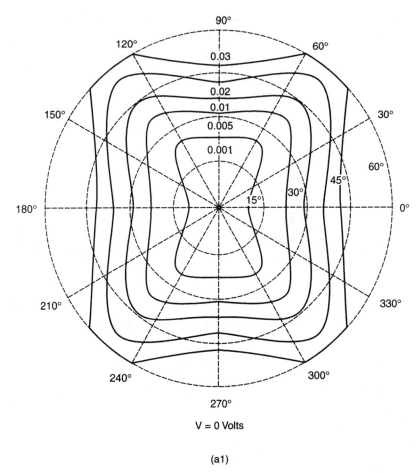

(a1)

Figure 5.12. (*a*) Isotransmittance viewing diagrams of a NB TN-LCD at three different applied voltages ($V = 0$, 2.10, 5.49 V).

wavelength λ and an applied voltage V, it is possible to display the transmission $T(\theta,\phi)$ in polar diagrams by using either density plots or contour plots (curves of equal transmission). In the polar diagrams, each point corresponds to a certain viewing angle (θ,ϕ). The center corresponds to normal incidence (on-axis viewing) with $\theta = 0$ and $\phi = 0$. The viewing angle characteristics can also be displayed by plotting $T(\theta,\phi)$ for vertical viewing ($\phi = \pi/2$) or horizontal viewing ($\phi = 0$) as a function of θ.

To illustrate the angular viewing properties of LCD, we consider the case of a TN cell with $\Delta nd = 0.48\,\mu\text{m}$. Figure 5.11 shows the horizontal and vertical viewing characteristics of the TN-LCD at three different voltages. In the normally black (NB) mode, the LCD is in a black state when the applied voltage is zero. Referring to Figure 5.11*a*, we note that the transmittance is indeed zero at normal incidence. In the horizontal viewing, we note that leakage occurs for

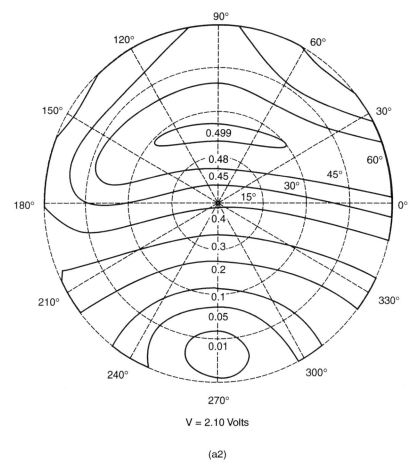

V = 2.10 Volts

(a2)

Figure 5.12. (*Continued*)

viewing angles greater than 20°. The leakage becomes severe at large viewing angles. In the vertical viewing, similar leakage occurs at large viewing angles. As we increase the applied voltage, the transmittance increases with the applied voltage. This is true for normal incidence as shown in Figure 5.11*a*. We note that the transmittance again depends on the angle of viewing. The variation of the transmittance with the viewing angle may degrade the performance of the display. An ideal LCD would have a transmittance that is insensitive to the variation of the viewing angles, and the transmittance varies monotonically with the applied voltage at all angles of viewing. Figure 5.11*b* shows the same viewing angle characteristics of a normally white TN-LCD. We note that the transmittance curves of NW TN-LCDs exhibit a complete left–right viewing symmetry. The symmetry is a result of the symmetry of the LCD system which exhibits a 180° rotation around the *x* axis (see Fig. 5.1*a* and Problem 5.10). It is important to note that normally black TN-LCDs, however, do not exhibit an

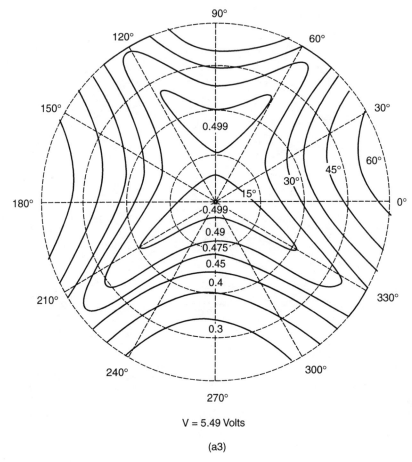

V = 5.49 Volts

(a3)

Figure 5.12. (*Continued*)

exact left–right viewing symmetry. The issue of viewing angle characteristics will be further discussed in Chapter 9.

As mentioned earlier, the viewing angle characteristics of TN-LCDs can also be presented by using iso-transmittance contours in polar diagrams. This is illustrated in Figure 5.12 with three different applied voltages. These curves are contours of equal transmittance in the polar coordinate. As an example, we now examine the iso transmittance curves of a NB TN-LCD as shown in Figure 5.12a. In the black state when the applied voltage is zero, we note that the transmission exhibits an approximate left–right viewing symmetry and up–down viewing symmetry. It can be shown that the approximate symmetry disappears at large viewing angles. The leakage of light in the black state is illustrated by the contours of equal transmittance at $T = 0.001, 0.005. 0.01, 0.02,$ and 0.03. We note that the angular window of viewing with a leakage of <0.005 covers a range of $-20°$ to $+20°$ in the horizontal direction and $-30°$ to $+30°$

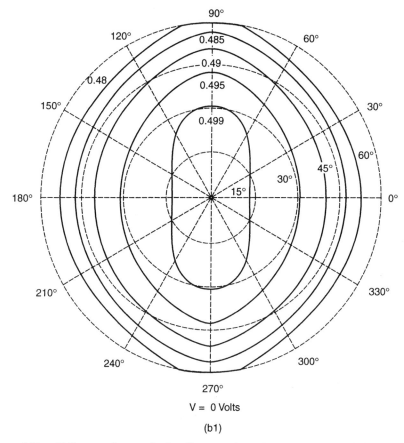

Figure 5.12. (*b*) Isotransmittance viewing diagrams of a NW TN-LCD at three different applied voltages ($V = 0$, 2.10, 5.49 V).

in the vertical direction. At an applied voltage of 2.1 V, the approximate viewing symmetries disappear. The TN-LCD exhibits a brighter transmission in the upper viewing directions. This is in agreement with the vertical viewing characteristics shown in Figure 5.11a. At an higher applied voltage of 5.49 V, the transmission of the LCD is higher at most angles of viewing as expected. However, the angular dependence still exists, as illustrated by the contours. It is important to note that the isotransmittance contours in the polar diagrams reveal more information about the viewing angle characteristics as compared with the simple vertical and horizontal plots. On the other hand, the simple vertical and horizontal plots are easier in recognizing the angular variation of the transmittance function. Figure 5.12b shows isotransmittance contours in polar diagrams for the normally white mode of the TN-LCD. Here, we note that the isotransmittance curves exhibit an exact left–right viewing symmetry at all voltages. The transmittance curves in Figures 5.11 and 5.12 are obtained by using the extended Jones matrix method, which will be discussed in Chapter 8.

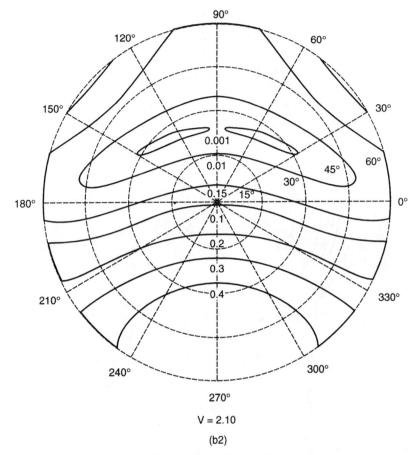

$V = 2.10$

(b2)

Figure 5.12. (*Continued*)

As we know, the standard Jones matrix method is no longer adequate for the range of viewing angles shown in the figures.

The issue of contrast was discussed in the section entitled "Contrast at Normal Incidence" for normal incidence only. Since the transmittance is a function of $(\theta, \phi, \lambda, V)$, the contrast ratio is, strictly speaking, also a function of $(\theta, \phi, \lambda, V)$. We first investigate the dependence of the contrast ratio on (θ, ϕ) for a TN-LCD illuminated with a monochromatic source (e.g., a laser beam). The contrast ratio of a TN-LCD can be defined as

$$C_{\mathrm{NB}}(\theta, \phi, \lambda, V) = \frac{T(\theta, \phi, \lambda, V)}{T(\theta, \phi, \lambda, 0)} \qquad (5.1\text{-}24)$$

$$C_{\mathrm{NW}}(\theta, \phi, \lambda, V) = \frac{T(\theta, \phi, \lambda, 0)}{T(\theta, \phi, \lambda, V)} \qquad (5.1\text{-}25)$$

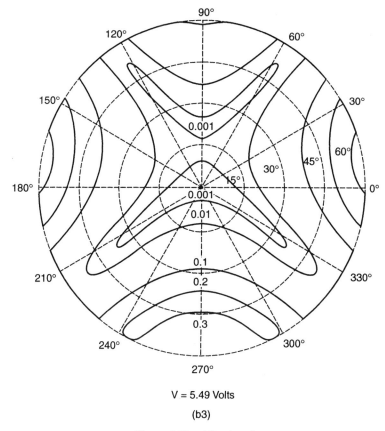

$V = 5.49$ Volts

(b3)

Figure 5.12. (*Continued*)

where NB stands for normally black and NW denotes normally white and V is the voltage of operation. The dependence on the applied voltage V was discussed earlier. We now focus our attention on the dependence on the viewing angles (θ, ϕ) at a given wavelength of illumination λ. Figure 5.13 and 5.14 show the angular dependence of the contrast ratio for a beam of monochromatic green light. As we know, the transmittance of a TN-LCD depends on the wavelength λ. It is important to note that the contrast ratio also depends on the wavelength of illumination.

Figure 5.13*a* shows the contrast ratio as a function of the horizontal and vertical viewing angle θ for a normally black mode of TN-LCD. In the horizontal viewing, we note that the contrast ratio is a symmetric function of θ with a peak of around 1000 at normal incidence. The contrast ratio drops monotonically as a function of θ, reaching ~ 50 for viewing angles over 30°. In the vertical viewing, the contrast ratio is no longer a symmetric function of θ. Sharp peaks of high

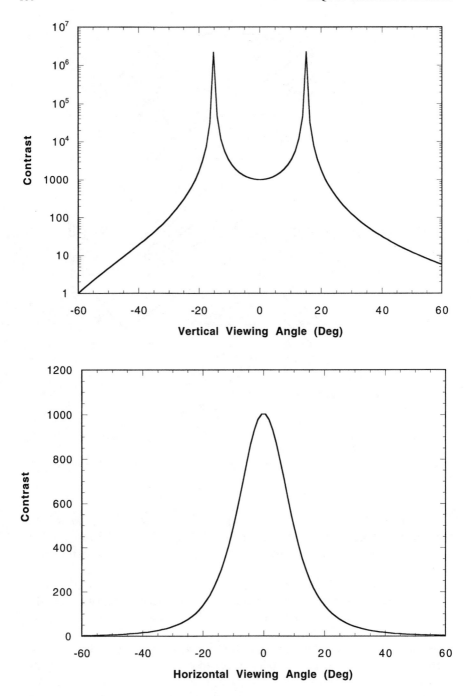

Figure 5.13. (a) Horizontal and vertical viewing contrast of a NB TN-LCD at $V = 5.49$ V and $\lambda = $ green.

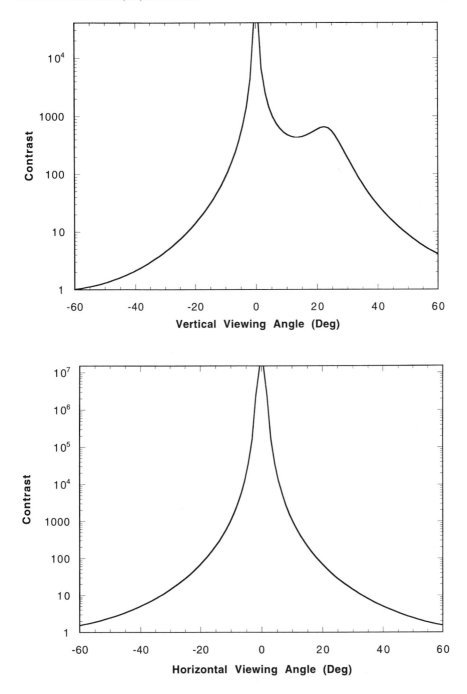

Figure 5.13.   (*b*) Horizontal and vertical viewing contrast of a NW TN-LCD at $V = 5.49$ V and $\lambda =$ green.

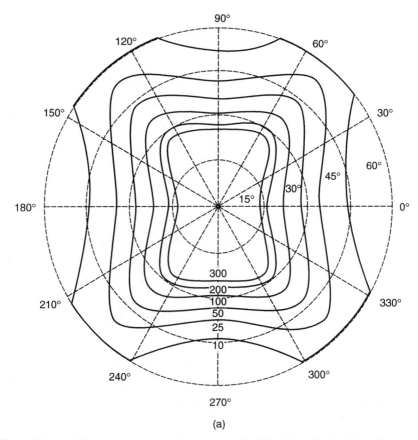

(a)

Figure 5.14. (*a*) Isocontrast viewing diagrams of a NB TN-LCD at $V = 5.49$ V and λ = green.

contrast ratio exist at some particular viewing angles. The contrast ratio drops to less than 100 for viewing angles over 30°. It is important to note that the presence of the peaks of high contrast is a result of the monochromatic illumination, which can lead to a near zero transmission of the black state at some particular angle of incidence. Figure 5.13*b* shows the contrast ratio as a function of the horizontal and vertical viewing angle θ for a normally white mode of TN-LCD. Generally speaking, these two modes of operation (NB and NW) have different contrast ratios. As most of the display systems are designed to provide the optimum viewing at normal incidence, the contrast ratio of both modes of operation drops as a function of the viewing angle θ. We note that the high contrast ratio at normal incidence is a result of the monochromatic illumination. The contrast ratio as a function of the viewing angles (θ, ϕ) can also be displayed by using isocontrast contours in the polar diagrams. Figure 5.14 shows the isocontrast viewing diagrams of the TN-LCD with $V = 5.49$ V and $\lambda = \lambda_{\text{green}}$. Each contour defines a field of viewing with a given contrast ratio.

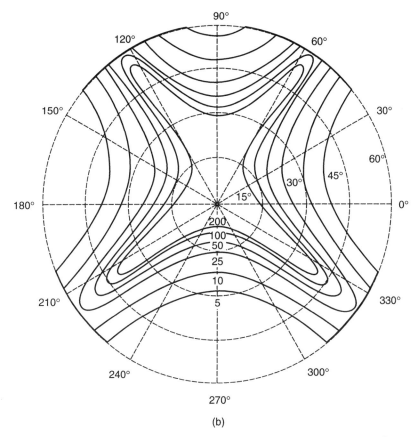

Figure 5.14.   (*b*) Isocontrast viewing diagrams of a NW TN-LCD at $V = 5.49$ V and $\lambda =$ green.

In the discussion above, we described the transmission properties at large viewing angles by using only one color (e.g., green). For displays with white light illumination, the integrated luminance given by Eq. (5.1-10) is the most important parameter. Using the integrated luminance, the contrasts are given by

$$C_{\text{NB}}(\theta, \phi, V) = \frac{L(\theta, \phi, V)}{L(\theta, \phi, 0)} \tag{5.1-26}$$

$$C_{\text{NW}}(\theta, \phi, V) = \frac{L(\theta, \phi, 0)}{L(\theta, \phi, V)} \tag{5.1-27}$$

where the luminance $L(\theta, \phi, V)$ is obtained by integrating the transmittance function $T(\theta, \phi, \lambda, V)$ over all wavelengths using Eq. (5.1-10). Figures 5.15 and 5.16 describe the isocontrast curves in the viewing plane. Each data point in Figure 5.15 is obtained by integrating over 50 wavelengths from 380 to 780 nm

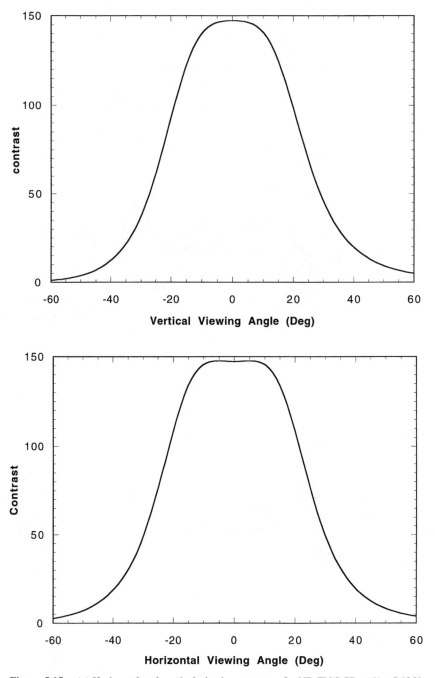

Figure 5.15. (*a*) Horizontal and vertical viewing contrast of a NB TN-LCD at *V* = 5.49 V.

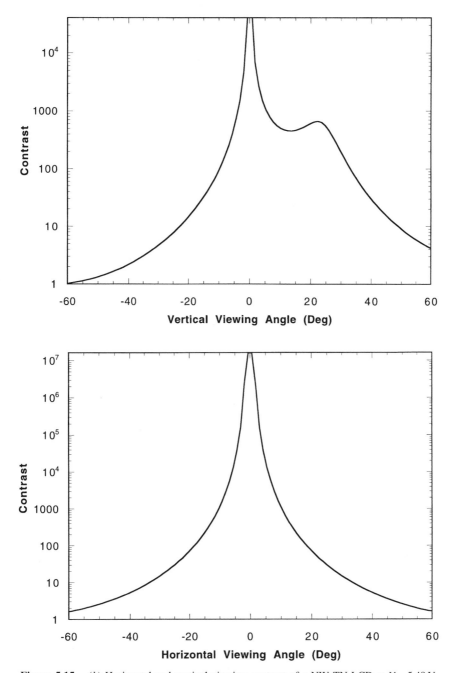

Figure 5.15. (*b*) Horizontal and vertical viewing contrast of a NW TN-LCD at $V = 5.49$ V.

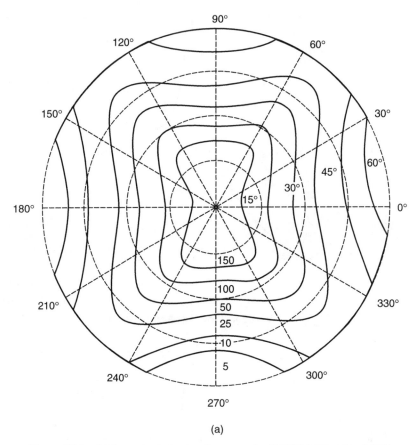

(a)

Figure 5.16. (*a*) Isocontrast viewing diagrams of a NB TN-LCD at $V = 5.49$ V.

using Eq. (5.1-10). Here we used the emission characteristics of a blackbody radiation at $6500\,\text{K}$ as the illuminant spectral distribution $D(\lambda)$. For different illumination sources, the isocontrast curves will look slightly different. Figure 5.16 is obtained by averaging over the three primary colors (red, green and blue).

Figure 5.15*a* shows the contrast ratio of a normally black TN-LCD as a function of the vertical and horizontal viewing angles with an applied voltage of $V = 5.49$ V at the white state. We note that the contrast ratio is around 150 at normal incidence. The contrast ratio remains around 150 for viewing angles within a field of view of $\pm 10°$. The contrast ratio then drops to below 100 for viewing angles greater than 20°. We also note that the contrast ratio remains above 10 for viewing angles up to $\pm 40°$.

Figure 5.15*b* shows the contrast ratio of a normally white TN-LCD as a function of the vertical and horizontal viewing angles with an applied voltage of 5.49 V at the black state. The contrast ratio exhibits a rather asymmetric dependence on the vertical viewing angle. We note that the contrast ratio is very

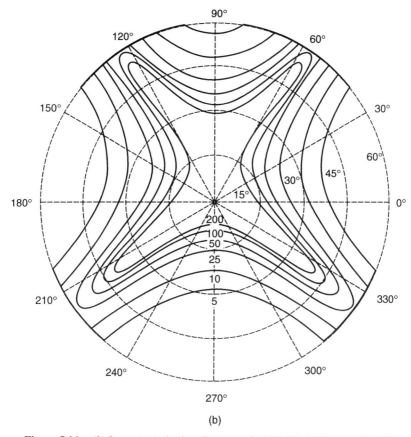

Figure 5.16. (*b*) Isocontrast viewing diagrams of a NW TN-LCD at $V = 5.49$ V.

high ($>10^4$) at normal incidence. This is in agreement with our earlier discussion (see Fig. 5.10). In the vertical viewing, the contrast ratio remains above 100 in the angular range of $(-10°, 30°)$. In the horizontal viewing, the contrast ratio remains above 100 in the angular range of $\pm 20°$. We note that the normally white mode of TN-LCD exhibits an exact left–right viewing symmetry.

Figure 5.16*a* shows the isocontrast contours of a normally black TN-LCD at an applied voltage of 5.49 V in the white state. Such a contour plot shows the angular viewing region in which the contrast ratio is above a specified value. Figure 5.16*b* shows a similar polar diagram for the normally white TN-LCD. The iso-contrast contours in the polar diagram give a better description of the angular dependence of the contrast ratio.

In the preceding discussion of the transmittance of LCD at large viewing angles, we assumed a zero pretilt angle for the purpose of illustrating the effect of off-axis incidence. In practice, a small pretilt angle ($\leqslant 1°$) is present in most LCD's for the stability of the liquid crystal when voltages are applied.

5.2. SUPERTWISTED NEMATIC (STN) DISPLAYS

As we discussed earlier, when a voltage is applied to a LC cell, the directors of
the LC molecules are forced to align along the direction of the electric field. A
balance between the elastic energy and the electrostatic energy leads to a
distribution of the director orientation in the LC cell. The redistribution of the
director leads to a change in the transmittance of the LC cell. Typical panels for
information displays requires a 2D array (say $M \times N$) of LC pixels. Each of the
pixels consists of a small LC cell that can be turned ON or OFF electrically. Thus,
each of the individual LC cells must be addressed electrically for information
displays. In principle this can be achieved by using $M \times N$ electrical connections. However, for practical purposes, the electrical addressing is achieved by
using multiplexing techniques that require only $M + N$ electrodes. The subject
of multiplexing and other addressing schemes will be discussed in the next
chapter.

   As a result of the multiplexing, the voltages applied at one cell cannot be
arbitrarily changed without affecting the applied voltages at other cells. This is
known as the *crosstalk*. As a result of the crosstalk, the ON and OFF voltages
applied at any of the cells cannot be too different. This severely limits the
contrast in display applications.

   A high contrast display requires a significant distortion (or redistribution) of
the LC directors in the cell at a reasonable applied voltage. Supertwisted nematic
(STN) liquid crystals offer a solution to this problem. This will be explained by
using the electrooptical distortion curves discussed in this section. Figure 5.17
shows examples of STN-LC cells with total twist angles of 180° and 270°. In
what follows, we discuss the principle of operation of STN cells.

5.2.1. Steepness of Electrooptical Distortion Curves of STN Cells

Referring to Figure 5.17b, we considered a TN cell with orthogonal rubbing
directions on the surfaces of the cell. The arrows indicate the direction of the LC
directors. It is important to note that the LC molecules are usually nonpolar. The
dashed arrows indicate the rubbing directions at the boundaries. It is also
important to note that the liquid crystal in this cell can have either a right-handed
270° twist or a left-handed 90° twist. Both configurations satisfy the boundary
conditions determined by the rubbing directions. Based on the argument in
Appendix D, the left-handed 90° would have a lower elastic energy. Thus, when
an ordinary nematic liquid crystal is filled in the cell, the liquid crystal would
maintain a left-handed 90° twist that has a lower elastic energy.

   To maintain a cell twist angle greater than 90° requires a nematic liquid
crystal with an intrinsically twisted nematic structure known as a *chiral nematic*.
Chiral nematics are ordinary nematic liquid crystals doped with a small
percentage of other materials that exhibit intrinsic optical rotatory power. As we
know, these optically active materials exhibit intrinsic "handedness" in the
rotation of the plane of polarization of a beam of incident light. The handedness

180° STN-LC

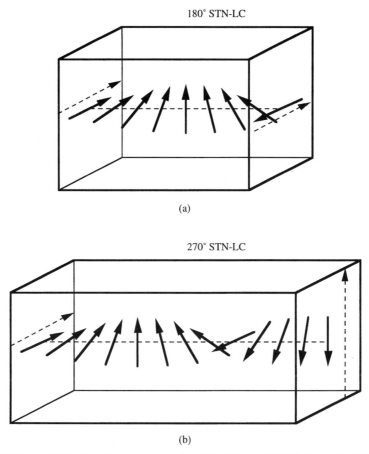

(a)

270° STN-LC

(b)

Figure 5.17. (*a*) STN cells with total twist angles of (*a*) 180° and (*b*) 270°. The dashed lines with arrows show the rubbing directions.

of the dopant molecules imparts a macroscopic twist to the whole nematic structure. The amount of twisting is characterized by a pitch length p, which is defined as the distance measured along the helical axis for the twist of the director to rotate by a full 360°. In an infinite medium of helically twisted nematic structure, the pitch is exactly the period of the medium along the axis of the helix. The structure is identical to that of the cholesteric phase described in Chapter 1. When the chiral nematic liquid crystal is filled into the cell, the director at the substrate planes anchor along the alignment directions, whereas the director in the cell twist according to the handedness of the chiral molecules. The pitch length is, in general, slightly different from that of the infinite medium, as a result of the boundary conditions. In an infinite medium, the pitch length is inversely proportional to the concentration of the chiral dopant. Thus, we can easily modify the pitch length by adjusting the chiral dopant concentration to suit the cell spacing d. The pitch p, the cell spacing d, and the cell twist angle Φ are related by the following condition:

$$\frac{d}{p} = \frac{\Phi}{2\pi} \qquad (5.2\text{-}1)$$

This condition provides a precise match of the intrinsic pitch length of the infinite medium to the pitch defined by the cell spacing and the alignment directions of a cell with a zero pretilt.

To account for the natural twist due to the presence of chiral molecules, the elastic energy density is written

$$U_{EL} = \tfrac{1}{2}k_1(\nabla\cdot\mathbf{n})^2 + \tfrac{1}{2}k_2(\mathbf{n}\cdot\nabla\times\mathbf{n}+q_0)^2 + \tfrac{1}{2}k_3(\mathbf{n}\times\nabla\times\mathbf{n})^2 \qquad (5.2\text{-}2)$$

where $\mathbf{n}$ is a unit vector representing the director distribution in the cell, k_1, k_2, k_3 are the three principal elastic constants, and $q_0 = 2\pi/p$ is the natural twist rate in the absence of the boundaries. We adopt a sign convention of positive q_0 for right-handed twist and negative q_0 for left-handed twist.

Consider the special case of a supertwisted nematic LC cell with a zero tilt angle and a uniform twist rate of q. Using Eq. (5.2-2), we obtain the following expression for the elastic energy density:

$$U_{EL} = \tfrac{1}{2}k_2(q-q_0)^2 \qquad (5.2\text{-}2a)$$

We note that, in the absence of boundaries, the lowest energy configuration occurs when the actual twist rate is exactly the same as the natural twist rate (i.e., $q = q_0$). Any deviation from the natural twist rate can lead to a higher elastic energy density. Thus, when the boundaries are present with prescribed rubbed directions, the actual twist rate is often slightly different from the natural twist rate in order to comply with the boundary conditions. The condition of d/p ratio in Eq. (5.2-1) corresponds to $q = q_0$.

The presence of the additional term q_0 in Eq. (5.2-2) for STN cells leads to more complicated electrodistortional curves (see Appendix E). The extent of the electrodistortion is often measured by examining the midlayer tilt angle as a function of the applied voltage. Figure 5.18 shows the electrodistortional curve for a TN (or STN) cell with various cell twist angles. We note that the steepness of the curve depends on the cell twist angle. By increasing the cell twist angle from 90° to 180–270°, we find that the steepness of the electrodistortional curve can be dramatically increased. As we mentioned earlier, a steep electrodistortional curve is a precondition to achieve a high contrast in a highly multiplexed display system. According to Figure 5.18, we find that the steepness increases as the cell twist angle increases. An infinite slope is obtained at a certain critical cell twist angle of about 270°. At higher cell twist angles, the curves become

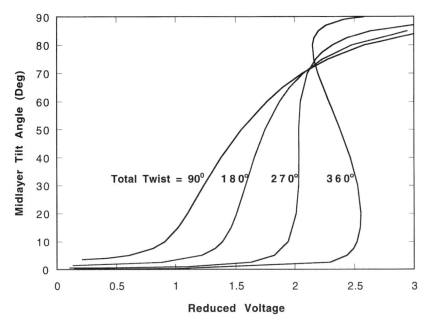

Figure 5.18.   Electrodistortional curve for a TN (or STN) cell with various cell twist angles.

doubled-valued, indicating a region of instability, bistability and hysteresis. It is important to note that the onset of the bistability occurs at a specific twist angle that depends on several device and material parameters, including d/p ratio, pretilt angle dielectric and elastic constants. Depending on these parameters, the steepest slope may occur at other twist angles, such as 240° or 210°.

Effect of Other Device and Material Parameters

In addition to the cell twist angle, the pretilt angle Θ at the boundaries can also affect the shape and the steepness of the electrodistortional curve. As we know, the electrodistortional curve is a result of the balance between the elastic restoring force and the electrostatic force. Thus, it is natural that the shape of the electrodistortional curve may depend on the elastic coefficients k_1, k_2, k_3, as well as the dielectric constants $\varepsilon_{\parallel}, \varepsilon_{\perp}$. The d/p ratio is both a device and material parameter. In what follows, we will discuss the dependence of the shape of the electrodistortional curve on these device and material parameters. The following discussion is based on a reference STN cell used in previous works [2,3].

Pretilt Angle

As we mentioned earlier, a small pretilt angle is important in TN cells to ensure a single domain under the influence of an applied electric field. The pretilt angle is even more important in STN cells. As a result of the larger twist angles, unwanted striped textures may occur in STN cells without a pretilt angle. In fact, the pretilt angles are often larger in STN cells, ranging from 5° to as large as 60°.

Generally speaking, increasing the pretilt angle causes the electrodistortional curve to shift toward lower voltages with a slight increase in the steepness of the curve. For the special case of a zero pretilt angle, there exists, a true threshold voltage below which there is uniform twist and zero tilt throughout the layer. When a pretilt angle exists, the tilt angle distribution $\theta(z)$ exhibits a minimum at the center of the cell (see Fig. 5.21). A uniform tilt angle $\theta(z) = $ constant may exist provided an appropriate voltage is applied to the cell (see Section 5.2.3).

Thickness Pitch Ratio d/p

As a result of the presence of the chiral molecules, a nature twist exists without the boundaries. Minimum elastic energy density occurs when the d/p ratio is exactly $\Phi/2\pi$. Any deviation from this value will cause an increase in the elastic energy density. This affects the balance between the electrostatic and the elastic energy, leading to a different distribution of the director. Thus, we expect to have a different electrodistortional curve. Many STN cells for display applications have a d/p ratio that is within the Grandjean zone defined by the following inequality [4]: $\Phi/2\pi - 0.25 < d/p < \Phi/2\pi + 0.25$. We note that the variation of the cell gap is within a quarter of a pitch ($p/4$). Generally speaking, lower d/p ratios shift the curves to lower voltages and make them steeper.

Bend / Splay Elastic Constant Ratio (k_3/k_1)

Increasing the k_3/k_1 ratio in an STN cell increases the steepness of the electrodistortional curves. Known nematic LC materials have k_3/k_1 ratios in the range 0.5–2.0, but most are in the range 1.2–1.8 (see Table 1.2).

Twist/Splay Elastic Constant Ratio (k_2/k_1)

The twist / splay elastic constant ratio lies between 0.5 and 0.6 for practically all LC materials. Decreasing the ratio shifts the upper portions of the electrodistortional curves to lower voltages, thereby increasing their steepness.

Dielectric Parameter γ

Generally speaking, increasing the dielectric parameter γ, defined as $\gamma = (\varepsilon_\parallel - \varepsilon_\perp)/\varepsilon_\perp$, decreases the steepness of the electrodistortion curve, as it does in the TN case.

5.2.2. Transmission Properties of STN-LCD in the Field-OFF State

We now consider the transmission properties of STN-LCDs. In a manner similar to TN-LCDs, there are various modes of operation in STN-LCDs. These include normally white (NW) and normally black (NB) operations. As a result of the relatively faster twist rate, the effect of waveguiding is no longer present. Thus, the polarizers in STN-LCDs are, in general, not parallel or perpendicular. In fact, the transmission axes of the polarizers are no longer parallel or perpendicular to

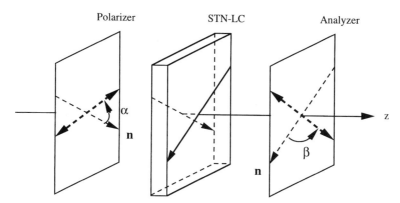

Figure 5.19. Schematic drawing of an STN-LCD, where α is the azimuth angle of the polarizer measured from the local director $\mathbf{n}(z = 0)$ and β is the azimuth angle of the analyzer measured from the local director $\mathbf{n}(z = d)$, where d is the cell thickness. The arrows on the STN cell indicate the director orientations at the surfaces. The double arrows indicate the transmission axes of the polarizers.

the local directors (rubbed directions) of the STN-LC. Generally speaking, two normal modes are excited by the incidence of a beam of polarized light. These two modes interfere after exiting the LC cell to produce either a bright state or a dark state. The general results obtained in Chapter 4 can now be employed to investigate the transmission of STN-LCDs in the field-OFF state.

Referring to Figure 5.19, we consider an STN-LCD that consists of a supertwisted nematic LC cell sandwiched between a pair of polarizers. We assume that the twist is right-handed. The azimuth angles of the transmission axes of the polarizers are denoted as α and β as shown in Figure 5.19.

If we assume a uniform tilt angle (either zero or small tilt angle) throughout the cell, then the transmission of the STN-LCD in the field-OFF state can be obtained by using Eq. (4.3-29). The polarizer's azimuth angles in Eq. (4.3-29) are related to α and β of Figure 5.19 by the following relationships:

$$\Phi_{\text{ent}} = \alpha, \qquad \Phi_{\text{exit}} = \phi + \beta \qquad (5.2\text{-}3)$$

where ϕ is the total twist angle. Note that the azimuthal angular separation between the polarizers is $(\Phi_{\text{exit}} - \Phi_{\text{ent}}) = \phi + \beta - \alpha$.

Using the new definition for the polarizer azimuth angles, we obtain the following transmission formula, according to Eq. (4.3-29):

$$T = \frac{1}{2}\left\{ \cos^2(\alpha - \beta) - \sin^2 X \sin 2\beta \sin 2\alpha \right.$$
$$+ \frac{\phi}{2X} \sin 2X \sin 2(\alpha - \beta) \qquad (5.2\text{-}4)$$
$$\left. - \phi^2 \frac{\sin^2 X}{X^2} \cos 2\beta \cos 2\alpha \right\}$$

where we recall that X is given by

$$X = \sqrt{\phi^2 + \frac{\Gamma^2}{4}} = \phi\sqrt{1 + u^2} \qquad (5.2\text{-}5)$$

where Γ is the phase retardation

$$\Gamma = \frac{2\pi}{\lambda}d(n_e - n_o) \qquad (5.2\text{-}6)$$

and u is the Mauguin parameter

$$u = \frac{\Gamma}{2\phi} \qquad (5.2\text{-}7)$$

It is important to note that if a uniform tilt angle exists ($\theta \neq 0$) in the cell, then n_e in Eq. (5.2-6) should be replaced by $n_e(\theta)$given by

$$\frac{1}{n_e^2(\theta)} = \frac{\cos^2\theta}{n_e^2} + \frac{\sin^2\theta}{n_o^2} \qquad (5.2\text{-}8)$$

where θ is the tilt angle. The transmission formula Eq. (5.2-4) can also be written

$$T = \frac{1}{2}\left\{\cos^2(\alpha + \beta) + \cos^2 X \cos 2\alpha \cos 2\beta \left(\tan 2\alpha - \frac{\phi}{X}\tan X\right) \right.$$
$$\left.\left(\tan 2\beta + \frac{\phi}{X}\tan X\right)\right\} \qquad (5.2\text{-}9)$$

We also note that the transmission formula is simpler by using the new definition for the polarizer azimuth angles. The transmission formula exhibits a 90° rotation symmetry. In other words

$$T\left(\alpha + \frac{\pi}{2}, \beta + \frac{\pi}{2}\right) = T(\alpha, \beta) \qquad (5.2\text{-}10)$$

Thus, if we rotate both the polarizer and the analyzer by 90° relative to the STN-LC cell, the transmission remains unchanged.

Although Eq. (5.2-4) or Eq. (5.2-9) can be employed to design the various modes of operation of STN-LCD, the techniques described in Chapter 4 offer some useful insight for this purpose. In Eqs. (4.3-38)–(4.3-40), we derived the input linear polarization states that can be transformed into linear polarization states (output beams) by an arbitrarily twisted LC cell with a uniform tilt. Using these formulas, we can obtain the azimuth angle of the polarizers for various modes of operation.

Before we consider the general cases, let us examine a special case of interest. According to Eq. (5.2-4), the transmission can be written

$$T = \tfrac{1}{2}\cos^2(\alpha - \beta) \tag{5.2-11}$$

provided

$$X = \phi\sqrt{1 + u^2} = \pi, 2\pi, 3\pi, \ldots \tag{5.2-12}$$

Thus, in this special case, the transmission is 100% if $\alpha = \beta$, and zero if $\alpha - \beta = \pi/2$. For a 270° twist cell, $\phi = 1.5\pi$, solutions for $X = 2\pi$, 3π, ... are possible. These correspond to a Mauguin parameter of $u = \sqrt{7}/3, \sqrt{3}, \ldots$ for the 270° STN cell. For $\lambda = 0.55$ μm, these correspond to $\Delta nd = 0.728$ μm, 1.429 μm, and so on. For applications in this special case, the relative azimuth angle between the polarizers is fixed by either $\alpha = \beta$, or $\alpha - \beta = \pi/2$, whereas the absolute azimuth angle can be arbitrarily chosen.

These numbers $\Delta nd = 0.728$ μm, 1.429 μm, and so on for a 270° twist cell are unfortunately too large for some LCD applications. In the following, we consider the general case when the parameter X is not an integral number of π. In the general cases, the azimuth angles (α, β) of both the polarizer and the analyzer must be properly chosen to obtain either a unity transmission or zero transmission for a specific wavelength. As we recall the transmission of a TN-LCD is somewhat insensitive to wavelength. This is a result of the waveguiding phenomenon in the Mauguin regime ($1 \ll u$). The transmission of an STN-LCD is no longer insensitive to the wavelength as a result of the higher twist rate. The following discussions are based on a single wavelength only.

Normally White Mode I (NW-I)

In the normally white operation, the transmission is 0.5 in the field-OFF state at a specific wavelength. Unlike the case in TN, the transmission of light at other wavelengths is much lower than $\tfrac{1}{2}$ in STN cells. So the white state is not quite white. The azimuth angle of the entrance polarizer is given by, according to Eq. (4.3-41)

$$\tan \alpha_1 = \sqrt{1 + v^2} - v \tag{5.2-13}$$

or equivalently

$$\tan 2\alpha_1 = \frac{1}{v} = \frac{\phi \tan X}{X} \tag{5.2-14}$$

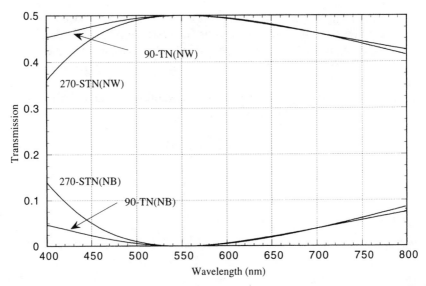

Figure 5.20. Transmission spectra of a 270° STN cell in the field-OFF state with $\alpha = -22°$, $\beta = 22°$, $\Delta nd = 0.48\,\mu m$, and a 90° TN cell in the field-OFF state with $\alpha = 0°$, $\beta = 0°$, $\Delta nd = 0.48\,\mu m$. The polarizer azimuth angles are chosen to have maximum (or zero) transmission in the field-OFF state for $\lambda = 550\,nm$.

It is important to remember that these angles α_1, β_2 are measured from the local director as indicated in Figure 5.19. And the azimuth angle of the analyzer (exit polarizer) is given by, according to Eq. (4.3-43)

$$\beta_1 = -\alpha_1 \qquad (5.2\text{-}15)$$

We note that the transmission is $T = \frac{1}{2}$ for these polarizer azimuth angles, according to Eq. (5.2-9). It is important to note that $T = \frac{1}{2}$ is applicable only for a specific design wavelength. A lower transmission will be obtained for other wavelengths (see Fig. 5.20). By examining Eq. (5.2-9), we find that the dependence on the wavelength is due to the $1/\lambda$ factor in Γ [Eq. (5.2-6)] and X. Thus, the deviation from $T = \frac{1}{2}$ depends on Δnd of the cell. Generally speaking, lower values of Δnd are desirable. If the STN cell is designed for a maximum transmission at around $\lambda = 0.55\,\mu m$, the field-OFF state (bright state) will have a yellowish green appearance, and the field-ON state (dark state) will have a black appearance. This is often referred to as the yellow mode [2].

Normally White Mode II (NW-II)

This mode is obtained by rotating both the polarizer and analyzer by 90° relative to the STN-LC cell. The transmission properties at normal incidence is identical to those of NW-I, according to Eq. (5.2-10). Specifically, the azimuth angle of the entrance polarizer is given by, according to Eq. (4.3-41)

$$\tan \alpha_2 = -\sqrt{1 + v^2} - v \qquad (5.2\text{-}16)$$

or equivalently

$$\tan 2\alpha_2 = \frac{1}{v} = \frac{\phi \tan X}{X} \qquad (5.2\text{-}17)$$

Note that although the two azimuth angles (α_1, α_2) are different, they have the same value for $\tan 2\alpha$. In fact, these two values of α are related by

$$\alpha_2 = \alpha_1 + \frac{\pi}{2} \qquad (5.2\text{-}18)$$

The azimuth angle of the analyzer (exit polarizer) is given by, according to Eq. (4.3-43)

$$\beta_2 = -\alpha_2 \qquad (5.2\text{-}19)$$

Normally Black Mode I

This mode can be obtained by keeping the azimuth angle of the entrance polarizer unchanged, while rotating the analyzer (exit polarizer) by 90°. Zero transmission is obtained for a specific wavelength. Leakage of light at other wavelengths is relatively stronger in STN cells than in TN cells. This is a result of the lack of waveguiding in STN cells. Thus, the black state is actually not quite black. Specifically, the azimuth angle of the entrance polarizer is given by, according to Eq. (4.3-41)

$$\tan \alpha_1 = \sqrt{1 + v^2} - v \qquad (5.2\text{-}20)$$

or equivalently

$$\tan 2\alpha_1 = \frac{1}{v} = \frac{\phi \tan X}{X} \qquad (5.2\text{-}21)$$

where X is given by Eq. (5.2-5). And the azimuth angle of the analyzer (exit polarizer) is given by

$$\beta_1 = \frac{\pi}{2} - \alpha_1 \qquad (5.2\text{-}22)$$

Note that the transmission axis of polarizers has no polarity. Thus, azimuth angles of ϕ and $\phi \pm \pi$ are the same orientation. We note that normally black mode I can be easily obtained from normally white mode I by rotating the analyzer (exit polarizer) by 90° in either direction. The zero transmission ($T = 0$) for these polarizer azimuth angles can also be obtained, according to Eq. (5.2-9). It is important to note that $T = 0$ is only for a specific design wavelength. A leakage of light occurs for other wavelengths, due to the dependence on the wavelength in Γ [Eq. (5.2-6)] and X (see Fig. 5.20). Thus, if the STN cell is designed for a zero transmission at around $\lambda = 0.55\,\mu m$, the field-OFF state (dark state) will have a blue-purple appearance, and the field-ON state (bright state) will have a white appearance. This is often referred to as the *blue mode* [2].

Normally Black Mode II

This mode is obtained by rotating both the polarizer and the analyzer in the NB-I mode by 90° relative to the STN-LC cell. The transmission properties at normal incidence is identical to those of NB-I, according to Eq. (5.2-10). Specifically, the azimuth angle of the entrance polarizer is given by, according to Eq. (4.3-41)

$$\tan \alpha_2 = -\sqrt{1 + v^2} - v \tag{5.2-23}$$

or equivalently

$$\tan 2\alpha_2 = \frac{1}{v} = \frac{\phi \tan X}{X} \tag{5.2-24}$$

$$\alpha_2 = \alpha_1 + \frac{\pi}{2} \tag{5.2-25}$$

where X is given by Eq. (5.2-5). The azimuth angle of the analyzer (exit polarizer) is given by

$$\beta_2 = \frac{\pi}{2} - \alpha_2 \tag{5.2-26}$$

Example 5.2. Consider a 270° STN-LCD with a uniform tilt angle throughout the cell. Let us calculate the polarizer azimuth angles α and β for three wavelengths of operation: $\lambda = 0.45$, 0.55, and 0.65 μm. The azimuth angles for the polarizer and the analyzer depend on the birefringence–thickness product $\Delta nd = [n_e(\theta) - n_o]d$, where θ is the tilt angle of the LC director. These angles (α, β) for Δnd between 0.40 and 1.00 μm are listed in Table 5.2. ∎

Table 5.2.   Polarizer Azimuth Angles

λ (μm)	Δnd (μm)	NW-I		NB-I	
		α	β	α	β
0.45	0.40	− 20.9°	20.9°	− 20.9°	− 69.1°
0.45	0.45	− 15.3°	15.3°	− 15.3°	− 74.7°
0.45	0.50	− 9.8°	9.8°	− 9.8°	− 80.2°
0.45	0.55	− 4.6°	4.6°	− 4.6°	− 85.4°
0.45	0.60	0.5°	− 0.5°	0.5°	89.5°
0.45	0.65	5.4°	− 5.4°	5.4°	84.6°
0.45	0.70	10.5°	− 10.5°	10.5°	79.5°
0.45	0.75	16.3°	− 16.3°	16.3°	73.7°
0.45	0.80	23.3°	− 23.3°	23.3°	66.7°
0.45	0.85	32.7°	− 32.7°	32.7°	57.3°
0.45	0.90	45.0°	− 45.0°	45.0°	45.0°
0.45	0.95	− 31.7°	31.7°	− 31.7°	− 58.3°
0.45	1.00	− 20.6°	20.6°	− 20.6°	− 69.4°
0.55	0.40	− 28.6°	28.6°	− 28.6°	− 61.4°
0.55	0.45	− 24.4°	24.4°	− 24.4°	− 65.6°
0.55	0.50	− 19.9°	19.9°	− 19.9°	− 70.1°
0.55	0.55	− 15.3°	15.3°	− 15.3°	− 74.7°
0.55	0.60	− 10.8°	10.8°	− 10.8°	− 79.2°
0.55	0.65	− 6.5°	6.5°	− 6.5°	− 83.5°
0.55	0.70	− 2.3°	2.3°	− 2.3°	− 87.7°
0.55	0.75	1.8°	− 1.8°	1.8°	88.2°
0.55	0.80	5.9°	− 5.9°	5.9°	84.1°
0.55	0.85	10.0°	− 10.0°	10.0°	80.0°
0.55	0.90	14.6°	− 14.6°	14.6°	75.4°
0.55	0.95	19.9°	− 19.9°	19.9°	70.1°
0.55	1.00	26.4°	− 26.4°	26.4°	63.6°
0.65	0.40	− 33.3°	33.3°	− 33.3°	− 56.7°
0.65	0.45	− 30.2°	30.2°	− 30.2°	− 59.8°
0.65	0.50	− 26.7°	26.7°	− 26.7°	− 63.3°
0.65	0.55	− 23.0°	23.0°	− 23.0°	− 67.0°
0.65	0.60	− 19.2°	19.2°	− 19.2°	− 70.8°
0.65	0.65	− 15.3°	15.3°	− 15.3°	− 74.7°
0.65	0.70	− 11.5°	11.5°	− 11.5°	− 78.5°
0.65	0.75	− 7.8°	7.8°	− 7.8°	− 82.2°
0.65	0.80	− 4.2°	4.2°	− 4.2°	− 85.8°
0.65	0.85	− 0.7°	0.7°	− 0.7°	− 89.3°
0.65	0.90	2.7°	− 2.7°	2.7°	87.3°
0.65	0.95	6.2°	− 6.2°	6.2°	83.8°
0.65	1.00	9.7°	− 9.7°	9.7°	80.3°

It is important to note that these polarizer azimuth angles for maximum transmission ($T = \frac{1}{2}$) or complete extinction ($T = 0$) depend on the operation wavelength λ. To illustrate the dependence on the wavelength, we plot the transmission of an STN cell as a function of wavelength in Figure 5.20. We note that in the normally black (NB) mode, the STN-LCD has a much stronger leakage in the blue spectral regime, leading to a bluish appearance in the black state. In the normally white (NW) mode, the STN-LCD has a much lower transmission in the blue spectral regime, leading to a yellowish green appearance in the white state. The transmission spectra of a TN cell are also shown in the same figure. We note that the TN cell is relatively insensitive to the wavelength variation as compared with the STN cell. Although the azimuth angles of the polarizers can be obtained for either zero or $\frac{1}{2}$ transmission for a given wavelength, the actual azimuth angles for STN-LCDs should be determined by optimizing the integrated transmission over the spectrum of the illuminating light source.

The colors in the field-OFF state of STN-LCDs are due to leakage of light, which is a result of the absence of waveguiding in the LC cell. These colors can be completely removed by using a double-layer STN cell, leading to true white or black in the field-OFF states. The double layer STN cells consist of an electrically switchable STN cell and a fixed STN cell with a reversed twist for color compensation in the field-OFF states.

5.2.3. Transmission Properties of STN-LCD in the Field-ON State

When an electric field of sufficient strength is turned ON, the LC molecules will be aligned parallel to the electric field. Similar to the situation in TN-LCD, the actual director distribution depends on the balance between the elastic energy density and the electrostatic energy density in the cell. Before we consider the general case of director distribution, let us examine a special case of approximately homeotropic alignment of the LC molecules in the cell due to the action of the electric field. The transmission for this special case is particularly simple. In the approximation of homeotropic alignment, the phase retardation Γ becomes zero. In this case, $X = \phi$, according to Eq. (5.2-5), and the transmission formula Eq. (5.2-4) becomes

$$T = \tfrac{1}{2}\cos^2(\beta + \phi - \alpha) \qquad (5.2\text{-}27)$$

which is exactly the transmission of a pair of polarizers with an azimuth angle of $(\beta + \phi - \alpha)$ between them. Equation (5.2-27) is a good estimate of the transmission of the STN cell in the field-ON state. However, in general, the director is not exactly in homeotropic alignment with the electric field. Jones matrix method and numerical techniques must be employed to calculate the transmission of the cell. However, we must obtain the director distribution $\theta(z)$ and $\phi(z)$ first before we can calculate the transmission.

Field-Induced Director Reorientation

According to Appendix E, the total electrostatic and elastic energy in the cell can be written

$$U = \frac{1}{2} \int_0^d \left(F_1 \theta'^2 + F_2 \phi'^2 + F_3 \phi' + \frac{D_z^2}{\varepsilon_\| \sin^2\theta + \varepsilon_\perp \cos^2\theta} \right) dz \qquad (5.2\text{-}28)$$

where $F_1, F_2,$ and F_3 are as follows:

$$F_1 = k_1 \cos^2\theta + k_3 \sin^2\theta \qquad (5.2\text{-}29)$$
$$F_2 = [k_2 \cos^2\theta + k_3 \sin^2\theta]\cos^2\theta \qquad (5.2\text{-}30)$$
$$F_3 = -2k_2 q_0 \cos^2\theta \qquad (5.2\text{-}31)$$

where θ' and ϕ' are the derivatives of the tilt and twist angles with respective to z, as defined in Eqs. (E-9) and (E-10), and q_0 is the natural twist rate of the LC material. Using the variational technique described in Appendix E, we obtain the following expression for the threshold voltage needed to achieve a uniform tilt angle throughout the cell:

$$V_{th} = \frac{1}{\sqrt{\varepsilon_\| - \varepsilon_\perp}} \left[4\pi k_2 \frac{d}{p_0} \Phi - \Phi^2 [k_3 - 2(k_3 - k_2)\cos^2\theta_0] \right]^{1/2} \qquad (\theta_0 \neq 0)$$

$$(5.2\text{-}32)$$

where θ_0 is the nonzero pretilt angle, $p_0 = 2\pi/q_0$ is the natural pitch, and Φ is the total twist angle. For the special case of zero pretilt angle $(\theta_0 = 0)$, the minimum voltage needed to create a small distortion in the cell is given by, according to Appendix E

$$V_{th} = V_{cl} \left[1 + \frac{k_3 - 2k_2}{k_1} \left(\frac{\Phi}{\pi}\right)^2 + 4 \frac{k_2}{k_1} \frac{d}{p_0} \frac{\Phi}{\pi} \right]^{1/2} \qquad (\theta_0 = 0) \quad (5.2\text{-}33)$$

where V_{cl} is the critical voltage needed to produce a finite tilt angle in a parallel nematic cell:

$$V_{cl} = \pi \frac{\sqrt{k_1}}{\sqrt{\varepsilon_\| - \varepsilon_\perp}} \qquad (5.2\text{-}34)$$

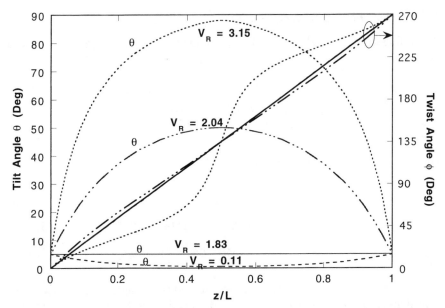

Figure 5.21. Director distribution $\theta(z)$ and $\phi(z)$ in an STN cell with 270° total twist angle at different reduced voltages. Reduced voltage $V_R = V/V_{c1}$.

It is important to remember that Eq. (5.2-33) is valid for the special case of zero pretilt angle ($\theta_0 = 0$). A discontinuity exists between Eqs. (5.2-32) and (5.2-33) at $\theta_0 = 0$. In other words, Eq. (5.2-32) does not reduce to Eq. (5.2-33) when θ_0 approaches 0. We note that Eq. (5.2-33) reduces to the result obtained previously for TN-LC by putting $p_0 = $ infinity:

$$V_T = V_{c1}\left[1 + \frac{k_3 - 2k_2}{k_1}\frac{\Phi^2}{\pi^2}\right]^{1/2} \tag{5.2-35}$$

The director distribution $\theta(z)$ and $\phi(z)$ can be obtained by solving the equations in Appendix E using numerical techniques. Figure 5.21 shows a typical director distribution in an STN cell at different applied voltages. In this case, a small pretilt angle of $\theta = 5°$ exists at the cell boundaries. We note that in the field-OFF ($V = 0$) state, the LC tilt angle $\theta(z)$ decreases as z increases from $z = 0$ to $z = d/2$ (midlayer), reaching a minimum value of the tilt angle at the midlayer point. At an applied voltage of $V_R = 1.83$ (reduced voltage $V_R = V/V_{c1}$), a uniform tilt $\theta(z) = 5°$ exists throughout the cell. For applied voltages V_R greater than 1.83, the tilt angle $\theta(z)$ is an increasing function of z between $z = 0$ and $z = d/2$, reaching a maximum at the midlayer point. The midlayer tilt angle $\theta(d/z)$ approaches 90° at applied voltages $1 \ll V_R$.

Transmission at Normal Incidence

Using the director distribution $\theta(z)$ and $\phi(z)$, we can now employ the Jones method to calculate the transmission spectrum of the STN cell. Here again, we can divide an STN cell into 20 or 40 sublayers of equal thickness. Each of these layers will be treated as a homogeneous uniaxially birefringent medium.

Figure 5.22 shows the transmission spectra of two different STN-LCDs at various applied voltages. The polarizers' transmission axes of these STN-LCDs are obtained from Table 5.2. These polarizers' azimuth angles are designed to yield a maximum transmission of 50%, for unpolarized light, at $\lambda = 0.55\ \mu m$ in the field-OFF state. Figure 5.22b plots the transmission spectrum of a different STN-LCD which offers a better contrast ratio. This is a result of the near-orthogonal orientation of the transmission axes of the polarizers. As we recall, the transmission in the field-ON state with $1 \ll V_R$ is given by Eq. (5.2-27). Thus a cell configuration with $(\beta + \phi - \alpha)$ near 90° or 270° will provide a better contrast ratio.

Contrast at Normal Incidence

As we indicated earlier, high-information-content displays with multiplexing addressing techniques suffer from the poor contrast ratios as a result of crosstalk between the addressing rows. Although the STN cells offer a solution for this problem, by virtue of their steep electrodistortional curves, the contrast ratio is still relatively poor for displays with high information content because of the absence of waveguiding. In TN-LCD, the transmission is relatively insensitive to the wavelength. In other words, if a TN cell is designed to have a zero transmission for the green light, the leakage is relatively low for the red and blue lights as well. This is no longer true in STN-LCDs due to the absence of waveguiding. In other words, if an STN cell is designed to have a zero transmission for green light, the leakage is relatively high for the red and blue lights. This can be seen from Figure 5.20. The leakage of light is a main source of the poor contrast.

The definitions of contrast ratios for an individual STN cell are the same as those of the TN cells. These contrast ratios for individual cells are only useful for low level multiplexed displays (alphanumeric) such as those used in automobiles panels, gas stations, watches, and calculators. For high-information-content displays with multiplexed addressing, the selection ratio of the applied voltage is limited by the following law of multiplexing [5]:

$$\frac{V_{ON}}{V_{OFF}} = \sqrt{\frac{\sqrt{N} + 1}{\sqrt{N} - 1}} \qquad (5.2\text{-}36)$$

where N is the number of rows in the displays panel (see Chapter 6 for more detailed discussion). For $N = 100$, the ratio is only 1.105.

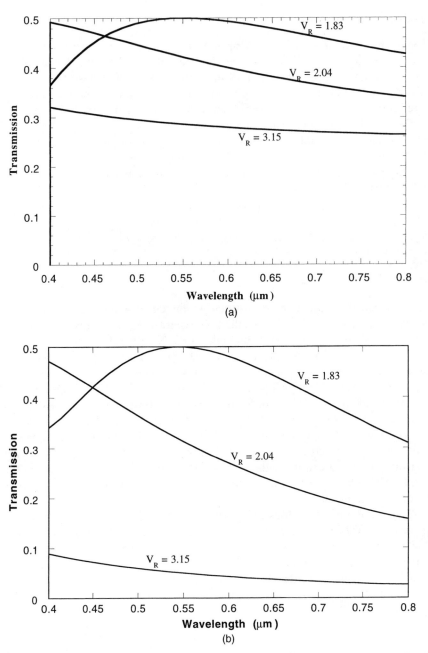

Figure 5.22. Transmission spectra (normal incidence) of a 270° STN-LCD at different reduced voltages, with (a) $\alpha = -22°$, $\beta = 22°$, $\Delta nd = 0.48\ \mu m$, reduced voltage $V_R = V/V_{c1}$ (b) $\alpha = 1.8°$, $\beta = -1.8°$, $\Delta nd = 0.75\ \mu m$, $V_R = V/V_{c1}$.

The contrast ratios for STN-LCDs are defined as follows:

$$C_{NW} = \frac{T(V_{OFF})}{T(V_{ON})} \qquad \text{(normally white)} \qquad (5.2\text{-}37)$$

$$C_{NB} = \frac{T(V_{ON})}{T(V_{OFF})} \qquad \text{(normally black)} \qquad (5.2\text{-}38)$$

As mentioned earlier, the leakage of light in the black state is relatively higher in STN-LCDs. This leads to a poor contrast ratio in STN display systems. In the case of TN-LCDs with multiplexed addressing, the voltages for ON and OFF states are governed by Eq. (5.2-36). For TN displays with a large number of rows N, V_{ON} is very close to V_{OFF}. This leads to $T(V_{ON}) \approx T(V_{OFF})$ and thus a very poor contrast ratio. In an STN cell with the proper twist angle, a very steep (near vertical) electrodistortional curve can be obtained (see Fig. 5.18 or 5.23). This is a very desirable property of the STN cell, allowing the possibility of high contrast ratios with a small voltage selection ratio. Thus, the applied voltages V_{ON} and V_{OFF} are often selected on both sides of the steepest slope (see Fig. 5.23). We note that the midlayer tilt angle is around 10° for a reduced voltage of

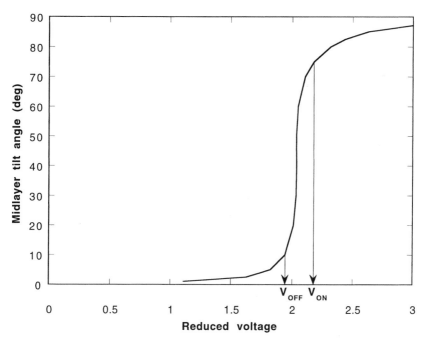

Figure 5.23. Electrodistortional curve of a 270° STN cell. The vertical lines indicate the nonselect voltage V_{OFF} and the select voltage V_{ON}.

$V_{OFF} = 1.9$. The midlayer tilt angle becomes about 75° at a reduced voltage of $V_{ON} = 2.2$.

As we know, a high contrast ratio requires a near zero denominator in Eqs. (5.2-37) and (5.2-38). As a result of the fast twist rate, the transmission of the field-OFF state $T(V_{OFF})$ varies with the wavelength (see Fig. 5.20). A near-zero integrated transmission of $T(V_{OFF})$ over the visible wavelengths is difficult to achieve. On the other hand, if the LC molecules can be homeotropically aligned because of an applied field, the transmission of the STN cell is simply $\cos^2 (\beta + \phi - \alpha)$, which can be zero if the polarizers are oriented at the appropriate azimuth angles. It can be shown that $\beta + \phi - \alpha = \pm \pi/2$ corresponds to a pair of crossed polarizers. Therefore the NW mode (yellow mode) is expected to have a higher contrast ratio.

Viewing Angles

Referring to Figures 5.24 and 5.25, we consider the viewing characteristics of a 270° STN-LCD at three different applied voltages. Figure 5.24 shows the transmission as a function of the horizontal viewing angle. We note that the display exhibits a left–right viewing symmetry. A severe leakage of light in the black state occurs for viewing angles greater than $\pm 30°$. The viewing exhibits no up–down symmetry. A significant leakage of light in the black state occurs for vertical viewing angles greater than $\pm 40°$.

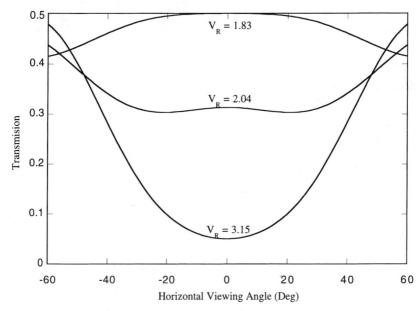

Figure 5.24. Horizontal viewing characteristics of a 270° STN-LCD at different reduced voltages, with $\alpha = 1.8°$, $\beta = -1.8°$, $\Delta nd = 0.75\,\mu m$, $\lambda = 0.55\,\mu m$. Reduced voltage $V_R = V/V_{cl}$, where V_{cl} is as given by Eq. (5.2-34).

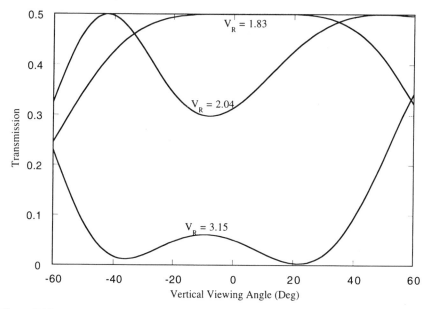

Figure 5.25. Vertical viewing characteristics of a 270° STN-LCD at different reduced voltages, with $\alpha = 1.8°$, $\beta = -1.8°$, $\Delta nd = 0.75$ μm, $\lambda = 0.55$ μm. Reduced voltage $V_R = V/V_{c1}$, where V_{c1} is as given by Eq. (5.2-34).

For NW-STN cells, in order to achieve a higher contrast ratio, the polarizers must be crossed. In this case, a high contrast can be achieved by using a high applied voltage in the black state. From Table 5.2, we realize that the cell gap d must be properly chosen.

5.3.   NEMATIC LIQUID CRYSTAL DISPLAY (N-LCD) MODES

In both TN and STN, the LC director **n** undergoes a twist in the cell. The twisting of the director allows the phenomenon of waveguiding to occur in TN cells. As a result of the twist, the displays show asymmetric viewing characteristics (in the vertical direction). Birefringence phase compensation for improving the viewing characteristics (Chapter 9) in these cells is more difficult to realize because of the complicated director distribution. We now consider the use of nematic LC cells for display applications. In this section, we consider three important cases of interest. These are homogeneously parallel aligned cell, vertically aligned (VA) cell, and bend-aligned (BA) cell.

5.3.1.   Parallel Aligned (PA) Cells

The LC cell in this case is effectively an a plate of uniaxially birefringent material in the field-OFF state. An electric field can be applied to reorient the LC director in the cell to control the phase retardation. The discussion is divided into the two following cases depending on the direction of the applied electric field.

Vertical Switching (E Field Perpendicular to LC Layer)

Referring to Figure 5.26, we consider a nematic liquid crystal display (N-LCD) consisting of a homogeneously parallel aligned cell sandwiched between a pair of crossed polarizers. The c axis (director) of the cell is oriented at 45° relative to the transmission axes of the polarizers. For display applications, the LC cell is coated with transparent electrodes so that an electric field along the z axis can be applied to the cell. In the field-OFF state, the LC cell exhibits a phase

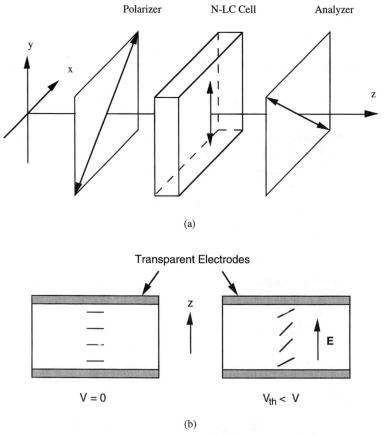

(a)

(b)

Figure 5.26. (a) Schematic drawing of a nematic liquid crystal display (N-LCD) using a parallel aligned cell; (b) director distribution in the field-OFF and field-ON states (V=applied voltage; V_{th} = threshold voltage).

retardation of

$$\Gamma = \frac{2\pi}{\lambda}(n_e - n_o)d \qquad (5.3\text{-}1)$$

where n_e, n_o are the principal refractive indices of the LC and material and d is the thickness of the LC layer. In a normally white (NW) operation, Γ is chosen to be an odd integral number of π (i.e., $\Gamma = \pi$, 3π, 5π, 7π, . . .). Usually, the lowest order ($\Gamma = \pi$) is chosen for the best viewing characteristics at large viewing angles. The transmission of the LCD is given by

$$T = \frac{1}{2}\sin^2\frac{\Gamma}{2} \qquad \text{(crossed polarizers)} \qquad (5.3\text{-}2)$$

When an electric field is applied to the LC cell, the LC director is reoriented (tilt) toward the direction of the electric field. This leads to a decrease of the phase retardation, which is now written

$$\Gamma = \frac{2\pi}{\lambda}\int[n_e(\theta) - n_o]dz \qquad (5.3\text{-}3)$$

where the integral is from $z = 0$ to $z = d$, $n_e(\theta)$ is given by

$$\frac{1}{n_e^2(\theta)} = \frac{\sin^2\theta}{n_o^2} + \frac{\cos^2\theta}{n_e^2} \qquad (5.3\text{-}4)$$

where θ is the tilt angle of the LC director. It is important to note that $\theta(z)$ is a function of position z, and the tilt angle distribution $\theta(z)$ depends on the applied voltage. In the extreme case of homeotropical alignment ($\theta = 90°$) by applying a strong electric field, the phase retardation reduces to zero. This leads to a zero transmission in a crossed polarizer configuration, according to Eq. (5.3-2). This mode of switching is sometimes referred to as the *electrically controlled birefringence mode* (ECB-LCD).

In principle, the same structure shown in Figure 5.26a can also be employed for a normally black (NB) operation. In this case Γ is chosen to be an integral number of 2π. (i.e., $\Gamma = 2\pi$, 4π, 6π, 8π, . . .). Again, for viewing characteristics at large viewing angles, Γ is usually chosen to be 2π ($\Gamma = 2\pi$). The white state is obtained by applying a voltage so that the integrated phase retardation in Eq. (5.3-3) becomes π. This operation is, however, not practical for two reasons: (1) the LC cell, with $\Gamma = 2\pi$, must be two times thicker, which is very undesirable in LCDs; and (2) leakage of light occurs when the phase retardation becomes less than π, as the applied voltage becomes too high.

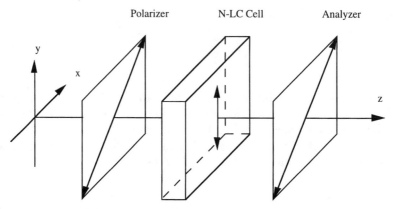

Figure 5.27.   Schematic drawing of a normally black vertical switching mode of N-LCD.

Referring to Figure 5.27, we consider a more practical normally black (NB) operation that involves the use of parallel polarizers. In the field-OFF state, the LC cell behaves like a half-wave plate, where Γ is an odd integral number of π (i.e., $\Gamma = \pi, 3\pi, 5\pi, 7\pi, \ldots$). Again, the lowest order ($\Gamma = \pi$) is usually chosen for the best viewing characteristics at large angles of viewing. The transmission of the LCD shown in Figure 5.27 is given by

$$T = \frac{1}{2} \cos^2 \frac{\Gamma}{2} \qquad \text{(parallel polarizers)} \qquad (5.3\text{-}5)$$

When an electric field is applied to the LC cell, the phase retardation Γ is as given by Eq. (5.3-3). In the extreme case of homeotropic alignment ($\theta = 90°$) by applying a strong electric field, the phase retardation reduces to zero. This leads to a maximum transmission in a parallel-polarizer configuration, according to Eq. (5.3-5).

Electrodistortion (Electrooptical Distortion) in the Vertical Switching Mode. When an electric field is applied to the cell along the z axis, the LC molecules are aligned toward the z axis. The distribution of the director tilt angle $\theta(z)$ is a result of the balance between the elastic restoring force and the electric force, as well as the boundary conditions. According to Appendix B, the elastic energy density can be written

$$U_{\text{EL}} = \frac{1}{2} (k_1 \cos^2 \theta + k_3 \sin^2 \theta) \left(\frac{d\theta}{dz} \right)^2 \qquad (5.3\text{-}6)$$

where θ is the tilt angle and k_1, k_3 are the elastic constants. The change of the electrostatic energy due to the reorientation of the directors can be written, according to Eq. (B-2) in Appendix B

$$\Delta U_{EM} = \frac{1}{2} \frac{D_z^2}{(\varepsilon_{\parallel} \sin^2 \theta + \varepsilon_{\perp} \cos^2 \theta)} - \frac{1}{2} \frac{D_z^2}{\varepsilon_{\perp}} \qquad (5.3\text{-}7)$$

where D_z ($=$ constant) is the z component of the displacement field vector. We note that the second term is a constant independent of the director orientation $\theta(z)$. According to the discussion in Appendix B, distortion occurs when the applied voltage is higher than a critical voltage:

$$V_{c1} = E_{c1} d = \pi \frac{\sqrt{k_1}}{\sqrt{\varepsilon_{\parallel} - \varepsilon_{\perp}}} \qquad \text{(tilt mode)} \qquad (5.3\text{-}8)$$

A numerical analysis can be carried out to obtain the director distribution $\theta(z)$ by using the equations in Appendix B. Figure 5.28 shows the director distribution $\theta(z)$ at various applied voltages. We note that the tilt angle $\theta(z)$ of the LC director is an increasing function of the applied voltage. In addition, the tilt angle $\theta(z)$ is a symmetric function with respect to the midlayer $z = d/2$. At an applied voltage of 5.71 V, the nematic cell is virtually homeotropically aligned with a tilt angle $\theta(z)$ of about 90°.

Figure 5.29 shows the transmission of the N-LCD at normal incidence as a function of the applied voltage. Using the green light as the design wavelength,

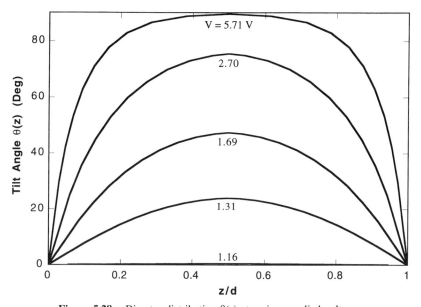

Figure 5.28. Director distribution $\theta(z)$ at various applied voltages.

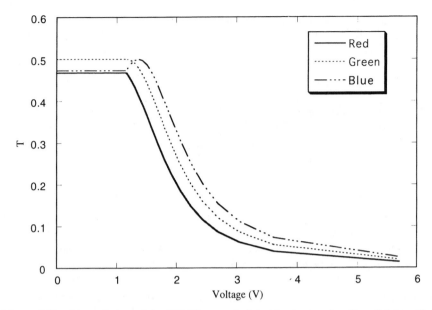

Figure 5.29. Transmission of the N-LCD at normal incidence as a function of the applied voltage.

the cell exhibits a maximum transmission of 0.5 for the green light. The transmission for the green light is a monotonically decreasing function of the applied voltage for $V_{th} < V$. The cell exhibits a lower transmission for the blue and red lights in the field-OFF ($V = 0$) state. As the applied voltage increases, the transmission for the blue light first increases to a maximum of 0.5, and then drops monotonically. As we know, the phase retardation $\Gamma(\lambda_{blue})$ is greater than π in the field-OFF state. As the LC director tilts, $\Gamma(\lambda_{blue})$ decreases. The maximum transmission of 0.5 occurs when $\Gamma(\lambda_{blue}) = \pi$. For the red light, the transmission is a monotonically decreasing function of the applied voltage. The phase retardation $\Gamma(\lambda_{red})$ is less than π and is a decreasing function of the applied voltage.

Figure 5.29 resembles Figure 5.9 for a NW TN-LCD. However, further comparison show that the transmission of TN-LCDs is less sensitive to wavelength variation than that of N-LCDs. This is a result of the waveguiding in the field-OFF state of TN-LCD's. In N-LCDs the transmission is based on polarization interference, which is often sensitive to wavelength.

In-Plane Switching (IPS)

Referring to Figure 5.30, we consider another switching mode of the homogeneously parallel aligned nematic cell for LCDs. This is known as the

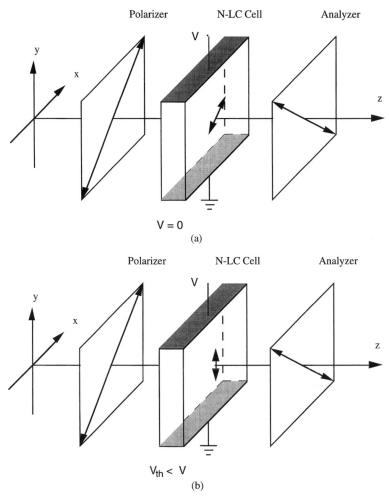

Figure 5.30. Schematic drawing of the in-plane switching mode of the N-LCD with normally black operations, where the shaded areas represent electrodes (V=applied voltage; V_{th} = threshold voltage). (a) In the field-OFF state, the LC director is parallel to the transmission axis of the polarizer. The transmission is zero, due to the crossed polarizers. (b) In the field-ON state, the director is aligned parallel to the electric field. Maximum transmission occurs with a proper choice of the cell thickness.

in-plane switching (IPS) [6,7] mode. In this case, the applied electric field is in the xy plane parallel to the LC layer surface. In the normally black (NB) operation, the transmission axis of the polarizer is parallel to the LC director at the input plane. The optical wave traversing through the LC cell is an extraordinary wave whose polarization state remains unchanged. This leads to a zero transmission due to the crossed polarizer configuration.

When an electric field is applied to the LC cell, the LC molecules are aligned toward the electric field (along the y axis). This leads to a new director distribution with a twist $\phi(z)$ in the xy plane. As a result of the boundary condition due to the anchoring of LC molecules with the rubbed directions, the twist angle ϕ is a function of position z. Except for a small pretilt, the tilt angle θ is zero since the electric field is in plane. Unlike the TN-LC cells, the twist here is not a simple function of z. Thus, a simple transmission formula is not available for this case. The Jones matrix method must be employed by dividing the LC cell into N thin layers of equal thickness (with N in the range of 20). However, if the applied field is sufficiently strong enough to ensure alignment of most of the LC molecules along the direction of the field, the LC cell is approximately in a homogeneously parallel alignment, by neglecting the thin boundary layers at the surfaces. In this approximation, the transmission can be written

$$T = \frac{1}{2}\sin^2 2\Delta\phi \sin^2 \frac{\Gamma}{2} \qquad (5.3\text{-}9)$$

where $\Delta\phi$ is the twist angle relative to the transmission axis of the polarizer due to the applied field, and Γ is the phase retardation of the LC cell. In the field-OFF $(V = 0)$ state, $\Delta\phi = 0$, Eq. (5.3-9) yields a transmission of $T = 0$. Selection of a proper cell thickness such that $\Gamma = \pi$, and a proper reorientation of the LC director such that $\Delta\phi = 45°$, Eq. (5.3-9) yields a transmission of $T = \frac{1}{2}$.

If we rotate the analyzer in Figure 5.30 by 90°, a normally white (NW) operation with parallel polarizers can be obtained. Several other modes of operation can be devised by choosing proper orientations of the LC cell relative to the transmission axes of the polarizers. Focusing on the operation shown in Figure 5.30, we now consider the electrodistortion of the LC cell under the application of an in-plane electric field.

Electrodistortion in the In-plane Switching (IPS) Mode. When an electric field is applied to the cell along the y axis, the LC molecules are aligned toward the y axis. The distribution of the director twist angle $\phi(z)$ is a result of the balance between the elastic restoring force and the electric force, as well as the boundary conditions. According to Appendix C, the elastic energy density can be written

$$U_{\text{EL}} = \frac{1}{2}k_2\left(\frac{d\phi}{dz}\right)^2 \qquad (5.3\text{-}10)$$

where ϕ is the twist angle measured from the x axis and k_2 is the twist elastic constant.

By taking into account the work done by the battery (or power supply), the net change in the electromagnetic energy density due to the reorientation of the LC director can be written, according to Eq. (C-4) in Appendix C

$$\Delta U_{\text{EM}} = \tfrac{1}{2}\varepsilon_\perp E^2 - \tfrac{1}{2}(\varepsilon_\parallel \sin^2 \phi + \varepsilon_\perp \cos^2 \phi)E^2 \qquad (5.3\text{-}11)$$

where ϕ is the twist angle, which depends on z. We note that the first term is a constant independent of the twist angle.

According to Appendix C, there exists a critical field E_{c1}, which is needed to create any twist of the director in the LC medium. For applied electric field $E < E_{c1}$, the midlayer twist angle $\Delta\phi$ is zero, with $\phi = 45°$ throughout the cell. The critical field can be written

$$E_{c1}d = \pi\frac{\sqrt{k_2}}{\sqrt{\varepsilon_\| - \varepsilon_\perp}} \qquad (5.3\text{-}12)$$

where k_2 is the twist elastic constant, and d is the cell gap. The critical voltage (threshold voltage) needed is thus given by

$$V_{\text{th}} = E_{c1}w = \pi\frac{\sqrt{k_2}}{\sqrt{\varepsilon_\| - \varepsilon_\perp}}\frac{w}{d} \qquad (5.3\text{-}13)$$

where w is the distance between the electrodes of the LC cell. We note that the critical voltage is proportional to the square root of the twist elastic constant and is inversely proportional to the square root of dielectric anisotropy. Figure 5.31 shows the twist angle distribution $\phi(z)$ at various applied voltages. We note that

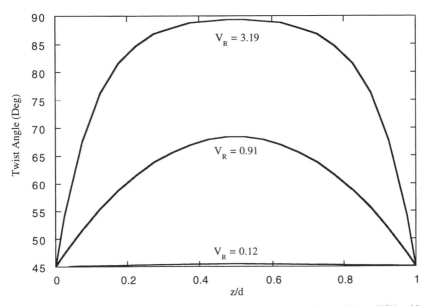

Figure 5.31. The twist angle distribution $\phi(z)$ at various reduced voltages, $V_R = V/V_{c1}$. Note that ϕ is measured from the x axis. So, $\phi = 45°$ in the field-OFF state.

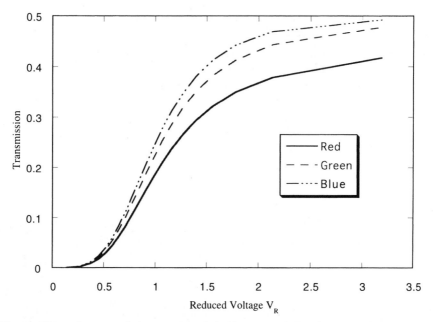

Figure 5.32. The transmission at normal incidence as a function of reduced voltage, $V_R = V/V_{cl}$.

the twist angle $\phi(z)$ is an increasing function of the applied voltage. The twist angle $\phi(z)$ is also symmetric with respect to the midlayer $z = d/2$. At a higher reduced voltage of $V_R = 3.19$, the twist angle $\Delta\phi$ due to the applied field is approximately 45° throughout the cell. Figure 5.32 shows the transmission at normal incidence as a function of applied voltage. We note that the transmission is an increasing function of the applied voltage. This is in agreement with Eq. (5.3-9), where $\Delta\phi$ is an increasing function of the applied voltage.

The cell gap used in Figure 5.32 is chosen so that the phase retardation for the green light is π when the LC director is oriented at 45°. Because of the boundary conditions, only molecules near the center of the cell are twisted by 45° (see Fig. 5.31). Therefore, the LC cell is slightly different from a half-wave plate and the transmission is not 50%. In addition, we note that the transmission of IPS-LCDs is also sensitive to wavelength variation. For IPS-LCDs to respond uniformly to all wavelengths, pixels with different colors need to have different cell gaps.

5.3.2. Vertically Aligned (VA) Cells

In a vertically aligned LC cell, the LC director is perpendicular to the surfaces of the cell (see Fig. 1.8). The LC cell in this case is effectively a c plate of uniaxially birefringent material in the field-OFF state $(V = 0)$. An electric field can be applied to reorient the LC director in the cell to control the phase retardation.

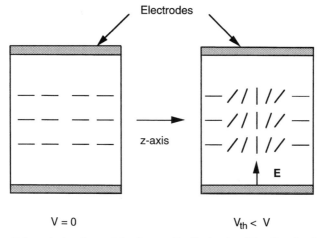

Figure 5.33. Schematic drawing and the director distribution in a VA cell under the influence of an applied electric field **E** that is perpendicular to the vertical alignment.

This mode of operation is also called *electrically controlled birefringence* (ECB). Figure 5.33 shows a schematic drawing of the director distribution in a VA cell under the influence of an applied electric field. The cells are enlarged for the illustration of the director distribution. In the VA-LCDs, the light is propagating along the z axis, whereas the electric field is perpendicular to the z axis.

Referring to Figure 5.34, we consider a normally black (NB) operation of the vertically aligned (VA) LCD using a pair of crossed polarizers. For normally incident light, the transmission of the LCD is, again, given by

$$T = \frac{1}{2} \sin^2 \frac{\Gamma}{2} \qquad \text{(crossed polarizers)} \qquad (5.3\text{-}14)$$

In the field-OFF state $(V = 0)$, the LC molecules are homeotropically aligned along the z axis. For normally incident light, the propagation is along the c axis of the cell with a zero phase retardation $(\Gamma = 0)$. Thus, the transmission is zero due to the crossed polarizers. When an electric field is applied to the LC cell, the LC director is reoriented (tilted) toward the direction of the electric field (see Fig. 5.34b). This leads to a distribution of the director tilt angle $\theta(z)$ and an increase of the phase retardation, which is now written

$$\Gamma = \frac{2\pi}{\lambda} \int [n_e(\theta) - n_o] dz \qquad (5.3\text{-}15)$$

where the integral is from $z = 0$ to $z = d, n_e(\theta)$, is as given by Eq. (5.3-4). The

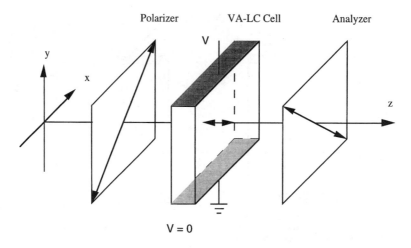

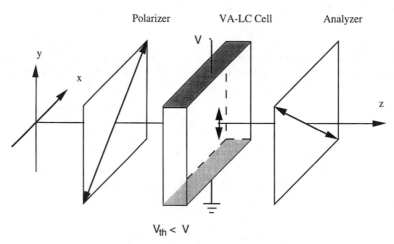

Figure 5.34. Normally black operation of a VA-LCD using a pair of crossed polarizers: (*a*) field-OFF state (black state); (*b*) field-ON state (white state).

phase retardation is an increasing function of the applied voltage. Under the appropriate conditions, a phase retardation of π can be achieved, which leads to a maximum transmission. The director distribution $\theta(z)$ in a vertically aligned cell can be obtained by using the technique described in Appendix B.

5.3.3. Bend-Aligned (BA) Cells

In a bend-aligned LC cell (also known as the *Pi cell*), the LC director has a bend distribution with a total bend of 180° (π) from surface to surface. This cell configuration was originally proposed by Bos and Koehler/Beran for an electrically controllable wave plate [8]. An electric field can be applied to

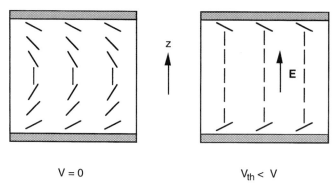

Figure 5.35. A schematic drawing of the director distribution in a bend-aligned cell under the influence of an applied electric field. The shaded areas are transparent electrodes.

reorient the LC director in the cell to control the phase retardation. Figure 5.35 shows a schematic drawing of the director distribution in a BA cell under the influence of an applied electric field. The cells are enlarged for the illustration of the director distribution. In the BA-LCDs, transparent electrodes are needed as the light is propagating along the z axis, which is also the direction of the electric field.

The BA cells, which have a phase retardation that is electrically controllable, can be employed for display applications. Figure 5.36 shows an example of BA-LCDs with a normally white (NW) operation using a pair of crossed polarizers. For normally incident light, the transmission of the LCD is, again, given by

$$T = \frac{1}{2}\sin^2\frac{\Gamma}{2} \qquad \text{(crossed polarizers)} \qquad (5.3\text{-}16)$$

where Γ is the phase retardation given by

$$\Gamma = \frac{2\pi}{\lambda}\int[n_e(\theta) - n_o]dz \qquad (5.3\text{-}17)$$

where the integral is from $z = 0$ to $z = d$, $n_e(\theta)$ is given by Eq. (5.3-4). In a (NW) operation, Γ is chosen to be an odd integral number of π (i.e., $\Gamma = \pi$, 3π, 5π, 7π, ...) in the field-OFF state ($V = 0$). Usually, the lowest order ($\Gamma = \pi$) is chosen for the best viewing characteristics at large viewing angles. This can be achieved by a proper choice of the cell thickness and the LC birefringence. When an electric field is applied to the LC cell, the LC director is reoriented (tilted) toward the direction of the electric field (see Fig. 5.35). This leads to a decrease of the phase retardation. When the field is sufficiently strong, the cell becomes homeotropically aligned, except at the boundaries. For normally incident light,

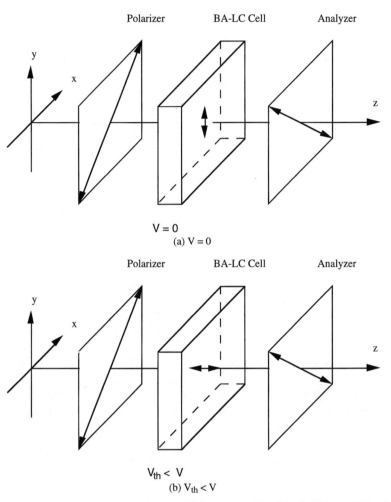

Figure 5.36. A schematic drawing of a BA-LCD in the normally white operation. (*a*) $V = 0$: The plane of bending (*yz*-plane) is oriented at 45° relative to the transmission axes of the polarizers. (*b*) $V_{th} < V$: In the field-ON state, the liquid crystal director is homeotropically aligned along the z axis, leading to a dark state.

Eq. (5.3-17) yields a zero phase retardation ($\Gamma = 0$). Thus, the transmission is zero as a result of the crossed polarizers. The director distribution $\theta(z)$ in a BA cell (or Pi cell) can be obtained by using the technique described in Appendix B.

Symmetry Property of Pi Cell

The BA cell has a unique property of viewing characteristics. As a result of the bend alignment, the phase retardation in the field-OFF state is a symmetrical function of the vertical viewing angles whose bend plane is in the yz plane. This viewing symmetry can be explained as follows with the aid of Figure 5.37.

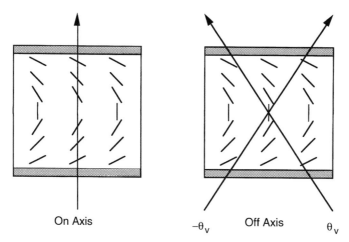

Figure 5.37. On-axis and off-axis viewing through a bend-aligned cell in the field-OFF state $(V = 0)$.

As a result of the bend alignment, the phase retardation for viewing in the bend plane has a self-compensating nature. The increase in the phase retardation in the first half of the cell due to off-axis viewing is compensated by the decrease in the phase retardation in the second half of the cell. This can be seen by decomposing the phase retardation integral into two parts:

$$\Gamma = \Gamma_1 + \Gamma_2 \tag{5.3-18}$$

where

$$\Gamma_1 = \frac{2\pi}{\lambda} \int_0^{d/2} [n_e(\theta) - n_o]dz \tag{5.3-19}$$

$$\Gamma_2 = \frac{2\pi}{\lambda} \int_{d/2}^{d} [n_e(\theta) - n_o]dz \tag{5.3-20}$$

In other words, Γ_1 is the phase retardation due to the first half of the cell, whereas Γ_2 is that of the second half of the cell. Both Γ_1 and Γ_2 are functions of the vertical viewing angle θ_v. As a result of the symmetric bend alignment relative to the center of the cell, it can be easily shown, by examining Figure 5.37, that the partial retardations, Γ_1, Γ_2, satisfy the following relation

$$\Gamma_1(\theta_v) = \Gamma_2(-\theta_v) \tag{5.3-21}$$

It follows immediately that, according to Eqs. (5.3-18)

$$\Gamma(\theta_v) = \Gamma(-\theta_v) \tag{5.3-22}$$

which leads to a vertical viewing symmetry. The symmetry in the vertical viewing direction is also responsible for the wide viewing angle of LCDs based on BA cells. It is important to note that the viewing will remain symmetric in the field-ON state provided the bend alignment remains symmetric relative to the center of the cell. The same result [Eq. (5.3-22)] can also be easily obtained using the principle of reciprocity and the symmetry of the cell [See Problem 5.15].

5.4. POLYMER DISPERSED LIQUID CRYSTAL DISPLAYS (PD-LCDs)

We recall that, in Chapter 1, we discussed the milky appearance of nematic liquid crystal in a glass bottle. The milky appearance is a result of the severe scattering of light due to the discontinuities of the refractive index at the boundaries of nematic LC droplets or domains whose physical dimensions are in the micrometer range. The milky appearance disappears when the sample is (1) heated beyond the clearing point so that the liquid crystal becomes optically isotropic or (2) poled into a single domain by using physical or chemical means. The scattering phenomenon of nematic LC droplets can be employed for display applications.

Referring to Figure 5.38, we consider a suspension of nematic LC droplets in a host medium (polymer) of matched refractive index n_p. The refractive indices of the nematic LC droplets depend on the order parameter S of the LC molecules in the droplets. Let the effective refractive indices be written

$$n'_e = n'_e(S), \qquad n'_o = n'_o(S) \tag{5.4-1}$$

where S is the order parameter. In the case of a homogeneously parallel alignment, the order parameter S is 1. In this case the refractive indices are written

$$n'^2_e(S = 1) = n^2_e \qquad n'^2_o(S = 1) = n^2_o \tag{5.4-2}$$

where n_e, n_o are the principal refractive indices of the nematic liquid crystal. In the extreme case of disorder (or a uniform distribution of **n** over all angles) with $S = 0$, the droplet becomes optically isotropic with the refractive index given by

$$n^2 = \frac{n^2_e + 2n^2_o}{3} \qquad (S = 0) \tag{5.4-3}$$

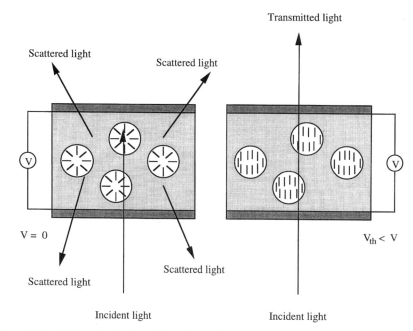

Figure 5.38. Schematic drawing of a polymer dispersed liquid crystal display (PD-LCD). The liquid crystal droplets are dispersed in a polymer host medium with a matching refractive index. Scattering occurs in the field-OFF state $(V = 0)$ when the liquid crystal molecules are not fully aligned in the droplets.

In most general cases, the LC molecules in the droplets are partially aligned because of the boundary conditions and the intermolecular forces. The effective refractive indices are somewhere between n_o and n_e. In other words

$$n_o < n'_e(S) < n_e, \qquad n_o < n'_o(S) < n_e \qquad (5.4\text{-}4)$$

The refractive index of the host medium is chosen to be n_o (i.e., $n_p = n_o$). Thus there exists a discontinuity of the refractive index at the droplet's boundary. As a result, scattering of light occurs. Consider now the incidence of a beam of unpolarized light into the LC cell. In the field-OFF state, the LC molecules within the droplets are only partially aligned. In addition, the averaged director orientation **n** of the droplets (if $S \neq 0$) exhibits a random distribution of orientation within the cell. The beam is significantly attenuated as a result of the energy loss via scattering.

When a sufficiently strong field ($\approx 1\ V_{rms}/\mu m$) is applied to the cell, all the LC molecules are aligned parallel to the electric field. If the light is also propagating in the direction parallel to the electric field, then the beam of light "sees" a refractive index of n_o, which is matched with the refractive index of the surrounding medium. In this state, the whole cell appears to be optically homogeneous (like a thin sheet of glass) for normally incident light. Thus the beam of light transmits through the cell with no scattering and no attenuation. This is the principle of operation of the PD-LC.

When the electric field is turned OFF, the droplets return to their original orientations with a partial alignment of the LC molecules within the droplets. This requires a restoring force. Polymer materials are ideal for this purpose. In most PD-LCDs, encapsulation procedures are employed to create micrometer-size capsules of LC droplets inside a polymer. By virtue of the rodlike or chainlike nature of the polymers, LC molecules within the droplets are likely to anchor in the directions parallel to the local polymer chains. Without an appropriate boundary to provide the restoring force, the switching from the ordered state to a disordered state must rely on the Brownian motion, which can take a very long time. It is important to note that there are several different configurations of director distribution in the droplets in the filed-OFF state, depending on the specific boundary conditions and the shape of each droplet.

The contrast ratio of the PD-LCDs is defined as

$$C(V) = \frac{T(V)}{T(0)} \tag{5.4-5}$$

where V is the applied voltage, $T(V)$ is the transmission of the field-ON state and $T(0)$ is the transmission of the field-OFF state. A high contrast ratio requires a high transmission in field-ON state and a near-zero transmission in the field-OFF state. High transmission in the field-ON state requires a good matching of the refractive index n_p of the host medium with the ordinary refractive index n_o of the liquid crystal. In other words, $n_p(\lambda) = n_o(\lambda)$ for most wavelengths λ in the spectral regime of operation.

The residual transmission (leakage) $T(0)$ depends on the attenuation of the incident beam due to scattering loss in the field-OFF state. Assuming a spherical shape of radius a, the scattering cross section σ can be written, according to Mie scattering theory [9–11]

$$\sigma = 2\pi a^2 \left[1 - \frac{\sin 2ka\, \delta n}{ka\, \delta n} + \frac{\sin^2 ka\, \delta n}{(ka\, \delta n)^2} \right] \tag{5.4-6}$$

where k is the wavenumber of light in vacuum ($k = 2\pi/\lambda$) and δn is a measure of the index discontinuity

$$\delta n = \left(\frac{n'}{n_p} \right) - 1 \tag{5.4-7}$$

where n' is the refractive index of the LC droplet according to Eq. (5.4-1), depending on the polarization state, and n_p is the index of refraction of the host medium. It is important to note that Eq. (5.4-6) is derived by assuming a sphere of isotropic medium with a refractive index of n', and is valid only when the index discontinuity is very small, e.g., $(\delta n' \ll 1)$.

In the limit of large spheres with $1 \ll ka\,\delta n$, the scattering cross section becomes, according to Eq. (5.4-6)

$$\sigma = 2\pi a^2 \qquad\qquad (5.4\text{-}8)$$

which is exactly twice the geometric cross section of the sphere. This result can also be obtained by assuming that the portion of light falling in the geometric area of the sphere is refracted and deflected into other directions. In addition, diffraction of light due to the removal of the amount of light in the geometric cross section also causes a scattering loss of the same amount of energy. This leads to the factor of 2 in Eq. (5.4-8). It is important to note that Eq. (5.4-8) is valid only when the index discontinuity is large enough $(1 \ll ka\,\delta n)$ to justify the complete removal of light falling into the geometric area via refraction. In most PD-LCDS with micrometer-size particles $(a \approx 1\ \mu m)$, where $ka\,\delta n$ is near unity, the scattering cross section must be obtained by using Eq. (5.4-6). In case when $ka\,\delta n \ll 1$, the scattering cross section can be written approximately as

$$\sigma = 2\pi a^2 (ka\,\delta n)^2 \qquad\qquad (5.4\text{-}9)$$

Figure 5.39 plots the normalized scattering cross section $(\sigma/2\pi a^2)$ as a function of the parameter $ka\,\delta n$. We note that the normalized cross section increases as a quadratic function according to Eq. (5.4-9) when $ka\,\delta n$ is small. The normalized cross section oscillates around 1 when $ka\,\delta n$ exceeds 2. This figure can be employed to determine the optimum droplet size for maximum scattering cross section. According to this figure, maximum scattering cross section occurs at around $ka\,\delta n = 2$.

The attenuation of a beam of light in a scattering medium is given by

$$I(z) = I(0)\exp\{-\alpha z\} \qquad\qquad (5.4\text{-}10)$$

where $I(z)$ is the intensity at position z; the attenuation coefficient α is related to the scattering cross section by

$$\alpha = N\sigma \qquad\qquad (5.4\text{-}11)$$

where N is the number of droplets per unit volume. The transmission in the

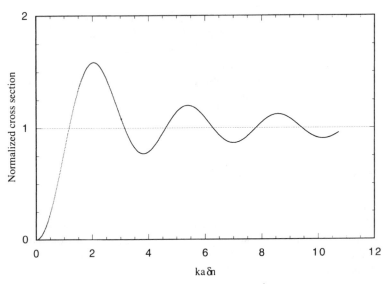

Figure 5.39. The normalized scattering cross section ($\sigma/2\pi a^2$) as a function of the parameter $ka\,\delta n$.

field-OFF state can thus be written

$$T(0) = \exp\{-\alpha d\} \tag{5.4-12}$$

where d is the thickness of the cell.

Example 5.3: 1-µm Droplets. Consider a droplet size of $a = 1\,\mu m$, with a concentration of $N = 0.1\,\mu m^{-3}$. This corresponds to a fill factor of about 40%. Using $\lambda = 0.55\,\mu m$, $\delta n = 0.1$, and Eqs. (5.4-6), we obtain a scattering cross section of $\sigma = 2\pi a^2(0.97) = 6.09\,\mu m^2$ and an attenuation coefficient of $\alpha = 0.609\,\mu m^{-1}$, which yields a transmission of $\exp(-6.09)$ in a cell $10\,\mu m$ thick, or $\exp(-12.2)$ in a cell $20\,\mu m$ thick.   ∎

Example 5.4: 2-µm Droplets. Consider a droplet size of $a = 2\,\mu m$, with a concentration of $N = 0.0125\,\mu m^{-3}$. This corresponds to the same fill factor of about 40%. Using $\lambda = 0.55\,\mu m$, $\delta n = 0.1$, and Eq. (5.4-6), we obtain a scattering cross section of $\sigma = 2\pi a^2(1.54) = 38.7\,\mu m^2$ and an attenuation coefficient of $\alpha = 0.484\,\mu m^{-1}$, which yields a transmission of $\exp(-4.84)$ in a cell $10\,\mu m$ thick, or $\exp(-9.68)$ in a cell $20\,\mu m$ thick.   ∎

So far, we have considered the transmission only at normal incidence (along the direction of the applied electric field). This is the situation when the beam of light is propagating along the c axis of the nematic LC droplets. For off-axis transmission, the LC droplets are optically birefringent. Thus, one polarization component will see a refractive index of n_o, while the other polarization component will see a refractive index of $n_e(\theta)$ given by

$$\frac{1}{n_e^2(\theta)} = \frac{\sin^2\theta}{n_e^2} + \frac{\cos^2\theta}{n_o^2} \qquad (5.4\text{-}13)$$

where θ is the angle between the c-axis and the direction of propagation in the droplet. Since the refractive index of the host medium is matched with n_o, the extraordinary polarization component of the incident beam will suffer from scattering loss at the boundary of the droplets as a result of the difference between n_p and $n_e(\theta)$ in the field-ON state. This leads to a decrease in the transmission and a degradation of the display at large viewing angles.

PD-LCDS offer several unique advantages, including high throughput due to the absence of polarizers, and simple cell construction due to the absence of alignment layers. They are particularly well suited for use in projection displays where high intensity operations are needed. The residual scattering in the bright state due to the extraordinary polarization component at large viewing angles is undesirable for direct-view displays. In the projection displays, the beam of light is often collimated. Thus, the scattering for off-axis light in the bright state is not a problem in projection displays. The PD-LCDs are also particularly useful for applications such as a variable light blocker, or a smart window whose transmission (from dark to clear) can be controlled electrically. Further discussion on the properties of PD-LCDs can be found in Doane [12].

5.5. REFLECTIVE LCDs

So far we have discussed LC displays only in the transmission configuration. In reflective LCDs, a mirror or reflector is placed at the back of the LC cell. Thus, the incident beam of light passes through the LC cell twice before exiting the device. Figure 5.40 shows a schematic drawing of reflective and transmissive modes of LCDs. We note that a rear mirror is an essential part in the reflective LCDs. Generally speaking, all transmissive LCDs discussed earlier can be employed for reflective LCDs. As a result of the double passage, only one polarizer is needed to obtain polarization interference to display the bright or dark state of reflection. On the other hand, reflective LCDs have one less degree of freedom in designing the normally white or normally black state. As a result, zero or unity reflectivity occurs only at specific values of Δnd in TN- or STN-based reflective LCDs. For example, normally black reflective LCDs based on a 45° TN cell with $R = 0$ in the field-OFF state is possible only when $\Delta nd/\lambda = 0.29$, 0.68, 1.23, and higher and with a polarizer azimuth angle of $-15.3°$, $34.3°$, $-39.1°$, and higher respectively. This issue was discussed in Section 4.6.

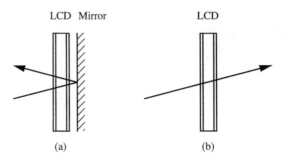

Figure 5.40. (*a*) A reflective LCD; (*b*) a transmissive LCD.

5.5.1. Reflection Properties at Normal Incidence

We now consider the reflectivity of a reflective LCD as a function of the applied voltage. Under the influence of an applied field, the LC director in the TN-LC cell undergoes a redistribution. The techniques described in Section 5.1 (or Appendix D) can be applied to obtain the director distribution functions $\theta(z)$ and $\phi(z)$. Once these functions are obtained, the Jones matrix method can then be employed to calculate the reflection properties. Again, we need to divide the LC layer into several thin sublayers (20–100). Each of these sublayers will be treated as a thin homogeneous birefringent plate with its own c axis orientation. Figure 5.41 shows the reflectivity $R(\lambda, V)$ of a reflective LCD based on a 45° TN-LC cell as a function of wavelength and applied voltage. The cell is designed to have a zero transmission at $\lambda = 0.55\,\mu m$ in the field-OFF state ($V = 0$). A slight leakage of light occurs at other wavelengths. As the applied voltage increases, the zero transmission occurs at a shorter wavelength (blue shift). Other than that, the transmission in the visible spectrum increases with the applied voltage. In the limiting case of a high applied voltage, the cell becomes homeotropically aligned with a zero phase retardation. As it has only one polarizer, the cell transmits all wavelengths of light.

5.5.2. Reflection Properties at Oblique Incidence

We now consider the reflection properties of reflective LCDs at large viewing angles (θ, ϕ). In the case when two polarizers are employed in the reflective LCDs (see Fig. 5.42b), the reflectivity can be easily obtained as follows:

$$R(\theta, \phi) = T(\theta, \phi)T(\theta, \phi + \pi) \tag{5.5-1}$$

where $T(\theta, \phi)$ is the transmission of the LCD without the mirror. The transmission $T(\theta, \phi)$ and $T(\theta, \phi + \pi)$ can be obtained by using the Jones matrix method.

The reflection properties of TN-based reflective LCDs using only one polarizer have been discussed in Section 4.6 for normal incidence. For a general

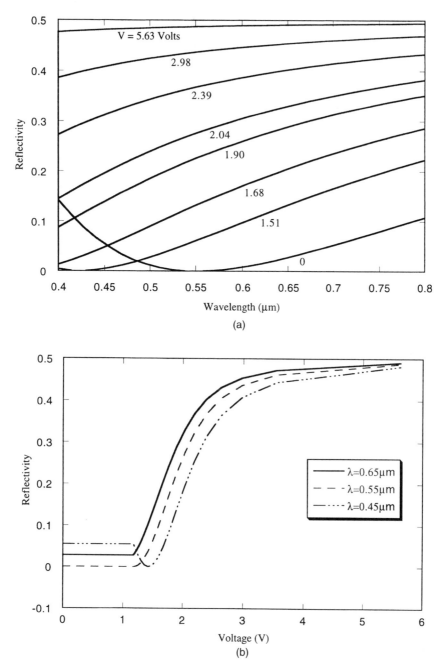

Figure 5.41. (a) The reflectivity spectrum $R(\lambda,V)$ of a reflective LCD at various applied voltages. (b) The reflectivity $R(\lambda, V)$ versus applied voltage V of a reflective LCD based on a 45° TN-LC cell, at $\lambda = 0.45$, 0.55, and 0.65 μm.

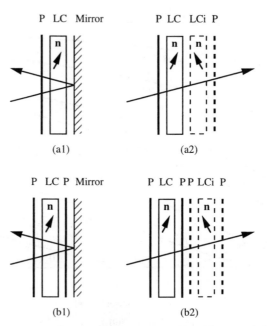

Figure 5.42. (*a*1) Schematic drawing of a reflective LCD with one polarizer (P = polarizer, LC = liquid crystal cell). (*a*2) An equivalent circuit for the calculation of the reflection properties of the display in (*a*1). LCi is the mirror image of the LC cell. Note that the mirror image of the director **n** is reversed in direction. In addition, the twist direction is also reversed in Lci. (*b*1) A reflective LCD with two polarizers. (*b*2) An equivalent circuit for the calculation of the reflection properties of the display in (*b*1).

angle of incidence (θ, ϕ), an equivalent circuit such as that shown in Figure 5.42 can be employed in conjunction with the Jones matrix method to calculate the reflectivity of the displays.

Using the general properties of the Jones matrix under mirror reflection derived in 4.1.3, we can express the reflection for the reflective LCD as

$$R(\theta, \phi) = |\mathbf{V} \cdot \tilde{M}(\theta, \phi + \pi)M(\theta, \phi)\mathbf{V}|^2 \qquad (5.5\text{-}2)$$

where M is the Jones matrix of the LC cell and $\mathbf{V}$ is the Jones vector of the incident beam as defined by the polarizer.

It is known that the transmission of NW TN-LCDs exhibits a left–right symmetry in its viewing angle characteristics. In other words

$$T(\theta, \phi) = T(\theta, \pi - \phi) \qquad \text{(NW TN-LCD)} \qquad (5.5\text{-}3)$$

The left–right symmetry in viewing is a desirable property of the displays. This symmetry, however, does not exist in a general reflective LCD based on TN or STN cells. In a reflective LCD an incident beam along the direction (θ, ϕ) is reflected into the direction $(\theta, \phi + \pi)$. The principle of reciprocity yields the following symmetry (twofold rotation symmetry):

$$R(\theta, \phi) = R(\theta, \phi + \pi) \tag{5.5-4}$$

This is a general property of reflective LCDs (i.e., a 180° rotation symmetry). The absence of left–right symmetry in a general reflective LCD is undesirable for direct-viewing applications. This is not a problem in projection displays. In a manner very similar to BA-LCDs, the reflective LCDs also exhibit a self-phase-compensation property. This leads to a wide viewing angle characteristics. By properly selecting the twist angle and the polarizer orientation, it is possible to design a reflective LCD which exhibits a left–right viewing symmetry. In fact, a display such as this will also exhibit a up–down viewing symmetry, according to Eq. (5.5-4).

5.6. PROJECTION DISPLAYS

In general, all LCDs described so far can be employed for projection displays. Most projection displays, especially for large screen projectors, require high brightness and high contrast ratios. Generally speaking, three LCDs are need for the three spectral regimes red, blue, and green in the projection displays. With the use of properly designed dielectric mirrors, the fundamental energy loss of $\frac{2}{3}$ due to color filters in direct-view displays can be eliminated or minimized. Referring to Figure 5.43, we consider a projection display system that involves the use of reflective mode LCDs. For projection displays, the incidence angles are usually relatively small (i.e., near normal incidence). Low loss polarizing beamsplitters (PBS) are employed to improve the optical energy throughput (brightness). The incident beam of light is split into three spectral components— R, G, and B—using spectral reflectors (dichroic mirrors). In fact, two high quality spectral filters for the blue and green spectral regimes and a mirror are adequate for the purpose. All the spectral components are polarized by the polarizing beamsplitters. Each of the polarizing beamsplitters is oriented at the proper azimuth angle as determined by the recipe described in Section 4.6. For the purpose of description, let us assume that the LC cells are 45° TN-LC and the RGB wavelengths are 0.65, 0.55, and 0.45 μm. All the TN-LC cells are located at the focal plane of the projection lens. In the normally white transmission (NWT) operation ($T = 0.5$, or $R = 0$) for a 45° TN-LC cell, consider the example in the first row of Table 4.3 with a polarizing beamsplitter azimuth of $-15.26°$ and a Mauguin parameter of $u = 1.16$. To achieve a normally white operation with $T = 0.5$, the Δnd for the LC cells must be 0.19 μ, 0.16 μ, and 0.13 μm for the three wavelengths $\lambda = 0.65$, 0.55, and 0.45 μm, respectively.

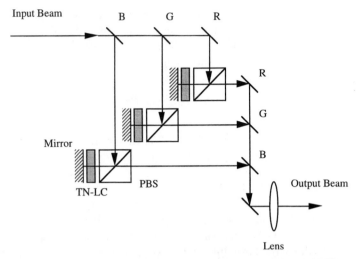

Figure 5.43. A projection display system based on reflective TN-LCDs.

If 90° TN-LC cells are employed, the normally white transmission (NWT with $R = 0$) requires a polarizer azimuth angle of $-32°$ and a Mauguin parameter of $u = 2.35$. To achieve a NWT with $T = 0.5$, and Δnd for the LC cells must be 0.765, 0.647, and 0.53 μm for the three wavelengths $\lambda = 0.65$, 0.55, and 0.45 μm, respectively.

Transmissive LCDs can also be employed for projection displays. Figure 5.44 shows a projection display system involves the use of transmissive TN-LCDs. All the TN-LCDs are located at the focal plane of the projection lens. The sheet polarizers can now be replaced with low loss polarizing beamsplitters (PBS) to improve the brightness. Dichroic mirrors (spectral filters B, G, R, B′, G′, R′) in the blue, green and red spectral regimes, respectively, are used to reflect light

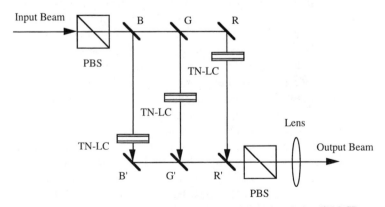

Figure 5.44. A projection display system based on transmissive TN-LCDs.

with corresponding wavelengths. In practice, some of the spectral reflectors (R and B$'$) can be replaced by mirrors to eliminate or minimize the energy loss due to dichroic filters.

5.7. OTHER DISPLAY SYSTEMS

Up to this point, all LCDs described in this book involve the use of nematic liquid crystals as the electrooptical medium to modulate the intensity of light. Actually, several other phases of the liquid crystals can also be employed for display applications. In the following, we discuss ferroelectric LCDs and cholesteric LCDs.

5.7.1. Ferroelectric LCD

The smectic $C(S_c)$ phase of LC materials can be employed for display applications. Figure 5.45 shows the molecular arrangement in the S_c phase. We note that the S_c phase consists of layers of LC molecules. The thicknesses of the layers are less than the molecular length. As a result, the molecules must tilt at an angle relative to the layer normal. Within each layer, the positions of the molecules are random with the long molecular axis $\mathbf{n}$ inclined at an angle θ with respect to the layer normal ($\mathbf{k}$). Because the angle θ is fixed, the molecular orientation is confined in a cone with a half-apex angle of θ. If the azimuth angle ϕ is fixed within a layer and remains the same throughout all the layers, the structure has a symmetry of the monoclinic class (see Table 3.1b), with a plane of symmetry formed by the vector $\mathbf{n}$ and the unit vector $\mathbf{k}$. The structure also exhibits a twofold rotational symmetry C_2 with the axis of symmetry parallel to $\mathbf{n} \times \mathbf{k}$. In other words, the LC medium is invariant under a rotation of 180°

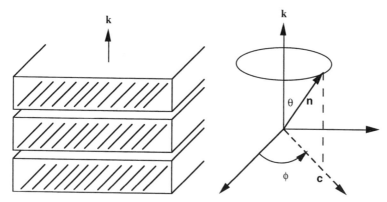

Figure 5.45.   The structure of smectic C phase of liquid crystals, where $\mathbf{k}$ is a unit vector normal to the molecular layer, θ is the angle between the molecular axis $\mathbf{n}$ and the layer normal $\mathbf{k}$, and ϕ is the azimuth angle of the molecular axis in the plane of the layer.

around the axis of symmetry along the direction $\mathbf{n} \times \mathbf{k}$. this symmetry is a result of an equal population distribution of molecules with axes in $\mathbf{n}$ and $-\mathbf{n}$.

Generally speaking, compounds exhibiting the S_c phase have a transverse component of the permanent electric dipole moments. In other words, the molecular dipole moment is not parallel to the molecular axis $\mathbf{n}$. However, the net dipole moment in a layer is zero due to the inversion symmetry and C_2 symmetry. The inversion symmetry here is a result of an equal population distribution of nonchiral molecules with axes in $\mathbf{n}$ and $-\mathbf{n}$. This is no longer true in a smectic C^* phase. If the liquid crystal in the smectic phase is doped with molecules with chirality, the structure distorts when the director precesses around the $\mathbf{k}$ axis leading to a helical configuration. The inversion symmetry and the mirror reflection symmetry are destroyed by the chirality. This is known as the smectic C^* phase. The pitch of the helix is often several thousand times the layer thickness. The optical properties of this structure are similar to those of a cholesteric liquid crystal where the local dielectric tensor in each layer is biaxial. In addition, a net dipole moment exists in each layer. As a result of the C_2 symmetry, the net dipole moment is always parallel to the C_2 axis that is in the direction of $\mathbf{n} \times \mathbf{k}$. Note that this net dipole moment is perpendicular to the molecular axis $\mathbf{n}$.

When an electric field is applied normal to the helical axis, the structure gets distorted. Above a critical field given by

$$E_c = \frac{\pi^4}{4} \frac{k_2}{Pp^2} \tag{5.7-1}$$

where p is the pitch of the undistorted helix, k_2 is the 'twist' elastic constant, and P is the polarization; the helix is completely unwound and the sample is poled. This leads to a homogeneously aligned LC cell with the director in a plane perpendicular to the electric field. When the field is reversed, the polarization vector is also reversed. The reversal of the polarization corresponds to a rotation of the molecular axis from (θ, ϕ) to $(\theta, \phi + \pi)$. It appears that the long molecular axis is rotated by an angle of 2θ.

Surface Stabilized Ferroelectric Liquid Crystal (SSF-LC) Cells

Clark and Lagerwall proposed the idea of using a very thin cell such that the boundary condition imposed by the alignment layers would be strong enough to suppress the helix [13]. The helical structure is indeed suppressed in a thin slab with a thickness less than the pitch of the helix, resulting in a homogeneously parallel alignment of the molecular axis. This is known as the "bookshelf"geometry, resembling a bookshelf with all the books tilting toward the same direction (see Fig. 5.46), with $\mathbf{k}$ parallel to the y axis). The addition of chirality destroys the inversion symmetry and the mirror reflection symmetry. Thus, if the molecules possess a transverse component of the permanent electric dipole moment, a net polarization may exist along the C_2 axis (z axis in

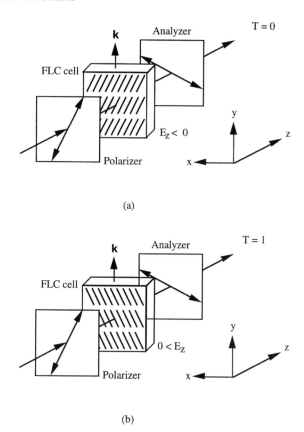

(a)

(b)

Figure 5.46. Schematic drawing of a ferroelectric LC cell for display applications. (*a*) When $E_z < 0$, the LC director is tilt toward upper right with an azimuth angle of π; (*b*) when $0 < E_z$, the LC director is tilt toward upper left with an azimuth angle of 0.

Fig. 5.46) in a layer. As a result of the C_2 symmetry, the net polarization is zero in the plane (*xy* plane in Fig. 5.46) perpendicular to the C_2 axis of symmetry. This is very similar to the spontaneous polarization in ferroelectric crystals. Note that this net polarization vector (dipole moment per unit volume) is perpendicular to the long molecular axis **n** and parallel to the layer. Thus, if the director **n** is in the *xy* plane (forced by the boundary condition), then the net polarization vector is along the *z* axis (see Fig. 5.46). This net polarization can point either up or down along the *z* axis, depending on whether the azimuth angle (measured from the *x* axis) is π or 0. In other words, the structure is bistable. An electric field along the *z* axis can be employed to switch the polarization. The switching rotates the director orientation around **k** from (θ, π) to $(\theta, 0)$ and vice versa. Here the angle θ is measured from the vector **k** that is parallel to the *y* axis in Figure 5.46. Although the molecular axes are actually rotated around the *y* axis (**k**) by an angle π, they appear like a rotation of the molecular axes around the *z* axis by an angle of 2θ.

For display applications, a surface stabilized F-LC cell is sandwiched between a pair of crossed polarizers. The transmission axis of the entrance polarizer is aligned parallel to the F-LC director in one of the bistable states. This corresponds to the black state (zero transmission; see Fig. 5.46a). When a proper electric field is applied, the polarization of F-LC is switched. The switching of the polarization leads to a reorientation of the molecular axis by an angle of 2θ relative to the polarizer. The transmission becomes

$$T = \frac{1}{2}\sin^2 4\theta \sin^2 \frac{\Gamma}{2} \tag{5.7-2}$$

where Γ is the phase retardation of the F-LC cell. By a proper choice of the cell thickness, it is possible to obtain a phase retardation of $\Gamma = \pi$. With a tilt angle of $\theta = 22.5°$, Eq. (5.7-2) yields a maximum transmission of $T = 0.5$. Thus, the transmission can be switched from 0 to 0.5 by reversing the polarity of the applied voltage (see Fig. 5.46).

It is important to note that F-LCD described in Figure 5.46 is a bistable device. In the operation, an applied electric field is employed to switch the molecular orientations from (a) to (b). Once the switching is done, the molecular orientation can be sustained without an applied electric field. Thus, the applied electric field can be removed without affecting the transmission of the device.

5.7.2. Cholesteric LCD

Cholesteric liquid crystals can also be employed for display applications. In Chapter 7 we will discuss the optical properties of cholesteric liquid crystals in the equilibrium planar state. In the equilibrium planar state (P) the medium exhibits a uniform twist of the director that is perpendicular to the helical axis. As a result of the uniform twist, the cholesteric liquid crystal is a periodic medium. For display applications, we consider a thin layer of cholesteric liquid crystal sandwiched between a pair of glass substrates. The helical axis of the liquid crystal is perpendicular to the substrates. It is known that selective reflection (Bragg reflection) occurs for light in the following wavelength range:

$$n_o p < \lambda < n_e p \tag{5.7-3}$$

where p is the pitch of the helix, and n_e, n_o are the principal refractive indices of the liquid crystal. The strong Bragg reflection may disappear if the helical structure in the cholesteric LC cell is destroyed or severely distorted.

In addition to the equilibrium planar state (P), there are several other states of the cholesteric liquid crystal in the cell. These include the homeotropic state (H), the focal conic state (FC), and some intermediate states (see Fig. 5.47). In a homeotropic state, the twist disappears and the LC director is uniformly aligned perpendicular to the substrates, resembling a vertically aligned (VA) nematic

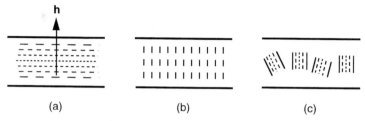

Figure 5.47. States (or textures) in cholesteric liquid crystals: (*a*) planar state; (*b*) homeotropic state; (*c*) focal conic state. The planar and focal conic states are stable at zero applied field. The homeotropic state can exist only with the presence of an applied electric field. Here **h** is a unit vector representing the helical axis.

cell. An applied electric field is needed to maintain the state. Thus, this state is often referred to as the field-induced nematic state. The Bragg reflection disappears completely in the homeotropic state. Thus, the cell appears optically like a plate of transparent medium. The focal conic state is similar to the planar state with the same intrinsic pitch of the helix. However, there are multiple domains in the focal conic state. Within each domain, the director is twisted with a helical axis approximately parallel to the substrates. At the domain boundary, the helical axis changes its azimuth angle abruptly. Forward scattering of light occurs in the focal conic state. If the rear substrate is painted black, then the focal conic state will appear dark. LCDs can be designed based on the different transmission and reflection properties of these states.

Transition from one state to another state can be achieved by applying an external electric field. For example, if an electric field is applied parallel to the substrates, a focal conic state can be transformed into a planar state. On the other hand, if an electric field is applied perpendicular to the substrates, a focal conic state can be transformed into a homeotropic state. These transitions can be employed for display applications. A cholesteric reflective display can be designed based on the transition between the focal conic state and the planar state. Since both of these states are stable, the device is bistable; in other words, the application of an external field can switch the states back and forth. In addition, the state can remain unchanged after the removal of the field. A cholesteric reflective or transmissive display can be designed based on the transition between the homeotropic state and the planar state. This would require an electric field perpendicular to the substrates. When the field is turned ON, the cholesteric liquid crystal cell is transformed into the homeotropic state, which transmits light. On removal of the field, the cholesteric LC cell returns to the planar state, which reflects light in selected spectral regimes.

5.8. SUMMARY

In summary, we have discussed the transmission properties of various liquid crystal displays, including TN-LCD, STN-LCD, N-LCD, VA-LCD, BA-LCD, reflective LCD, PD-LCD, ferroelectric LCD, and cholesteric LCD. It is important to note that LCD technology is developing very rapidly. New modes of operation and new cell configuration are constantly being invented and developed. The fundamental principles described in this chapter and the previous chapters will be very useful for the design and development of new LCDs. It is also important to note that there are other display systems that are not covered in this chapter, including plasma addressed LCDs. Interested readers are referred to the references and the proceedings of LC display conferences.

REFERENCES

1. C. H. Gooch and H. A. Tarry, *J. Phys. D: Appl. Phys.* **8**, 1575–1584 (1975).

2. T. Scheffer and J. Nehring, "Twisted nematic and supertwisted nematic mode LCDs," in *Liquid Crystals—Applications and Uses*, B. Bahadur, ed., World Scientific, 1993 Chapter 10, p. 231, and references cited therein.

3. T. J. Scheffer and J. Nehring, *J. Appl. Phys.* **58**, 3022 (1985).

4. F. Grandjean, *C. R. Acad. Sci.* (Paris) **172**, 71 (1921).

5. P. M. Alt and P. Pleshko, *IEEE Trans. Electron Devices* **ED-21**, 146–155 (1974).

6. M. Oh-e, M. Ohta, S. Aratani, and K. Kondo, *Digest Asia Display '95* 577 (1995).

7. M. Ohta, M. Oh-e, and K. Kondo, *Digest Asia Display '95* 68 (1995).

8. P. J. Bos and K. R. Koehler/Beran, *Mol. Cryst. Liq. Cryst.* **113**, 329–339 (1984).

9. See, for example, H. C. van de Hulst, *Light Scattering by Small Particles*, Wiley 1957, p. 176.

10. H. C. van de Hulst, *Recherches astron. obs. Utrecht* No. 11, Part 1 (1946).

11. R. Penndorf, "Total Mie scattering coefficients for spherical particles or refractive index $n \approx 1.0$," *J. Opt. Soc. Am.* **47**, 603 (1957).

12. J. William Doane, "Polymer dispersed liquid crystal displays," in *Liquid Crystals—Applications and Uses*, B. Bahadur, ed., World Scientific, 1993, Chapter 14, p. 361, and references cited therein.

13. N. A. Clark and S. T. Lagerwall, "Sub microsecond bistable electro-optic switching in liquid crystals," *Appl. Phys. Lett.* **36**, 899 (1980).

SUGGESTED READINGS

J. Dijon, "Ferroelectric LCDs," in *Liquid Crystals—Applications and Uses*, B. Bahadur, ed., World Scientific, 1993, Chapter 13, p. 305, and references cited therein.

T. Miyashita, Y. Yamaguchi, and T. Uchida, "Wide-viewing-angle display mode using bend-alignment LC cell," *Jpn. J. Appl. Phys.* **34**, Part 2, L177–179 (1995).

S.-T. Wu and C.-S. Wu, *Jpn. J. Appl. Phys.* **35**, Part I, 5349–5354 (1996).

C.-L. Kuo, C.-K. Wei, S.-T. Wu, and C.-S. Wu, "Reflective direct view display using a mixed-mode twisted nematic cell," *Jpn. J. Appl. Phys.* **36**, Part I, 1077 (1997).

S.-T. Wu and C.-S. Wu, *Appl. Phys. Lett.* **68**, 1455–1457 (1996).

E. Beynon, K. Saynor, M. Tillin, and M. Towler, *IDRC'97 Digest* L-34 (1997).

T. J. Scheffer and J. Nehring, *J. Appl. Phys.* **45**, 1021 (1984).

M. Schadt and F. Leenhouts, *Appl. Phys. Lett.* **50**, 236 (1987).

H. L. Ong, *J. Appl. Phys.* **64**, 4867 (1988).

H. S. Kwok, "Parameter space representation of liquid crystal display operating modes," *J. Appl. Phys.* **80**, 3687 (1996).

PROBLEMS

5.1. *Birefringent dielectrics between parallel electrodes*:

Consider a slab of unifrom anisotropic dielectric material sandwiched between a pair of parallel electrodes. Let ε be the dielectric tensor and d be the thickness of the slab. A voltage V is applied on the electrodes. Let z be the axis perpendicular to the electrodes, and assuming that the slab and the electrodes are of infinite extent.

(a) Show that $E_x = E_y = 0$ regardless of the orientation of the principal axes of the dielectric.

(b) Show that the displacement field vector **D** is, in general, not parallel to the z axis.

(c) Let the dielectric be a uniaxial crystal with its c axis oriented at angle θ relative to the z axis. Find the displacement field vector **D** in terms of E_z.

(d) Show that $E_x = E_y = 0$ even if $\varepsilon(z)$ is a function of z.

5.2. *Electrostatic energy of capacitors with birefringent dielectrics*:

Consider an empty capacitor made of a pair of electrodes. A slab of uniform anisotropic dielectric material (of dielectric tensor ε) is moved from infinity into the capacitor.

(a) If the electrodes are isolated from a power supply (or battery), the surface charge density σ on the electrodes is kept constant. Find the change in the electrostatic energy.

(b) Assume that the electrodes are connected to a power supply (or battery), so that the voltage drop between the electrodes is kept a constant. Find the net change in the electrostatic energy. (Note that there is an additional work done by the battery as the dielectric is moved into the capacitor.)

5.3. *Polarization ellipses in TN-LC*:

Consider the polarization states of transmitted wave through a TN-LC with a right-handed twist. Use $n_o = 1.5$, $n_e = 1.6$, $L = 10\,\mu m$, and $\lambda = 0.45,\ 0.55,\ 0.65\,\mu m$.

(a) Consider a linearly polarized incident light propagating along z axis. Letting θ be the angle between the **E** vector and the director (c axis) at

$z = 0$, find the ellipticity e of the transmitted wave as a function of θ. Plot e as a function of θ. Also find the sense of rotation of the transmitted wave as a function of θ.

(b) Consider the incidence of a right-handed circularly (RHC) polarized light along the z axis. Find the output polarization state. Discuss the possibility of converting RHC into LHC polarized light by using a TN-LC.

5.4. Consider a 90° TN-LC and a ($-90°$) TN-LC in series with a continuous c axis (director) at the interface. Find the output wave $\mathbf{E}_o$ as a function of the input wave $\mathbf{E}_i$ (propagating along the z axis). Use $n_o = 1.5$, $n_e = 1.6$, $L = 10\,\mu m$, and $\lambda = 0.45$, 0.55 0.65 μm. Compare the results with a 180° TN-LC.

5.5. Derive Eq. (A-16)—elastic energy density due to twist and tilt in a TN-LCD.

5.6. Derive Eqs. (B-7) and (B-22)—electrooptical distortion (tilt mode).

5.7. Derive Eq. (C-9)—electrooptical distortion (twist mode—IPS).

5.8. Derive Eqs. (D-20), and (D-25)—electrooptical distortion (TN-LCD).

5.9. Consider a 90° TN-LC and a ($-90°$) TN-LC in series with a continuous c axis (director) at the interface. The whole structure is sandwiched between a pair of parallel polarizers with the transmission axes parallel to the directors.

(a) Using $\Delta nd = 0.48\,\mu m$ for the cells, find the transmission for $\lambda = 0.45$, 0.5, 0.65 μm. Compare the results with a 180° TN-LC.

(b) Find the minimum cell thickness such that green light at $\lambda = 0.5\,\mu m$ has a zero transmission in the 180° TN-LCD. Find the residual transmission of blue and red light at $\lambda = 0.45$, 0.65 μm. Compare the result with the 90° + ($-90°$) TN-LCD.

5.10. Determine the symmetry of viewing in normally white TN-LCDs and STN-LCDs. The transmission of light through a TN-LCD as shown in Figure 5.1a is dependent on the incidence wavevector $\mathbf{k} = (k_x, k_y, k_z)$.

(a) Show that the TN-LCD shown in Figure 5.1a is invariant under a rotation of 180° around the x axis. Thus, the transmission exhibits the following symmetry: $T(k_x, k_y, k_z) = T(k_x, -k_y, -k_z)$

(b) Show that the principle of reciprocity can be written $T(k_x,k_y,k_z) = T(-k_x, -k_y, -k_z)$.

(c) Using the result in (a) and (b), show that $T(k_x,k_y,k_z) = T(-k_x,k_y,k_z)$, which implies that a left–right symmetry exists in the viewing.

(d) Show that the same symmetry in (c) can also exist in STN-LCD, provided the transmission axes of the polarizers are aligned symmetrically with respective to the bisector of the rubbing directions.

5.11. Calculate the leakage of light in a 90° TN-LCD at normal incidence. Consider a 90° TN-LCD in the normally black mode of operation. The Mauguin parameter should be $u = \sqrt{3}, \sqrt{15}, \sqrt{35}, \ldots$ to obtain a zero transmission (assuming ideal polarizers).

(a) Find the values of Δnd of the LC cell for the wavelength $\lambda = 0.55\,\mu m$.

(b) Find the transmission of the LCD for $\lambda = 0.45, 0.65\,\mu m$.

5.12. Determine the leakage of light in a 90° TN-LCD at normal incidence. Consider a 90° TN-LCD in the normally black mode of operation with a Mauguin parameter of $u = \sqrt{3}$. Let the transmission of the sheet polarizers be $T_1 = 0.8$ and $T_2 = 0.05$, where T_1 is the polarization parallel to the transmission axis of the polarizer, and T_2 is the polarization perpendicular to the transmission axis of the polarizer. Find the transmission of the TN-LCD.

5.13. *Minimum elastic energy density in TN-LC cells*:

(a) Show that in a TN cell with a uniform twist rate and zero tilt, the elastic energy density is

$$U_{EL} = \frac{1}{2}k_2\left(\frac{2\pi}{4d}\right)^2$$

where $d = $ cell gap.

(b) Show that, according to Eq. (5.1-11), in a TN cell with a uniform twist rate and non-zero-pretilt angle θ_0, the elastic energy density is minimum when $\theta(z) = \theta_0$.

5.14. *Difference between E and O modes of operation*:

(a) Show that for normal incidence the transmission of these two modes are identical.

(b) Show that, in general, the transmission of these two modes are different.

5.15. The BA-LCD shown in Figure 5.36 exhibits an exact rotation symmetry of 180° around the y axis C_{2y}. Using the procedure similar to those used in Problem 5.10, show that the transmission exhibits an exact vertical viewing symmetry. An approximate horizontal viewing symmetry also exists in the same BA-LCD. Show that the slight asymmetry in the horizontal viewing is due to the polarizers.

6

Matrix Addressing, Colors, and Properties of LCDs

In Chapter 5, we described the principle of operation of an individual LC cell for display applications. To control the information display, the LC cell must be electrically addressed. For the display of high content information (e.g., TV and computer monitors), we need a 2D array (say, $M \times N$) of LC pixels. Each pixel consists of a small LC cell that can be turned ON or OFF electrically. In principle, this can be achieved by using $M \times N$ electrical connections. However, for practical purposes, the electrical addressing is achieved by using multiplexing techniques, which requires only $M + N$ electrodes. The addressing of the display pixels can be achieved by two different methods. They are multiplexed addressing (passive) and the active matrix (AM) addressing. In this chapter, we will discuss these two methods of addressing, as well as some important properties of LCDs.

6.1. MULTIPLEXED DISPLAYS

As a result of the multiplexing, the voltages applied at one cell cannot be arbitrarily changed without affecting the applied voltages at other cells. This is known as *crosstalk*. As a consequence, the effective ON and OFF voltages applied at any of the cells cannot be too different. This severely limits the contrast ratio in display applications. In this section, we discuss the effect of the crosstalk.

Referring to Figure 6.1, we consider 2D array $(M \times N)$ LCD pixels that are addressed by N rows and M columns of electrodes. In the multiplexing, a positive voltage V_s is applied to one row at a time, starting from the first row. Focusing on the first row at the moment, we find that the M elements in the first row can be turned ON or OFF depending on the voltages applied to each element. Let V_d or $-V_d$ be the voltages applied to the elements (assume a positive V_d). The instantaneous voltage drops at the pixel electrodes in the selected row are

ON state: $\quad V = V_s - (-V_d)$
OFF state: $\quad V = V_s - V_d$

We note that a negative voltage $-V_d$ at the mth column and a positive voltage V_s at the nth row can produce an applied voltage of $V_s + V_d$ across the LC cell. This

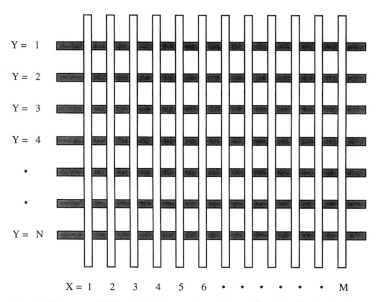

Y = 1
Y = 2
Y = 3
Y = 4
•
•
Y = N

X = 1 2 3 4 5 6 • • • • • • M

Figure 6.1. A 2D array of LC pixels that can be addressed by $M \times N$ electrical connections using N rows and M columns of electrodes.

creates an ON pixel at the (m,n) location. On the other hand, voltage drops also exist at other pixels. These include the voltage drop of $\pm V_d$ at all pixels in the nonselected row, and the voltage drop of $(V_s - V_d)$ at the OFF pixels of the selected row.

It is known that liquid crystals do not respond to the applied electric field instantaneously, because of the finite time needed in the reorientation of the directors. Thus, the actual molecular orientations depend on the root mean square (rms) average of the applied voltage. In each multiplexing cycle, each row is selected only during $1/N$ of the cycle time T. Thus, the rms voltage of the ON and OFF states are

$$V^2_{\text{ON}} = \frac{1}{N}(V_s + V_d)^2 + \frac{N-1}{N}(V_d)^2 \qquad (6.1\text{-}1)$$

$$V^2_{\text{OFF}} = \frac{1}{N}(V_s - V_d)^2 + \frac{N-1}{N}(V_d)^2 \qquad (6.1\text{-}2)$$

Subtracting Eq. (6.1-2) from Eq. (6.1-1), we obtain

$$\frac{4V_d V_s}{N} = V^2_{\text{ON}} - V^2_{\text{OFF}} \qquad (6.1\text{-}3)$$

Adding Eqs. (6.1-1) and (6.1-2), we obtain

$$V_{\text{ON}}^2 + V_{\text{OFF}}^2 = 2V_d^2 + \frac{2}{N}V_s^2 \qquad (6.1\text{-}4)$$

According to Eq. (6.1-3), we find that the difference of the rms voltages of the ON state and the OFF state is near zero when N is large. This may lead to a poor contrast in the display.

In what follows, we find the optimum value of N given the desired rms voltages V_{ON} and V_{OFF}. By eliminating V_s from Eqs. (6.1-3,4), we obtain

$$N = 8V_d^2 \frac{V_{\text{ON}}^2 + V_{\text{OFF}}^2}{\left(V_{\text{ON}}^2 - V_{\text{OFF}}^2\right)^2} - \frac{16V_d^4}{\left(V_{\text{ON}}^2 - V_{\text{OFF}}^2\right)^2} \qquad (6.1\text{-}5)$$

To find the optimum value for N, we differentiate Eq. (6.1-5) with respect to V_d while keeping V_{ON} and V_{OFF} constant. This leads to

$$V_d = \tfrac{1}{2}\sqrt{V_{\text{ON}}^2 + V_{\text{OFF}}^2} \qquad (6.1\text{-}6)$$

with

$$N_{\max} = \frac{\left(V_{\text{ON}}^2 + V_{\text{OFF}}^2\right)^2}{\left(V_{\text{ON}}^2 - V_{\text{OFF}}^2\right)^2} \qquad (6.1\text{-}7)$$

According to this equation, the maximum number of rows in multiplexing is limited by the difference between the desired rms voltages V_{ON} and V_{OFF}. Equation (6.1-7) can also be written

$$\frac{V_{\text{ON}}}{V_{\text{OFF}}} = \sqrt{\frac{\sqrt{N_{\max}} + 1}{\sqrt{N_{\max}} - 1}} \qquad (6.1\text{-}8)$$

where we note that the ratio of the rms voltages (also known as the *selection ratio*) approaches unity when $N_{\max}$ is very large, leading to a poor contrast. This equation was first obtained by Alt and Pleshko [1] in 1974, and is often referred to as the "iron law of multiplexing." For $1 \ll N_{\max}$, this equation can be written approximately as

$$\frac{V_{\text{ON}}}{V_{\text{OFF}}} = 1 + \frac{1}{\sqrt{N_{\max}}} \qquad (6.1\text{-}9)$$

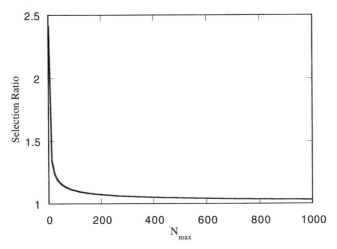

Figure 6.2. Selection ratio of multiplexed addressing versus column number [Eq. (6.1-8)].

For a column number of $N_{max} = 400$, Eq. (6.1-9) yields a selection ratio of only 1.05, which means that the rms voltage of the ON state is only 5% more than that of the OFF state. Figure 6.2 shows the selection ratio as a function of the row number. We note that the selection ratio approaches unity at large N_{max}.

The ratio V_d/V_s is often referred to as the *bias ratio*. At the optimum multiplexing, the bias ratio is given by, according to Eqs. (6.1-3), (6.1-6), and (6.1-7)

$$\frac{V_d}{V_s} = \frac{1}{\sqrt{N_{max}}} \qquad (6.1\text{-}10)$$

Example 6.1. Consider a multiplexed LC display of 600×400 pixels. Assuming $V_{OFF} = 2.0\,\text{V}$, we obtain the following rms ON voltage, according to Eq. (6.1-8): $V_{ON} = 2.10\,\text{V}$, and $V_s = 29.0\,\text{V}, V_d = 1.45\,\text{V}$. We note that the difference between V_{ON} and V_{OFF} is very small.   ■

For practical applications in LCDs, it is important that

$$V_{OFF} \le V_{th} \qquad (6.1\text{-}11)$$

$$V_{ON} \ge V_{th} + \Delta \qquad (6.1\text{-}12)$$

where V_{th} is the threshold voltage needed to induce a reorientation of the LC director and Δ is a transition voltage. From the operational point of view, it is desirable to define a device parameter

$$P \equiv \frac{\Delta}{V_{th}} \qquad (6.1\text{-}13)$$

which is a measure of the nonlinearity of the electrooptical characteristics of the device. This parameter completely specifies the maximum scanned lines and the corresponding voltages. According to Eqs. (6.1-9) and (6.1-11)–(6.1-13), we obtain

$$P \equiv \frac{\Delta}{V_{\text{th}}} \leq \frac{1}{\sqrt{N_{\text{max}}}} \qquad (6.1\text{-}14)$$

A technique known as "dual scan" is often employed in multiplexed displays. In a dual-scan operation, an upper column driver provides the voltages (V_d or $-V_d$) for the upper $N/2$ rows, and a lower column driver provides the voltages for the lower $N/2$ rows. The dual scan improves the bias ratios or the device parameter P by a factor of $\sqrt{2}$.

6.2.   ACTIVE-MATRIX DISPLAYS

As we discussed earlier, the multiplexing technique of addressing (also known as *passive matrix*) leads to a poor contrast as the number of rows N increases. Although the use of STN liquid crystals can provide a significant improvement in the contrast when N is large, severe limitations still exist in cell performance. These include response time, viewing angles, and gray-scale stability. These problems can be eliminated by using active-matrix addressing. The performance of active-matrix (AM) LC cells is compared to that of simple multiplexed LC cells in Table 6.1. As the manufacturing techniques continue to improve, it is now recognized that the ultimate solution for high quality, large area, high-information-content, color gray-scale display applications requires the use of AM addressing. The discussion that follows is based on a classic book chapter by F. C. Luo [2], to which the reader is referred to for more details.

6.2.1.   Principle of TFT Operation

An active-matrix liquid crystal display (AM-LCD) incorporates a 2D array (matrix) of circuits to provide the electrical addressing of the individual pixels.

Table 6.1.   Performance of AM LC Cells and Simple Multiplexed LC Cells

	Active-Matrix LC cells	Simple Multiplexed LC cells
LC mode	TN	STN
Contrast ratio (CR)	100 : 1	15 : 1
Viewing angles—horizontal	$(-60°, +60°)$	$(-30°, +30°)$
Viewing angles—vertical	$(-30°, +45°)$	$(-25°, +25°)$
Response time	30–50 m sec	150 m sec
Multiplexed lines	>1000	400
Gray scale	>16	8

The AM circuit has an active device in each pixel element defined by the crossover of the row and the column bus lines. The active devices in typical AM-LCDs are thin film transistors (TFTs). A simple matrix circuit can incorporate a transistor and a capacitor in each of the pixel elements for addressing of the display. Electrical charges can be stored in the capacitors to maintain a steady state voltage across the electrodes. Being isolated from other pixels by the transistors, the voltages remain constant while other pixel elements are being addressed electrically. Thus, the contrast ratio in AM-LCDs is not subject to the Alt-Pleshko limitation, Eq. (6.1-8). In what follows, we describe briefly the principle of operation of the active matrix (AM) LCDs, especially the TFT-LCDs.

Referring to Figure 6.3, we consider a color TFT-LCD that consists of a TFT matrix circuit on a glass substrate and a color filter glass plate with the LC layer sandwiched in between. Each pixel element is defined by the row and column bus lines in the TFT matrix circuit. Figure 6.4a is a schematic diagram of the TFT array circuit for the active-matrix addressing of LCD. A detailed electrical circuit diagram of each pixel element is shown in Figure 6.4b. We note that each pixel element has one TFT and a LC capacitor formed between the ITO transparent output electrode on the TFT matrix circuit and the ITO backplane electrode on the top glass substrate/color filter with the LC layer as the insulator (see also Fig. 6.5a).

Figure 6.5a is a schematic drawing of a side view of the TFT-LCD module. We note that the top glass plate contains pixel color filters and a black matrix, which are then covered with a passivation layer and the ITO common electrode. A thin layer of polyimide is then coated as the alignment layer. The main purpose of the black matrix is to improve the display quality by blocking the stray light passing through the gap between the pixel electrodes. The black matrix also protects the semiconductor layer from external optical illumination,

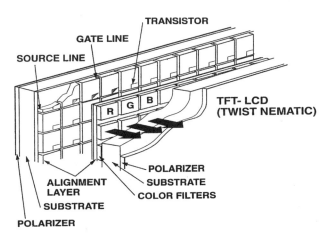

Figure 6.3. Schematic drawing of a TFT-LCD; R, G, and B stand for color filters (red, green, and blue).

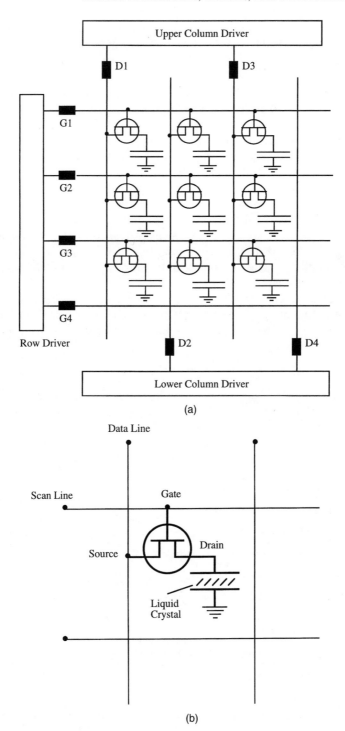

Figure 6.4. (*a*) Schematic drawing of TFT array circuit for LCD; (*b*) electrical circuit diagram of one pixel element.

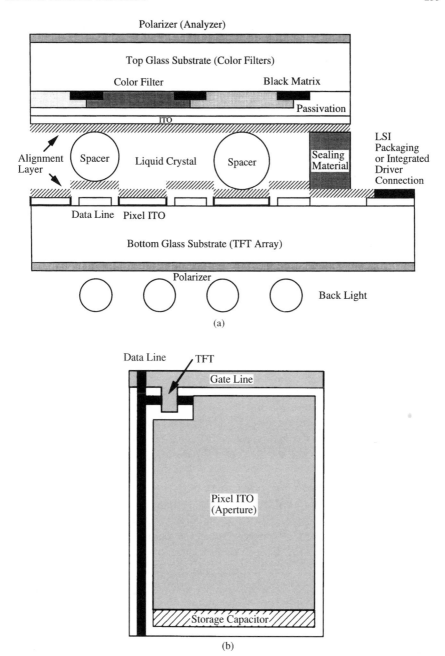

Figure 6.5. (*a*) Schematic drawing of a side view of AM-LCD module; (*b*) schematic drawing of a typical pixel layout in a-Si TFT-LCD.

which can affect the conductivity in the TFTs via photoconductivity of the semiconductor. We also note that the cell gap is kept uniform by using transparent spacer balls (or fibers). Figure 6.5b is a schematic drawing of a typical layout of a pixel in a-Si (amorphous silicon) TFT-LCD. We note that the useful area for light transmission (aperture) is only a fraction of the total pixel area, due to the real estate needed for the driving circuits, storage capacitor, and scan and data bus lines.

The principle of operation is described as follows. Again, the cells are addressed one line at a time. When a row (scan or gate line) is addressed, a positive voltage pulse (of duration T/N (where N is the number of rows and T is the frame time) is applied to the line turning on all the transistors along the row. The transistors act as switches transferring electrical charges to the LC capacitors from the respective columns (data or source line). When addressing other rows, a negative voltage is applied to the gate lines turning OFF all the transistors along the line and holding the electrical charges in the LC capacitors for one frame time T until the line is addressed again. If the LC material used in the cell is twisted nematic (TN), it is desirable to use AC voltage to drive the LC element. This is often achieved by switching the polarity of the data voltage in alternate frames. An example of the driving voltage waveforms of both the gate and source lines, and the voltage on the pixel electrode is shown in Figure 6.6 [3, 4].

The operation of the TFT in LCD can be divided into four steps in time [4]:

1. At time 1 (see Fig. 6.6a), during the addressing time of an odd frame (at the end of the frame), a positive voltage of duration T/N is applied to the gate line V_G turning on the TFT. Thus, the LC pixel electrode (ITO) is charged from $V_p = -0.9 V_{ON}$ at time 1 to greater than $V_p = 0.9 V_{ON}$ at time 2 within the time duration of T/N, due to the positive source voltage $V_{SD} = V_{ON}$.

2. At time 2, the gate voltage becomes negative, turning OFF the TFT and simultaneously the source voltage V_{SD} changes from $+V_{ON}$ to $-V_{ON}$. During the time period between 2 and 3 with a duration of $(N-1)T/N$, the voltage on the LC pixel electrode V_P remains about $0.9 V_{ON}$ as the LC capacitor is now isolated from the data lines.

3. At time 3 (the next addressing time), the TFT is turned ON again by applying a positive gate voltage of duration T/N. The LC capacitor now sees a negative source-to-drain voltage $V_{SD} = -V_{ON}$. Thus, the pixel electrode is charged (actually discharged) from $V_p = +0.9 V_{ON}$ at time 3 to less than $V_p = -0.9 V_{ON}$ at time 4 within the time duration of T/N.

4. At time 4, the TFT is turned OFF by the negative gate voltage and simultaneously the source voltage V_{SD} changes from $-V_{ON}$ to $+V_{ON}$.

This completes the cycle of the four steps in time. We note that the voltage on the LC pixel electrode V_P does not remain constant during the duration $(N-1)T/N$ because of a slight leakage of current at the LC cell. The leakage of

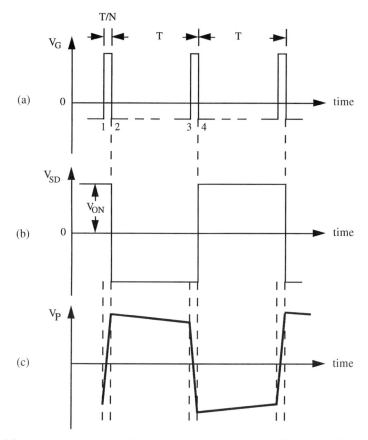

Figure 6.6. An example of the driving voltage waveforms: (*a*) applied to one of the gate bus lines; (*b*) applied to one of the source bus lines; (*c*) on the LC pixel electrode [3,4].

current is due to a very small residual conductivity of the LC material. We note that the pixel voltage V_p across the LC cell in the field-ON state is actually alternating in time. This is necessary to maintain the domain stability in the TN cell. The director of the liquid crystal responds to the root mean square of the pixel voltage V_p. To turn off a pixel, we simply apply a zero source-to-drain voltage $(V_{SD} = 0)$ to the bus line. This removes all the charges in the LC capacitor, leading to a zero pixel voltage $(V_p = 0)$.

The TFT device in AM-LCDs works like a metal oxide semiconductor field effect transistor (MOSFET). A typical MOSFET consists of a conducting channel (e.g., n-type semiconductor) between two ohmic contacts, called *source* and *drain*, respectively; the width of the conducting channel is controlled electrically by a rectifying contact called the *gate*; the gate voltage controls the current flowing between source and drain. The metal gate is separated from the conducting channel by an oxide insulting layer. This insulating oxide layer ensures an almost zero gate current. When an appropriate voltage is applied to the gate, an n-type conducting channel is induced. If a voltage V_{SD} is applied between the source and the drain, a current I_{SD} will flow from source to drain, and the magnitude of this current is controlled by the gate voltage V_G. The conducting channel can be completely cut off by reverse-biasing the gate. This feature is particularly useful for the operation of AM-LCDs. For detail characteristics of MOSFET, the reader is referred to Reference [5].

The current and voltage characteristics of an MOSFET based TFT can be written [2,5]

$$I_{SD} = \frac{W\mu C_G}{L} V_{SD} \left(V_G - V_T - \frac{V_{SD}}{2} \right) \qquad \text{when} \qquad V_{SD} < (V_G - V_T)$$

$$I_{SD} = \frac{W\mu C_G}{2L} (V_G - V_T)^2 \qquad \text{when} \qquad V_{SD} > (V_G - V_T)$$

$$(6.2\text{-}1)$$

where W = width
L = length of conducting channel of TFT
μ = effective mobility of semiconductor
C_G = gate capacitance per unit area
V_{SD} = source–drain voltage
V_G = gate voltage
I_{SD} = source–drain current
V_T = threshold voltage of the TFT.

In an MOSFET based TFT, a conducting channel is induced when the gate voltage is larger than the threshold voltage. We note that the TFT device is in the ohmic regime when $V_{SD} < (V_G - V_T)$. In this regime, the current is approximately proportional to the voltage. We also note that the current I_{SD} is practically independent of the voltage V_{SD} when $V_{SD} > (V_G - V_T)$. This is the so-called pinchoff or saturation regime, and the current is called the *saturation current*.

We now consider the requirement on the characteristics of the TFT for AM-LCD operations. The electrical current I_{ON} of the TFT to charge the LC cell, which acts like a capacitor, during a time period of T/N must satisfy the following relation

$$2K_1 C_{LC} V_{ON} < \frac{T}{N} I_{ON} \qquad (6.2\text{-}2)$$

where K_1 = an engineering operation constant
V_{ON} = ON voltage of the LC element
C_{LC} = capacitance of the LC element
N = number of addressed lines (rows) in the display
T = frame time.

For bilevel operations such as an alphanumeric or graphical displays, the requirement on the pixel voltage is not very tight since all voltages above the ON voltage of the LC element will be acceptable. The constant K_1 can be a value between 1 and 10 to allow for any nonuniformity in the TFTs in the panel and the fact that the ON current of the TFT is not a constant during the charging of the pixel. For a 6.25×6.25-in. panel with 1024×1024 pixel elements, I_{ON} should be greater than 1.09×10^{-6} A if we pick K_1 to be 5. For gray-scale operations, the RC (resistance $\times$ capacitance) time constant is required to be at least one third of the charging period to obtain less than 5% error; also K_1 should be greater than 10. In this case, I_{ON} should be greater than 2.18×10^{-6} A.

During the OFF period, the TFTs are turned off. The OFF currents (leakage) of the TFTs should be low enough so that the charge stored in the LC capacitors will not leak away to affect the appearance of the panel. The leakage current is generally attributed to thermionic field emission and residual thermal charge carriers. For bilevel display panels, there are two separate conditions. For OFF pixels, the voltage on the pixels should not increase to a level exceeding the threshold voltage of the LC material and turn the pixels partially ON. For the ON pixels, the voltage on the pixels should not drop below a voltage at which the pixels appear partially ON or OFF. For typical LC materials, this voltage change ΔV should be smaller than 1.5 V. The requirement on the OFF current (leakage current) of the TFT is thus

$$I_{OFF} T < K_2 \Delta V C_{LC} \qquad (6.2\text{-}3)$$

where T is the frame time, K_2 is another engineering operation constant that allows for the uniformity variations of TFT characteristics in a TFT array and the temperature variations of the OFF currents of the TFTs. For a typical a-Si TFT, the OFF current can increase by a factor of 10 when the temperature is raised from 20 to 70°C. Again using the 1024×1024 array of LC cells as an example, $\Delta V = 1.5$ V and $K_2 = 0.1$, I_{OFF} should be less than 3.28 pA. For gray-scale operations, ΔV should be less than one gray level, or 0.08 V for a 16-level display panels, I_{OFF} should be less than 0.175 pA.

The charge stored in the LC capacitors can also leak through the RC relaxation due to a small residual conductivity of the LC materials (i.e., the resistance R is not infinite). For a 5% error of the rms voltage on the LC capacitors, the RC time constant of the LC materials is required to be 10 times the frame time of addressing. For a display with a 16-ms frame time, the RC time constant should be longer than 160 ms. When considering the requirements of elevated temperature performance of the cell, the room temperature RC time constant should be over 1 s. Typical resistivities of TN-LC materials are in the range from 10^{11} to 10^{12} $\Omega \cdot$cm. The actual resistivity of TN-LC in a display cell may be affected by the alignment polyimide material, the buffing material, or the cleaning procedure after buffing and the sealing epoxy. It is important to note that great care in handling all these areas needs to be exercised to ensure optimized LCD performance.

In order to reduce the severe requirements on the TFT OFF current I_{OFF} and the LC cell's RC time constant, a storage capacitor can be added in parallel to each pixel element. This leads to a larger capacitance C_{LC} that relaxes the OFF current condition according to Eq. (6.2-3), and also increases the RC time constant. The storage capacitor can be either a general ground capacitor or a capacitor formed between the ITO output electrode and the following scan line. The additional capacitor leads to an increased fabrication complexity and a reduced yield. In addition, the storage capacitor takes up real estate of the pixel, leading to a smaller aperture ratio. For LCDs with moderate or low resolutions (e.g., ≤ 200 lines per inch), the storage capacitor is desirable but not required. For high resolution displays with 500 lines or more per inch, the storage capacitor is considered to be essential. In AM-LCD technology using polysilicon TFTs, the leakage current can also be reduced by replacing a single TFT with multiple TFTs in series connected to a common gate [6,7].

In a TFT circuit, there are many parasitic capacitances, including capacitance C_{GD} due to the physical overlap of the gate and the drain electrodes. There also exists a finite capacitance C_{GP} between the gate bus line and the ITO output electrode. Some of the more important parasitic capacitances are shown in Figure 6.7. In most TFTs for LCDs, $C_{GP} \ll C_{GD}$. During the addressing of a line, when the scan line voltage is dropped from V_{ON} to V_{OFF} as shown in Figure 6.8, a negative voltage shift of

$$\Delta V_P = \Delta V_G \frac{C_{GD}}{C_{\text{LC}} + C_{GD}} \qquad (6.2\text{-}4)$$

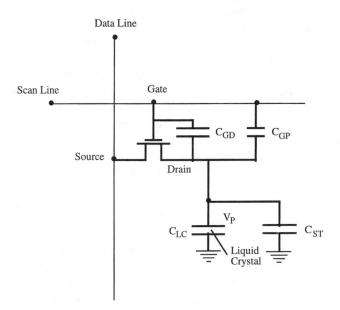

Figure 6.7. Parasitic capacitances in a TFT circuit.

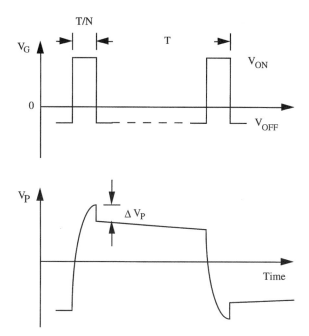

Figure 6.8. Voltage shift in the pixel voltage caused by the scan line voltage change and the presence of parasitic capacitance.

results, where $\Delta V_G = V_{ON} - V_{OFF}$. The voltage shift ΔV_P can lead to the problem of flicker and the degradation of the gray-scale performance. Unfortunately, this voltage shift cannot be compensated by using a simple voltage shift of the backplane electrode, because (1) C_{GD} is not always uniform throughout the TFT array and (2) C_{LC} is dependent on the voltage applied to the LC cell element due to the intrinsic dielectric anisotropy of the material. The voltage shift ΔV_P can, however, be reduced by the addition of a storage capacitor C_{ST} in each pixel. A voltage compensation scheme can be developed to achieve a symmetrical driving voltage across the LC cell. The discussion of the compensation scheme is beyond the scope of this book. Interested readers are referred to Reference 10. There are other parasitic capacitance (not shown in Fig. 6.7), including the source–drain capacitance C_{SD} of the TFT, the capacitance between the data line, and the ITO pixel electrode C_{SP}. The capacitance coupling from the data voltage can lead to gray scale voltage distortion. Again, electrical addressing techniques can be developed to reduce the distortion. Interested readers are referred to references 2, 8, and 9 and references cited therein.

6.2.2. Array Fabrication

Amorphous silicon (a-Si), polycrystalline silicon (poly-Si), and cadmium selenide (CdSe) are by far the most widely investigated semiconductor materials for application in AM-LCDs. Each of the materials has its own unique fabrication processes. For instance, CdSe TFTs have the longest history of technological development. Although CdSe TFTs have all the required characteristics to drive the matrix circuits and the peripheral circuits, large scale high resolution color displays have not yet been demonstrated. On the other hand, a-Si TFTs have been the most popular and widely developed materials, possibly because of their low OFF current (I_{OFF}) and adequate ON current (I_{ON}) characteristics and the convenience of using the same processing and fabrication equipment as Si-MOS devices. Polysilicon (poly-Si) TFTs are currently being actively developed for future AM-LCDs. By using poly-Si materials, the peripheral driving circuits can be integrated with the TFT matrix circuit on the same substrate. In addition, poly-Si materials offer a higher electron (and hole) mobility, potentially up to $300 \, cm^2/V \cdot s$. The higher mobility of poly-Si TFT translates into smaller transistors (higher pixel density) and storage capacitors, hence a higher aperture ratio than an a-Si TFT-LCD of comparable drive current and pixel size [11,12]. Crystalline-Si (c-Si) materials are also being investigated for future TFT-LCD applications. Higher resolution [>2000 dpi (dots/inch)] and electron (and hole) mobility are important features of c-Si. Table 6.2 summaries a comparison of the a-Si, poly-Si, and CdSe TFT materials.

TFT Structures

Generally speaking, two types of TFT structure are commonly used in display applications. Figure 6.9a shows a bottom gate (inverted staggered) structure, as the gate is at the bottom and the source–drain electrodes are on the top side of the semiconductor. This structure has been used extensively in a-Si TFT arrays. Figure 6.9b shows a top gate (coplanar) structure, as the gate is on the top and all electrodes (gate, source, drain) are on the same side of the semiconductor. This structure is quite popular for poly-Si TFT arrays. Both structures have been used in CdSe TFT arrays.

Table 6.2. Comparison of the Three TFT Materials

	Amorphous Si	Poly-Si	CdSe
Mobility ($cm^2/V \cdot s$)	0.3–1	10–300	25–150
Substrate	Hard glass / sodalime	Hard glass / quartz	Hard glass / sodalime
Processing temperature	<300°C	600–1000°C	<350°C
ON/OFF ratio	10^5–10^7	10^6	10^5–10^7
OFF current	<1 pA	1–10 pA	1–10 pA
Integrated circuit	Difficult	Product	Lab demo
Largest size	14 in.	9.5 in.	9.5 in.
Most pixels	1100×1440	480×960	400×600

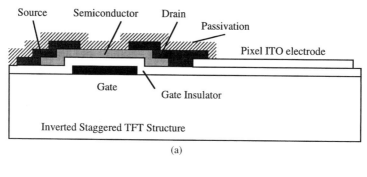

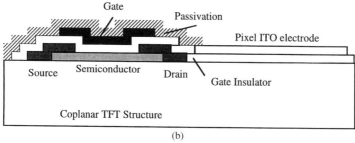

Figure 6.9. (*a*) An inverted staggered TFT structure; (*b*) a coplanar TFT structure.

TFT Types

a-Si TFT. The development of a-Si TFT for AM-LCDs has benefited significantly from the early advancement of silicon technology for a-Si solar cells in the 1970s. The extensive knowledge on the deposition techniques and electrical characteristics accumulated during these years can be readily applied to TFT-LCD applications. By virtue of its intrinsically high resistivity, a-Si TFT arrays have been very successful in AM-LCD applications. Although there have been rapid developments in the fabrication processes, they can be generally divided into two types: A and B. For TFT-LCDs, the substrate must be a thin and transparent flat plate with a strain point at least 100°C above the maximum processing temperature, alkaline-free and enough structure integrity, such as Corning 7059 glass. The substrate must be able to sustain the TFT-LCD processing cycles (different temperature, pressure, chemical environments) without significant chemical attack, shrinkage, or thermal mismatch to ensure the manufacturability. The gate material can be conducting metals such as Ti, Ti-Mo, Ta, and Mo–Ta, which can be deposited by DC magnetron sputtering. Both Ta and Mo–Ta gates have been widely used as the materials can be easily anodized to yield a pinhole-free gate insulator. These two fabrication processes for bottom gate structures are discussed as follows:

Type A fabrication process

1. Deposition of a pattern of gate metal for gate delineation.
2. Deposition of a sandwich of SiN_x (gate insulator), i-a-Si (semiconductor), and n^+ a-Si (source, drain) layers in one vacuum pump down by using the plasma-enhanced chemical vapor deposition (PECVD) process.
3. Etching of n^+ a-Si and i-a-Si layers into islands for individual TFTs.
4. Deposition and delineation of source, drain, and ITO electrodes.
5. Etching of n^+ a-Si layer to form the conducting channels for the TFTs using the source–drain electrodes as the mask.
6. Etching (timed or differential) of n^+ a-Si layer to reveal the intrinsic a-Si layer.
7. Deposition of a thin passivation layer.

Type B fabrication process

1. Deposition of a pattern of gate metal for gate delineation.
2. Deposition of a sandwich of SiN_x (gate insulator), i-a-Si (semiconductor), and SiN_x (passivation) layers in one vacuum pump down by using the same PECVD process.
3. Etching of top SiN_x and i-a-Si layers into islands for individual TFTs.
4. Etching of the top SiN_x layer to expose i-a-Si layer surface for source and drain regions.
5. Deposition of thin layers of n^+ a-Si and source–drain metal (Cr, Mo, Cr–Al, Ti) in sequence.
6. Etching and patterning of source–drain electrodes.
7. Deposition of a thin ITO layer by sputtering.
8. Etching to form the LC output electrodes.

We note that process A requires only four mask levels, whereas process B requires five mask levels. However, the etch stop of the n^+ a-Si layer (step 6) in process A is difficult to control. As a result, the i-a-Si layer cannot be reduced to a few hundred angstroms to minimize the photoconductivity effect. Thus, the photocurrent of the TFT obtained in this process is in general higher than that of process B. Since the passivation layer is not deposited in the same pump down as the i-a-Si layer, the OFF (leakage) current of the TFT obtained in this process is, in general, higher than those obtained in process B. We also note that process B requires a very careful cleaning of the interface before the deposition of the n^+ a-Si layer (step 5) to ensure good ohmic contact for the TFTs. An ON/OFF current ratio greater than 10^6 and mobility of $0.3–1\,cm^2/V\cdot s$ have been achieved.

Poly-Si TFT

The fabrication processes of poly-Si TFTs are generally divided into two paths according to the maximum processing temperatures and the substrate size [11]. The first path leads to high temperature (HT) devices for high resolution view finders and projection displays, with a maximum processing temperature above 900°C. The second path is aimed at low temperature (LT) devices to be processed on large glass substrate of about 40 cm [12,13], with a maximum processing temperature below 600°C. Currently, the LT process has been the focus of most development efforts for reasons of cost-effectiveness. As for the device configurations, there are again, top gate and bottom gate structures. The top gate configuration with the self-aligned ion implantation architecture based on low pressure chemical vapor deposition (LPCVD)-deposited Si thin films is popular among MOS manufacturers. The bottom gate configuration using crystallized PECVD a-Si:H as the active material is preferred among a-Si TFT-LCD manufacturers for the apparent compatibility with large area a-Si TFT technology. Poly-Si layers are typically based on solid-phase crystallization (SPC) of a-Si layers deposited by the low pressure chemical vapor deposition (LPCVD) at around 600°C. The substrate can be either quartz or hard glass depending on whether a thermally grown oxide or a deposited low temperature oxide is used as the gate insulator. As mentioned earlier, top gate (coplanar) structures are often used for poly-Si TFTs. An example of the fabrication process of a top gate low temperature (LT) poly-Si TFT structure is described as follows [2]:

1. Deposition of a 150 nm-thick-phosphorous-doped poly-Si layer (a-Si with solid-phase crystallization) on a hard glass substrate.
2. Etching and patterning to form the source–drain electrodes.
3. Deposition of a very thin (25-nm) layer of undoped poly-Si layer by LPCVD at 600°C.
4. Etching to form TFT.
5. Deposition of ITO layer.
6. Etching to form the ITO data lines and pixel ITO output electrodes.
7. Deposition of the gate insulator (150 nm SiO_2) using thermal CVD.
8. Sputtering and delineation of gate electrode (Cr).

For the high temperature poly-Si TFT, the gate insulator is thermally grown at around 1000°C and the source–drain doping is accomplished by self-aligned ion implantation. Self-aligned-gate processes have the advantage of minimizing the gate-to-drain electrode overlapping. Figure 6.10 shows an example of a self-aligned-gate process. In the self-aligned process, a poly-Si layer is deposited and patterned into islands for TFT. Then a layer of SiO_2 and a layer of poly-Si are deposited in sequence. The stack is then etched to expose the two ends of each poly-Si island. After the etching, the exposed poly-Si areas are treated with ion

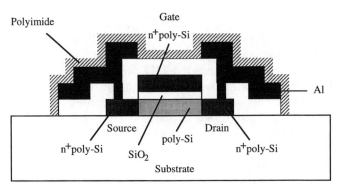

Figure 6.10.   A self-aligned gate poly-Si TFT fabrication process.

implantation, converting the poly-Si into n^+-doped poly-Si. The doping process by the ion implantation creates the gate, source, and drain, where the top poly-Si layer is the gate electrode. In addition, the top poly-Si layer (gate) along with the SiO_2 layer underneath act as a mask defining the source and drain regions of the TFT, thus achieving the process of self-alignment. After depositing another layer of SiO_2 and via hole etching, Al is deposited and delineated to serve as the ohmic contact for the source–drain electrodes. Finally, ITO is sputter deposited and etched as output electrodes.

As mentioned earlier, poly-Si materials offer the capability of integrating the TFT arrays with the driver circuits and the peripheral circuits. Another promising aspect of poly-Si TFT is the possibility of converting a-Si films into poly-Si films by using laser annealing (LA). By using this laser-assisted conversion, both poly-Si and a-Si can be processed on the same substrate where a-Si TFT is the matrix array and poly-Si is the driver circuits. The integration can significantly reduce the required display interconnections and increase the compactness and the reliability of the display. The significantly higher electron (and hole) mobility in poly-Si offers the possibility of a higher addressing speed. According to Eq. (6.2-1), the source–drain current I_{SD} is proportional to the electron mobility μ. With a higher current, it takes less time to charge the LC capacitor, leading to the capability of higher operation frequency for the column driver.

CdSe TFT. CdSe has the advantages of being a low temperature processing material and having a relatively higher electron mobility of up to $150\,\mathrm{cm^2/V \cdot s}$. In an all-photolithography CdSe fabrication process for an inverted staggered structure, all layers are delineated by liftoff processes except the ITO layer, which was patterned by etching. The process is described as follows.

1. Deposition of a pattern of gate metal (Ni) for gate delineation by sputtering.
2. Deposition of gate insulator layer (Al_2O_3) by sputtering.

3. Deposition of a pattern of CdSe for TFT's.

4. Delineation of In–Au source and drain electrodes.

5. Deposition of a thin passivation layer.

6. Delineation of ITO electrodes.

CdSe TFTs can have an ON current of greater than 1 μA and an OFF currents of less than 1 pA. One of the potential advantages of the CdSe TFT process is the possibility of integrating the row and column driver circuits with the matrix circuits.

6.2.3. Cell Assembly

Here we discuss the assembly of the display panel of TFT-LCDs. As described earlier, the display panel consists of mainly two glass plates with LC material sandwiched in between [2,14]. One glass plate contains the TFT array, and the other contains the color filters and the common electrode. In the assembly, the inner surfaces of both glass plates are first coated with a thin layer of polyimide with a thickness ranging from a few hundreds to one thousand angstroms. The polyimide layers are then rubbed in prearranged directions for the alignment of the LC director. Next, a narrow strip of epoxy is either dispensed or printed on the boarders of the color filter glass plate to form a ring with a narrow opening. To maintain a uniform spacing (4–10 μm) between the glass plates, fiber glass spacer rods or plastic spacer beads are dispensed uniformly on the TFT glass plate. The uniform spacing can also be maintained by delineating a dyed polyimide post on each TFT in the array. We note that the post not only maintains the proper spacing, but also serves a light-blocking function. The two plates are then assembled together with the pixels in the TFT array properly aligned with the pixels in the color filters. In addition, the two plates are also properly oriented so that a desired twist angle (TN-LC) is obtained. After pressing and curing, the assembly is vacuum-filled with LC and plugged by a quick-setting epoxy. After the completion of the LC cell assembly process, the final step is performed with external components. Plastic polarizers are carefully adhered to both sides of the LC cell with the transmission axes aligned in the predetermined directions depending on the director twist. Finally, the LC cell is mounted on the circuit board or rigid substrate connected with the driving and peripheral circuits of LSI (large scale integration).

In the discussion above we described several TFT structures and some of the fabrication processes for a-Si, poly-Si, and CdSe TFTs and the cell assembly process. These discussions were intended to illustrate the basic operation principles. It is important to note that the fabrication technologies, device architectures and the assembly process are advancing very rapidly. It is important to keep an eye on the latest developments in these areas, including new materials and new fabrication processes.

6.3.   OPTICAL THROUGHPUT OF TFT-LCDs

The presence of the TFT array circuits, scan and data bus, polarizers, transparent electrodes, and other elements will affect the brightness or light throughput of the displays. In color displays, the presence of the color filters leads to a significant decrease of the brightness due to the absorption of light in the filters. In what follows, we consider the energy loss at each of the components.

6.3.1.   Polarizers

A beam of unpolarized light suffers 50% energy loss due to a perfect polarizer. In practice, sheet polarizers are used for flat panel displays. Most commercially available polarizers consist of polarizer materials such as HN22, HN32, HN38S, and HN42HE sandwiched between two thin high quality glass substrates. These polarizer materials are usually made of stretched PVA (poly vinyl alcohol) films containing highly concentrated iodine dyes. The polarizers for LCD applications have typical transmission around 45% or less for visible wavelengths. For example, a polarizer material, HN42HE, manufactured by Polaroid corporation transmits roughly 42% of an unpolarized input beam. The deviation of 8% from perfect polarizers is due to residual absorption for the transmitted polarization component. For display applications, the polarizer materials are cemented to the outer surfaces of the glass plates of the LC cell.

Example 6.2.   Consider a LCD using polarizers with a maximum transmission of 42% for unpolarized light. The energy loss due to the polarizer is 58%, whereas the energy loss due to the analyzer is about 8%. Thus, the total energy loss due to the polarizers alone is 62% ($= 1 - 0.42 \times 0.92$).

The loss of 3 dB (50%) of energy at the first polarizer is very undesirable. Using sheet polarizers, the 50% of the energy is absorbed by the polarizer. There is no possibility of recovering the energy loss. However, if polarizing beamsplitters (PBS) are used, 50% of the energy is deflected out of the viewing field. The 50% of the energy deflected may be recycled by using optical techniques. This may improve the optical energy throughput. Unfortunately, ordinary polarizing beam splitters are often bulky and unsuitable for flat panel displays. However, for projection displays, these polarizing beamsplitters can be very important in improving the optical energy throughput. Fig. 6.11 describes an example of the use of polarizing beam splitters for projection displays [15].

                                                                                          ■

Referring to Figure 6.11, we consider a Sagnac ring interferometer consisting of a polarizing beamsplitter (PBS) and three mirrors (M). In practice, two mirrors are adequate. An incident beam of unpolarized light, entering at port A is split into two which are then directed toward each other via the mirrors. The beams recombine at the same PBS and then exit the interferometer at port B. A TN-LC cell without polarizers is inserted in the beam path between the mirrors. In the field-OFF state, the TN-LC cell with the proper orientation serves as a

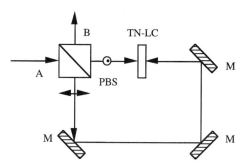

Figure 6.11. A schematic drawing of the use of polarizing beamsplitters in a Sagnac interferometer configuration for projection displays.

polarization rotator, rotating the polarization state by 90°. As a result, the beams recombine at the PBS and exit the interferometer through port A, in the direction opposite to the incident beam. In the field-ON state, the LC molecules are homeotropically aligned, and the cell exhibits no phase retardation to the beams. As a result the polarization states of the beams remain unchanged. Thus, the beams recombine at the PBS and exit the interferometer through port B. In this configuration, there is no energy loss due to the polarizer.

The use of cholesteric liquid crystal (CLC) as polarizers for LCDs has been proposed. A thin sheet of CLC can work as a polarizer that can separate right-handed circularly (RHC) polarized waves from left-handed circularly (LHC) polarized waves within a spectral regime of interest. Specifically, a CLC-based polarizer can reflect LHC polarized light and transmit RHC polarized light. The reflected light can be recycled by using a mirror behind the backlight, leading to a higher energy throughput. This concept is illustrated in Figure 6.12.

Unpolarized light emitted from the backlight is split into two parts of circularly polarized light. The RHC polarized light is transmitted through the CLC cell and then converted into linear polarized light by using a quarter-wave plate. The LHC polarized light is reflected at the CLC cell. The reflected LHC light is then reflected by the rear mirror. On reflection, the light becomes RHC polarized, and can then transmit through the CLC cell. This is the basic principle of recycling the light by using CLC polarizers. The polarizing properties of CLC will be discussed in detail in Chapter 7.

In addition to the energy loss due to the polarizers, the transmission of light through the LC cell suffers energy loss due to the presence of the black matrix and storage capacitor (aperture ratio), residual absorption of the ITO layers, and scan and data bus lines. As a result the typical light throughput is around 20% for monochrome and 5% for color LCD. Color filters are the major source of energy loss in color LCDs. Each color filter absorbs at least 67% (two-thirds) of the energy entering into that particular color pixel. The energy loss and the typical fraction of light transmission are listed in Table 6.3, where we note that the major source of energy loss in the color filters.

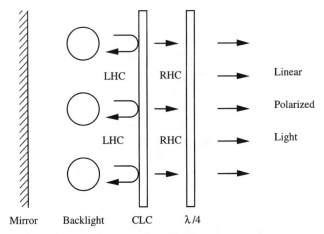

Figure 6.12. CLC polarizers for throughput improvement in LCD applications.

Table 6.3.   Transmission of Various LCD Components

Energy Loss	Transmission (%)
Polarizer	43
Color filters	25
TFT aperture ratio	80
Liquid Crystal	95
Analyzer	80
Overall (color LCDs)	5

6.3.2. Color Filters

Many types of color filters have been reported for the color LCDs. These include dyed gelatin, dyed polyimide, and colored inks. Dyed gelatin is by far the most widely used material for color filters. Currently, photolithography is employed to generate small color pixels. At first, a thin layer (2–3 μm thick) of base material (e.g., glue, gelatin, PVA, polyimide) that has dye acceptance property is coated on the glass substrate. The base is then covered with a thin layer of photoresist. Selected areas formed by photoexposure with photomasks are then opened for red dyeing by dipping into the red dye solution. After photoresist is removed, this procedure is repeated for green and blue dyes. Finally, the surface of the dyed base material is coated with a passivation layer (polyimide), thus isolating the LC molecules from the color filter layer and providing a smooth surface for the deposition of the ITO transparent electrode. Precision printing with RGB inks is a cost-effective way to eliminate the photolithography technique.

Generally speaking, three color filters are needed for the display of all colors: red, green, and blue filters. If we assume that the optical energy is evenly distributed in these three spectral regimes, then it is clear that each color filter can transmit only one-third of the total optical energy. In other words, two-thirds of the energy is absorbed by the color filters, if the filters are based on absorption. The transmission bandwidth of the color filter is usually very wide to allow a higher brightness at the expense of a lower color purity. In practice, the color filters have a bell-shape spectral transmission curve centered at their respective colors that match with the peaks of the RGB-enhanced backlighting. Thus the color filters can only transmit less than one-third of the total energy. This explains the typical transmission of only 20% as shown in Table 6.3.

The use of diffractive gratings or interference filters for color LCDs has been proposed. The basic principles of recycling the energy using either diffractive gratings or interference filters are depicted in Figure 6.13.

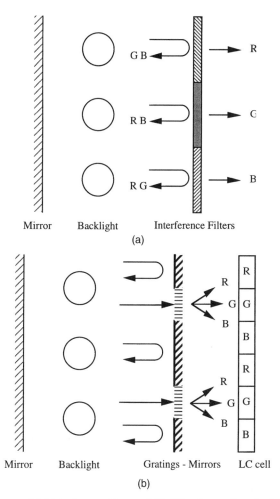

Figure 6.13. The use of (*a*) interference filters and (*b*) diffractive gratings for improving the light throughput in LCDs.

Figure 6.13*a* shows a schematic drawing of the use of interference filters for improving the light throughput in LCDs. Optical interference filters based on multilayer thin films can be designed to transmit the desired colors in R,G,B spectral regimes of interest. The optical energy outside the transmission band will be reflected by the interference filters. The reflected energy is then recycled by using a rear mirror.

Figure 6.13*b* shows a schematic drawing of the use of diffractive gratings for improving the light throughput in LCDs. By using a grating–mirror combination in which about one-third of the area is occupied by diffractive gratings. Thus, one-third of the incident light is diffracted into R,G,B beams, whereas two-thirds of the incident light is reflected. The one-third open aperture for the gratings is important to ensure the color separation in space. The reflected light is then recycled by using a rear mirror.

6.4. COLORS IN LCDs

From the fundamental point of view, any source of color light can be described by its spectral distribution. This can be easily obtained experimentally by using a spectrometer such as a diffraction grating or a prism. Human eyes, however, do not function like a spectrometer. It is generally believed that human eyes contain three different pigments (say, R,G,B) which receive the light with different absorption spectra, so that R pigments absorb strongly in the red region, G pigments absorb strongly in the green, and B pigments absorb strongly in the blue. Thus, when a beam of light is directed into the eyes, we perceive a color that is the result of the mixing of the amount of absorption received by the three different sets of pigments. As a result of the mechanism of color vision of human eyes, different spectral distribution can produce the same color. In addition, any color can be made from three different colors (e.g., red, green, and blue). Furthermore these fundamental laws of colors can be described mathematically [16].

The discussion in the previous chapters, suggests that TN-LCD can be employed to modulate the transmission of light. In black-and-white (B/W) displays, the transmission properties of the LCD must be insensitive to wavelength. This can be achieved by using TN-LCD with Mauguin parameters in the regime $1 \ll u$. For color displays, the easiest way to achieve colors is to use color filters in series with B/W displays. Using three spectral filters (red, green and blue) for each pixel, we are able to obtain different colors by controlling the intensity transmission of each of the three subpixels. As a result of the color mixing, we are able to generate almost all colors of interest.

In most color LCDs, each pixel of information consists of three subpixels with color filters. The color perceived by our eyes is a mixture (sum) of the three colors in the pixel. Letting the intensity transmission for each of the subpixels be t_1, t_2, t_3, and the transmission spectra of the three color filters be

$f_1(\lambda), f_2(\lambda), f_3(\lambda)$, the effective transmission spectrum of the pixel can be written

$$T(\lambda) = t_1 f_1(\lambda) + t_2 f_2(\lambda) + t_3 f_3(\lambda) \qquad (6.4\text{-}1)$$

Here we assume that t_1, t_2, t_3 are insensitive to the variation of the wavelength. Strictly speaking, t_1, t_2, t_3 are functions of wavelength. However, the spectral dependence of $T(\lambda)$ is dominated by the transmission of the color filters $f_1(\lambda), f_2(\lambda)$, and $f_3(\lambda)$. In color LCDs, t_1, t_2, t_3 can be independently controlled by applying different voltages to each subpixel. Once $T(\lambda)$ is obtained, it is possible to calculate the chromaticity coordinates of the color transmitted by the pixel. In what follows, we discuss the calculation of the chromaticity coordinates and the mixing of colors.

To calculate the chromaticity coordinates (x,y), we must first obtain the *tristimulus values* (X, Y, Z) by using the following formulas [17,18],

$$X = k \int S(\lambda) T(\lambda) \bar{x}(\lambda) d\lambda \qquad (6.4\text{-}2)$$

$$Y = k \int S(\lambda) T(\lambda) \bar{y}(\lambda) d\lambda \qquad (6.4\text{-}3)$$

$$Z = k \int S(\lambda) T(\lambda) \bar{z}(\lambda) d(\lambda) \qquad (6.4\text{-}4)$$

where $\bar{x}(\lambda)$, $\bar{y}(\lambda)$, and $\bar{z}(\lambda)$ are the *color matching functions*, $S(\lambda)$ is the source spectrum (backlight illuminant), and $T(\lambda)$ is the transmission spectrum of the information pixel according to Eq. (6.4-1); k is the normalization factor defined as

$$k = \frac{100}{\int S(\lambda) \bar{y}(\lambda) d\lambda} \qquad (6.4\text{-}5)$$

This definition of k makes the Y tristimulus value equal to 100 for a perfect transmission system with $T(\lambda) = 1$ for all λ. The color matching functions for 10° viewing angles (CIE 1964 *Standard Colorimetric Observer*) are written as $\bar{x}_{10}(\lambda), \bar{y}_{10}(\lambda), \bar{z}_{10}(\lambda)$ whose spectral tristimulus values are tabulated in Table 6.4, and plotted as functions of the wavelength in Figure 6.14. The tristimulus values X_{10}, Y_{10}, Z_{10} are obtained if the color matching functions $\bar{x}_{10}(\lambda), \bar{y}_{10}(\lambda)$, $\bar{z}_{10}(\lambda)$ are used. The color matching functions of CIE 1931 *Standard Colorimetric Observer* are denoted $\bar{x}(\lambda), \bar{y}(\lambda), \bar{z}(\lambda)$ for 1–4° viewing angles.

Table 6.4. Spectral Tristimulus Values of Color Matching Functions for 10° Viewing Angles $\bar{x}_{10}(\lambda), \bar{y}_{10}(\lambda), \bar{z}_{10}(\lambda)^a$

λ (nm)	$\bar{x}_{10}(\lambda)$	$\bar{y}_{10}(\lambda)$	$\bar{z}_{10}(\lambda)$	x	y
380	0.00015995	0.00001736	0.00070478	0.18133	0.019685
385	0.00066244	0.00007156	0.00292780	0.18091	0.019542
390	0.00236160	0.00025340	0.01048220	0.18031	0.019348
395	0.00724230	0.00076850	0.03234400	0.17947	0.019044
400	0.01910970	0.00200440	0.08601090	0.17839	0.018711
405	0.04340000	0.00450900	0.19712000	0.17712	0.018402
410	0.08473600	0.00875600	0.38936600	0.17549	0.018134
415	0.14063799	0.01445600	0.65675998	0.17323	0.017806
420	0.20449200	0.02139100	0.97254199	0.17063	0.017849
425	0.26473701	0.02949700	1.28250000	0.1679	0.018708
430	0.31467900	0.03867600	1.55348000	0.16503	0.020283
435	0.35771900	0.04960200	1.79849990	0.16217	0.022487
440	0.38373399	0.06207700	1.96728000	0.15902	0.025725
445	0.38672599	0.07470400	2.02729990	0.15539	0.030017
450	0.37070200	0.08945600	1.99480000	0.151	0.036439
455	0.34295699	0.10625600	1.90070000	0.14594	0.045217
460	0.30227301	0.12820099	1.74537000	0.13892	0.05892
465	0.25408500	0.15276100	1.55490010	0.12952	0.07787
470	0.19561800	0.18519001	1.31756000	0.11518	0.10904
475	0.13234900	0.21994001	1.03020000	0.095732	0.15909
480	0.08050700	0.25358900	0.77212501	0.072777	0.22924
485	0.04107200	0.29766500	0.57006001	0.045194	0.32754
490	0.01617200	0.33913299	0.41525400	0.020987	0.44011
495	0.00513200	0.39537901	0.30235600	0.007302	0.56252
500	0.00381600	0.46077701	0.21850200	0.005586	0.67454
505	0.01544400	0.53135997	0.15924899	0.021874	0.75258
510	0.03746500	0.60674101	0.11204400	0.04954	0.8023
515	0.07135800	0.68566000	0.08224800	0.085024	0.81698
520	0.11774900	0.76175702	0.06070900	0.12524	0.81019
525	0.17295299	0.82332999	0.04305000	0.16641	0.79217
530	0.23649099	0.87521100	0.03045100	0.20706	0.76628
535	0.30421299	0.92381001	0.02058400	0.24364	0.73987
540	0.37677199	0.96198797	0.01367600	0.27859	0.7113
545	0.45158401	0.98220003	0.00791800	0.31323	0.68128
550	0.52982599	0.99176103	0.00398800	0.3473	0.65009
555	0.61605299	0.99910998	0.00109100	0.38116	0.61816
560	0.70522398	0.99734002	0.00000000	0.41421	0.58579
565	0.79383200	0.98237997	0.00000000	0.44692	0.55308
570	0.87865502	0.95555198	0.00000000	0.47904	0.52096
575	0.95116198	0.91517502	0.00000000	0.50964	0.49036
580	1.01416000	0.86893398	0.00000000	0.53856	0.46144
585	1.07430010	0.82562298	0.00000000	0.56544	0.43456
590	1.11852000	0.77740502	0.00000000	0.58996	0.41004
595	1.13430000	0.72035301	0.00000000	0.6116	0.3884
600	1.12399010	0.65834099	0.00000000	0.63063	0.36937

Table 6.4. (*Continued*)

λ (nm)	$\bar{x}_{10}(\lambda)$	$\bar{y}_{10}(\lambda)$	$\bar{z}_{10}(\lambda)$	x	y
605	1.08910000	0.59387797	0.00000000	0.64713	0.35287
610	1.03048000	0.52796298	0.00000000	0.66122	0.33878
615	0.95073998	0.46183401	0.00000000	0.67305	0.32695
620	0.85629702	0.39805701	0.00000000	0.68266	0.31734
625	0.75493002	0.33955401	0.00000000	0.68976	0.31024
630	0.64746702	0.28349301	0.00000000	0.69548	0.30452
635	0.53511000	0.22825401	0.00000000	0.70099	0.29901
640	0.43156701	0.17982800	0.00000000	0.70587	0.29413
645	0.34369001	0.14021100	0.00000000	0.71025	0.28975
650	0.26832899	0.10763300	0.00000000	0.71371	0.28629
655	0.20430000	0.08118700	0.00000000	0.71562	0.28438
660	0.15256800	0.06028100	0.00000000	0.71679	0.28321
665	0.11221000	0.04409600	0.00000000	0.71789	0.28211
670	0.08126060	0.03180040	0.00000000	0.71873	0.28127
675	0.05793000	0.02260170	0.00000000	0.71934	0.28066
680	0.04085080	0.01590510	0.00000000	0.71976	0.28024
685	0.02862300	0.01113030	0.00000000	0.72002	0.27998
690	0.01994130	0.00774880	0.00000000	0.72016	0.27984
695	0.01384200	0.00537510	0.00000000	0.7203	0.2797
700	0.00957688	0.00371774	0.00000000	0.72036	0.27964
705	0.00660520	0.00256456	0.00000000	0.72032	0.27968
710	0.00455263	0.00176847	0.00000000	0.72023	0.27977
715	0.00314470	0.00122239	0.00000000	0.72009	0.27991
720	0.00217496	0.00086419	0.00000000	0.71565	0.28435
725	0.00150570	0.00058644	0.00000000	0.71969	0.28031
730	0.00104476	0.00040741	0.00000000	0.71945	0.28055
735	0.00072745	0.00028404	0.00000000	0.71919	0.28081
740	0.00050826	0.00019873	0.00000000	0.71891	0.28109
745	0.00035638	0.00013955	0.00000000	0.71861	0.28139
750	0.00025097	0.00009843	0.00000000	0.71829	0.28171
755	0.00017773	0.00006982	0.00000000	0.71796	0.28204
760	0.00012639	0.00004974	0.00000000	0.71761	0.28239
765	0.00009015	0.00003554	0.00000000	0.71724	0.28276
770	0.00006453	0.00002549	0.00000000	0.71686	0.28314
775	0.00004634	0.00001834	0.00000000	0.71646	0.28354
780	0.00003341	0.00001325	0.00000000	0.71606	0.28394

[a] Where x, y are the chromaticity coordinates of monochromatic light.

The chromaticity coordinates of any given color are defined as follows:

$$x = \frac{X}{X+Y+Z}, \qquad x_{10} = \frac{X_{10}}{X_{10}+Y_{10}+Z_{10}} \qquad (6.4\text{-}6)$$

$$y = \frac{Y}{X+Y+Z}, \qquad y_{10} = \frac{Y_{10}}{X_{10}+Y_{10}+Z_{10}} \qquad (6.4\text{-}7)$$

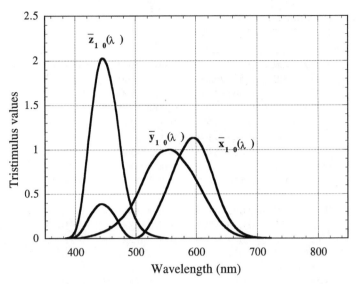

Figure 6.14. The spectral tristimulus values of the color matching functions for 10° viewing angles as functions of the wavelength.

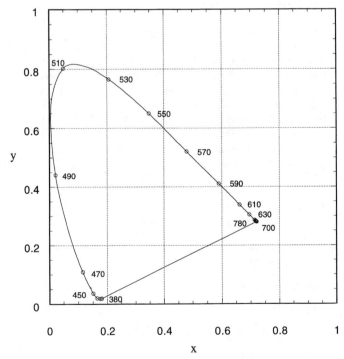

Figure 6.15. CIE 1964 (x,y) chromaticity diagram showing the spectrum locus of monochromatic stimuli from 380 to 780 nm, and the purple line.

Each color specification is represented by a point (x, y) in the chromaticity coordinate system. Figure 6.15 shows the curve by connecting all points of the chromaticity coordinate of monochromatic light, and the purple line, which is the straight line between 380 and 780 nm.

6.4.1.   Addition (or Mixing) of Colors and CIE 1976 Color Space ($L^*u^*v^*$)

Let (x_1, y_1) be the chromaticity coordinates of the color due to the color filter of the first subpixel, and similarly (x_2, y_2) for the color filter of the second subpixel, and (x_3, y_3) for the color filter of the third subpixel. Let (X_1, Y_1, Z_1), (X_2, Y_2, Z_2), (X_3, Y_3, Z_3) be the corresponding tristimulus values for the subpixels. It can be easily shown that the tristimulus values for the whole pixel can be written

$$X = t_1 X_1 + t_2 X_2 + t_3 X_3 \tag{6.4-8}$$

$$Y = t_1 Y_1 + t_2 Y_2 + t_3 Y_3 \tag{6.4-9}$$

$$Z = t_1 Z_1 + t_2 Z_2 + t_3 Z_3 \tag{6.4-10}$$

where t_1, t_2, and t_3 are the transmittance of the subpixels. We note that the resultant tristimulus values are weighted sums of the individual tristimulus values. The chromaticity coordinates for the whole pixel can thus be calculated by using Eqs. (6.4-6) and (6.4-7). If all the intensity transmission t_1, t_2, t_3 can be varied independently between 0 and 1, then all colors within the triangle formed by the three points (x_1, y_1), (x_2, y_2), and (x_3, y_3) can be obtained. If leakage of light exists at the dark state, then t_1, t_2, t_3 cannot be very close to zero. This may lead to the problem of color desaturation. When this happens, the color LCD cannot display some of the colors within the triangle formed by the three points (x_1, y_1), (x_2, y_2), and (x_3, y_3), especially the colors near the corners and the edges of the triangle.

In 1976, the CIE recommended CIE $L^*u^*v^*$ for applications dealing with additive mixtures of colored light such as the lighting and color TV industries. The same color space can be employed for LCDs. Given a chromaticity coordinate (x, y) for each gray-scale level of the pixel, we can calculate the coordinate (u', v') as follows:

$$u' = \frac{4x}{-2x + 12y + 3} \tag{6.4-11}$$

$$v' = \frac{9y}{-2x + 12y + 3} \tag{6.4-12}$$

The rectangular coordinates of CIE 1976 include L^* (metric lightness), u^*

(redness–greenness), and v^* (yellowness–blueness) scales, which are defined as follows;

$$L^* = 116\sqrt[3]{\frac{Y}{Y_0}} - 16 \qquad (6.4\text{-}13)$$

$$u^* = 13L^*(u' - u_0') \qquad (6.4\text{-}14)$$

$$v^* = 13L^*(v' - v_0') \qquad (6.4\text{-}15)$$

where u', v', Y and u_0', v_0', Y_0 are the (u', v') chromaticity coordinates and the luminance of the pixel under examination and those of a suitable reference (e.g., a chosen white background), respectively. For investigating the angular dependence of the color viewing in LCDs, u', v', Y are the (u', v') coordinates and luminance of the pixel under examination and u_0', v_0', Y_0 are the (u', v') coordinates and luminance of the backlight at normal incidence.

The polar coordinates of the CIE 1976 color space include S_{uv} (saturation), C_{uv} (metric chroma), and H_{uv} (hue angle) scales, which are calculated as follows:

$$S_{uv} = 13\sqrt{(u' - u_0')^2 + (v' - v_0')^2} \qquad (6.4\text{-}16)$$

$$C_{uv} = \sqrt{u^{*2} + v^{*2}} \qquad (6.4\text{-}17)$$

$$H_{uv} = \tan^{-1}\frac{v^*}{u^*} \qquad (6.4\text{-}18)$$

For the purpose of quantitatively comparing the difference between a pair of samples (L_2^*, u_2^*, v_2^*), (L_1^*, u_1^*, v_1^*), we define the following parameters in the CIE 1976 $(L^*u^*v^*)$ color space:

Color difference: $\quad \Delta E = \sqrt{(L_2^* - L_1^*)^2 + (u_2^* - u_1^*)^2 + (v_2^* - v_1^*)^2}$

$$(6.4\text{-}19)$$

Chromaticity difference: $\quad \Delta_{uv} = \sqrt{(u_2^* - u_1^*)^2 + (v_2^* - v_1^*)^2} \qquad (6.4\text{-}20)$

Chroma difference: $\quad \Delta C_{uv} = \sqrt{u_2^{*2} + v_2^{*2}} - \sqrt{u_1^{*2} + v_1^{*2}} \qquad (6.4\text{-}21)$

Hue angle difference: $\quad \Delta H_{uv} = \tan^{-1}\frac{v_2^*}{u_2^*} - \tan^{-1}\frac{v_1^*}{u_1^*} \qquad (6.4\text{-}22)$

For investigating the angular dependence of the color viewing in LCDs, u_2^*, v_2^*, Y_2 are the (u^*, v^*) chromaticity coordinates and luminance of the pixel under examination and u_1^*, v_1^*, Y_1 are the (u^*, v^*) chromaticity coordinates and luminance of the same pixel at normal incidence.

REFERENCES

1. P. M. Alt and P. Pleshko, "Scanning limitations of liquid crystal displays," *IEEE Trans. Electron Devices* **ED-21**, 146–155 (1974).

2. F. C. Luo, "Active matrix LC displays," in *Liquid Crystals—Applications and Uses*, B. Bahadur, ed., World Scientific, 1993, Chapter 15, page 397; and references therein.

3. F. C. Luo, I. Chen, and F. C. Genovese, *IEEE Trans. Electron Devices* **ED-28**, 740–743 (1981).

4. F. C. Luo, D. Pultorak, and E. Freeman, *IEEE Trans. Electron Devices* **ED-30**, 202–206 (1983).

5. See, for example, A. van der Ziel, *Solid State Physical Electronics*, Prentice-Hall, 1976.

6. S. Morozumi et al., *SID Digest '85* 278 (1985).

7. J. C. Sturm, I.-W. Wu, and M. Hack, *IEEE Trans. Eectron Devices* **ED-42**, 1561–1563 (1995).

8. F. C. Luo, "Active matrix liquid crystal displays: an overview," *Proc. SPIE*, Vol. 1815, *Display Technologies*, p. 50 (1992); see also *Opto News Lett.* (43), 1–4 (Oct. 1993).

9. I. W. Wu, *Opto News Lett.* (43), 5–8 (Oct. 1993); (44), 11–14 (Dec. 1993); (45), 32–35 (Feb. 1994).

10. T. Yanagisawa, K. Kasahara, and M. Kajimura, *Proc. Jpn. Display '86*, 192 (1986).

11. A. Chiang, "Polysilicon thin film transistor technology for AM-LCDs," *Proc. SPIE* **1815**, 128 (1992).

12. H. Oshima et al., *SSDM Ext. Abs.* **577** (1991).

13. K. Yoneda, *IDRC '97 Digest*, M40–M47 (1997).

14. S. Morozumi, "Materials and assembling process of LCDs," in *Liquid Crystals—Applications and Uses*, B. Bahadur ed., World Scientific, 1993, Chapter 7, p. 171, and references cited therein.

15. S. T. Wu and C. S. Wu, *Jpn. J. Appl. Phys.* Part 1, **35**, 5349–5354 (1996).

16. R. P. Feynman, R. B. Leighton, and M. Sands, *The Feynman Lectures on Physics*, Addison-Wesley, 1963, Chapter 35.

17. G. Wyszecki, "Colorimetry," in *Handbook of Optics*, G. Driscoll, ed., McGraw-Hill, 1978, Chapter 9.

18. G. Wyszecki and W. S. Stiles, *Color Science: Concepts and Methods, Quantitative Data and Formulas*, Wiley, 1967.

SUGGESTED READINGS

G. Wyszecki and W.S. Stiles, *Color Science: Concepts and Methods, Quantitative Data and Formula*, 2nd ed., Wiley, 1982.

M. Ishikawa, M. Sato, Y. Hisatake, and H. Hatoh, *J. SID* **2/4**, 169 (1994).

PROBLEMS

6.1. *Multiplexed addressing*:
   (a) Differentiate Eq. (6.1-5) with respect to V_d while keeping V_{ON} and V_{OFF} constant. Derive Eqs. (6.1-6) and (6.1-7).
   (b) Derive Eq. (6.1-8):

$$\frac{V_{ON}}{V_{OFF}} = \sqrt{\frac{\sqrt{N_{max}} + 1}{\sqrt{N_{max}} - 1}}$$

   (c) Find the selection ratio for $N = 400$.

6.2. Derive the following expressions Eq. (6.2-1) for the current in MOSFET:

$$I_{SD} = \frac{W\mu C_G}{L} V_{SD}\left(V_G - V_T - \frac{V_{SD}}{2}\right) \qquad \text{when} \qquad V_{SD} < (V_G - V_T)$$

$$I_{SD} = \frac{W\mu C_G}{2L}(V_G - V_T)^2 \qquad \text{when} \qquad V_{SD} > (V_G - V_T)$$

   where W = width
   L = length of conducting channel of TFT
   μ = effective mobility of the semiconductor
   C_G = gate capacitance per unit area
   V_{SD} = source–drain voltage
   V_G = is the gate voltage
   I_{SD} = source–drain current
   V_T = threshold voltage of the TFT

6.3. Derive Eq. (6.2-2):

$$2K_1 C_{LC} V_{ON} < \frac{T}{N} I_{ON}$$

   where K_1 = an engineering operation constant
   V_{ON} = ON voltage of the LC element
   C_{LC} = capacitance of the LC element
   N = number of addressed lines in the display
   T = frame time.

6.4. Derive Eq. (6.2-3):

$$I_{OFF} T < K_2 \Delta V C_{LC}$$

   where T is the frame time and K_2 is another engineering operation constant that allows for the uniformity variations of TFT characteristics in a TFT array and the temperature variations of the OFF currents of the TFTs.

6.5. Derive Eq. (6.2-4):

$$\Delta V_P = \Delta V_G \frac{C_{GD}}{C_{LC} + C_{GD}}$$

6.6. To calculate chromaticity coordinates of mixture of monochromatic light, consider a source of monochromatic light with $S(\lambda) = \delta(\lambda - \lambda_0)$, where λ_0 is the wavelength of the monochromatic source. Using $T(\lambda) = 1$, Table 6.4, and Eqs. (6.4-2)–(6.4-4) the tristimulus values (X, Y, Z) can be obtained. The chromaticity coordinates (x,y) can then be obtained by using Eqs. (6.4-6) and (6.4-7). It can be easily shown that the chromaticity coordinates for the following monochromatic light are:

$\lambda_0 = 450$ nm (blue): $x = 0.1510$, $y = 0.0364$
$\lambda_0 = 550$ nm (green): $x = 0.3473$, $y = 0.6501$
$\lambda_0 = 650$ nm (red) $x = 0.7137$, $y = 0.2863$

(a) Consider an equal mixture of the three sources of monochromatic light described above. Show that the resultant color has the following chromaticity coordinates:

$x = 0.2683$, $y = 0.2729$ $(R + G + B)$

This color appears to be sky white, which is somewhat different from the daylight white.

(b) Show that the chromatic coordinates of an equal mixture of any two of the monochromatic sources of light are

$x = 0.2262$, $y = 0.2716$ $(B + G)$
$x = 0.4197$, $y = 0.5782$ $(G + R)$
$x = 0.2257$, $y = 0.0696$ $(B + R)$

It is important to note that the chromaticity coordinates are not simple averages of the individual chromaticity coordinates of the color components.

6.7. To calculate equienergy white light, consider an artificial source of white light with $S(\lambda) =$ constant. Using $T(\lambda) = 1$, Table 6.4, and Eqs. (6.4-2)– (6.4-4), show that the tristimulus values (X, Y, Z) are the same, that is, $X = Y = Z$, and that the chromaticity coordinates are

$x = 0.3333$, $y = 0.3333$ (equienergy white)

These are quite close to the daylight, whose coordinates are

$x = 0.311$, $y = 0.338$ (daylight white)

In calculating the tristimulus values, the integrations in Eqs. (6.4-2)– (6.4-4) are approximated by summations.

7

Optical Properties of Cholesteric LCs (CLCs)

As we mentioned earlier, when molecules possessing chirality are added into nematic liquid crystal media, a twist of the directors occurs and the resulting structure is termed *cholesteric*. In this chapter, we investigate the optical properties of an ideal cholesteric liquid crystal (CLC) in which the twist rate of the director is uniform throughout the medium. Specifically, we will study the normal modes of propagation, their dispersion relationship $\omega(\mathbf{k})$ and the polarization states of these modes. Starting from the dielectric tensor of the CLC medium, exact solutions of the wave equation are obtained for propagation along the helical axis. The general results and some of the special cases are useful in the understanding of the propagation of polarized light in TN and STN cells. Various operation regimes, including the Bragg regime, the circular regime, and the Mauguin regime, will be discussed.

7.1. OPTICAL PHENOMENA IN CLCs

Cholesteric liquid crystals (CLC) exhibit several unique and interesting optical properties. These include strong optical rotatory powers, selective reflection of circularly polarized light. As a result of the helical arrangement of the directors, circular form birefringence occurs leading to a very strong optical rotatory power (in the range of several thousand degrees per millimeter). The selective reflection of light is a result of the spatially periodic variation of the dielectric tensor in a helical structure. For light propagating parallel to the helical axis, Bragg reflection occurs when the wavelength is in the following range

$$n_o p < \lambda < n_e p \tag{7.1-1}$$

where p is the pitch of the CLC structure and n_o, n_e are the principal refractive indices of the LC medium. In a CLC structure, the director completes a full 360° turn in a distance of p. The Bragg reflected light is circularly polarized if the incident wave propagates along the helical axis. In a right-handed CLC structure, only the right-handed circularly (RHC) polarized component is transmitted through the medium, whereas the left-handed (LHC) component is strongly reflected. Here, we recall that we define the handedness of circular polarization by using the sense of revolution of the electric field vector in time. Thus, for a

RHC polarized beam of light propagating along the direction of the right-hand thumb, the electric field revolves in time along the direction of the fingers. It is important to note that the instantaneous spatial electric field pattern of a LHC polarized light is a right-handed helical structure. Thus, we find that only the component of optical polarization for which the instantaneous spatial electric field pattern matches the helical structure of the director is strongly reflected. The other component is transmitted.

The enormously large optical activity is a result of the helically twisted structure. Generally speaking, there are two contributions to the optical rotatory power: (1) optical rotation due to molecular chirality and (2) optical rotation due to macroscopic structure chirality. The best examples are (1) sugar solutions and (2) quartz crystals. In sugar solutions, the optical rotation is due to the chirality of the sugar molecules that are randomly distributed with random orientations. In crystalline quartz, SiO_2 molecules are bonded in a helical structure along the c axis of the crystal, leading to a significant optical rotatory power. SiO_2 molecules has no chirality. Thus, it is expected that fused silica exhibits no optical rotatory power.

7.2.   DIELECTRIC TENSOR OF AN IDEAL CLC

Consider a right-hand twisted anisotropic medium with a twist angle given by

$$\psi = qz = \frac{2\pi}{p}z \qquad (7.2\text{-}1)$$

where p is the pitch and q is a constant. Figure 7.1 shows a schematic drawing of the right-handed helical structure. The local directors of the LC medium are twisted in the xy plane, with a twist angle given by Eq. (7.2-1). The dielectric constant in the local principal coordinate is given by

$$\varepsilon_{\text{local}} = \begin{pmatrix} \varepsilon_1 & 0 & 0 \\ 0 & \varepsilon_2 & 0 \\ 0 & 0 & \varepsilon_3 \end{pmatrix} \qquad (7.2\text{-}2)$$

where $\varepsilon_1, \varepsilon_2, \varepsilon_3$ are the principal dielectric constants. For most CLC media made of rodlike molecules, the principal dielectric constants can be written

$$\varepsilon_1 = \varepsilon_0 n_e^2, \qquad \varepsilon_2 = \varepsilon_0 n_o^2, \qquad \varepsilon_3 = \varepsilon_0 n_o^2 \qquad (7.2\text{-}3)$$

where n_e, n_o are the refractive indices of the rodlike LC molecules in the nematic phase, and ε_0 is the dielectric constant of the vacuum. We assume $n_o < n_e$ for LC materials with rodlike molecules.

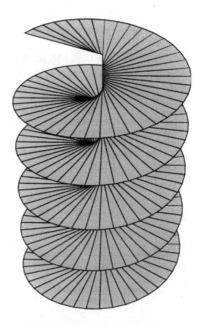

Figure 7.1. Schematic drawing of a right-handed
helical structure.

We now consider the propagation of optical waves along the helical axis z. In
the laboratory coordinate system xyz, the dielectric tensor can be written

$$\varepsilon(z) = R(-\psi)\varepsilon_{\text{local}}R(\psi) \qquad (7.2\text{-}4)$$

where $R(\psi)$ is the coordinate rotation matrix

$$R(\psi) = \begin{pmatrix} \cos\psi & \sin\psi & 0 \\ -\sin\psi & \cos\psi & 0 \\ 0 & 0 & 1 \end{pmatrix} \qquad (7.2\text{-}5)$$

Figure 7.2 shows the coordinate rotation and the definition of the twist angle ψ.
Using Eqs. (7.2-3)–(7.2-5), the dielectric tensor in the laboratory coordinate
system can then be written

$$\varepsilon(z) = \begin{pmatrix} \bar{\varepsilon} + \frac{1}{2}\Delta\varepsilon\cos 2qz & \frac{1}{2}\Delta\varepsilon\sin 2qz & 0 \\ \frac{1}{2}\Delta\varepsilon\sin 2qz & \bar{\varepsilon} - \frac{1}{2}\Delta\varepsilon\cos 2qz & 0 \\ 0 & 0 & \varepsilon_3 \end{pmatrix} \qquad (7.2\text{-}6)$$

where $\bar{\varepsilon} = \frac{1}{2}(\varepsilon_1 + \varepsilon_2)$ and $\Delta\varepsilon = \varepsilon_1 - \varepsilon_2 = \varepsilon_0(n_e^2 - n_o^2)$. Note that the dielectric
tensor is a periodic function of z with a period of $p/2$. For LC molecules with no
polarity, $\psi = 180°$ is identical to $\psi = 360°$. Thus, although the pitch is p, the

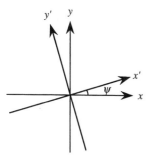

Figure 7.2. The coordinate rotation.

period is actually $p/2$. The dielectric tensor (7.2-6) can be written

$$\varepsilon(z) = \varepsilon_0 \begin{pmatrix} \beta + \alpha \cos 2qz & \alpha \sin 2qz & 0 \\ \alpha \sin 2qz & \beta - \alpha \cos 2qz & 0 \\ 0 & 0 & n_o^2 \end{pmatrix} \qquad (7.2\text{-}7)$$

where

$$\beta = \tfrac{1}{2}(n_e^2 + n_o^2) \qquad (7.2\text{-}8)$$

$$\alpha = \tfrac{1}{2}(n_e^2 - n_o^2) \qquad (7.2\text{-}9)$$

The new parameter β is a measure of the average refractive index, whereas α is a measure of the dielectric anisotropy. For CLC media with small birefringence ($n_e - n_o \ll n_e$ or n_o), these parameters are often written

$$\alpha = n \, \Delta n \qquad (7.2\text{-}10)$$

$$\beta = n^2 \qquad (7.2\text{-}11)$$

where n is the average refractive index and Δn is the birefringence

$$\Delta n = n_e - n_o \qquad (7.2\text{-}12)$$

In what follows, we will substitute the dielectric tensor Eq. (7.2-7) into the wave equation and find the solutions.

7.3. EXACT SOLUTIONS AT NORMAL INCIDENCE

We now investigate the normal modes of propagation in the medium, including Bloch waves, the dispersion relation, and polarization states. Exact solutions are available for propagation along the helical axis (z axis). The electric field vector of the optical wave must satisfy the following wave equation

$$\frac{d^2}{dz^2}E + \omega^2 \mu \varepsilon E = 0 \tag{7.3-1}$$

where ε is the dielectric tensor given by Eq. (7.2-7). The wave equation can be simplified by defining

$$E_u = E_x + iE_y \tag{7.3-2}$$

$$E_v = E_x - iE_y \tag{7.3-3}$$

Using Eqs. (7.3-2) and (7.3-3), we obtain

$$\frac{d^2}{dz^2}\begin{pmatrix} E_u \\ E_v \end{pmatrix} = -\frac{\omega^2}{c^2}\begin{pmatrix} \beta & \alpha e^{-2iqz} \\ \alpha e^{2iqz} & \beta \end{pmatrix}\begin{pmatrix} E_u \\ E_v \end{pmatrix} \tag{7.3-4}$$

This equation supports exact solutions of the following form:

$$\begin{pmatrix} E_u \\ E_v \end{pmatrix} = \begin{pmatrix} ae^{iqz} \\ be^{-iqz} \end{pmatrix}e^{-ikz} \tag{7.3-5}$$

where k is the wavenumber and a,b are constant (all these are to be determined). The solutions satisfy the Bloch–Floquet theorem, which requires that the normal modes be written in terms of the product of a periodic function (with a period of $p/2$) and an exponential function. Substitution of Eq. (7.3-5) into Eq. (7.3-4) yields two independent modes with the wavenumbers given by

$$k^2 = k_1^2 = \beta k_0^2 + q^2 - \sqrt{4q^2\beta k_0^2 + \alpha^2 k_0^4} \tag{7.3-6}$$

$$k^2 = k_2^2 = \beta k_0^2 + q^2 + \sqrt{4q^2\beta k_0^2 + \alpha^2 k_0^4} \tag{7.3-7}$$

where k_0 is the wavenumber in vacuum

$$k_0 = \frac{\omega}{c} = \frac{2\pi}{\lambda} \tag{7.3-8}$$

For each wavenumber k, a set of constants a and b can be obtained. The polarization states of the modes depend on the constants a and b. We now discuss the dispersion relation (ω vs. k) and the polarization states of the normal modes.

7.3.1. Dispersion Relation

The wavenumbers of the normal modes are denoted by k_1, k_2 with $k_1 < k_2$. We note that k_2 is always real for all frequencies. In Figure 7.3, we plot $-k_2$, $-k_1$, k_1, k_2 as functions of frequency ω for a right-handed CLC medium. In a right-handed twist CLC, the normal modes with $k = -k_2$, k_2 are always right-handed elliptically polarized. We note that $-k_2$, k_2 vs ω are almost straight lines. Because wavenumbers are real, they are propagating waves in the medium. The polarization ellipse depends on the frequency of the wave. The polarization ellipse will be discussed later in this section. In the following frequency regime, k_1 becomes pure imaginary:

$$\frac{q}{n_e} < \frac{\omega}{c} < \frac{q}{n_o} \tag{7.3-9}$$

or equivalently in terms of the wavelength

$$n_o p < \lambda < n_e p \tag{7.3-10}$$

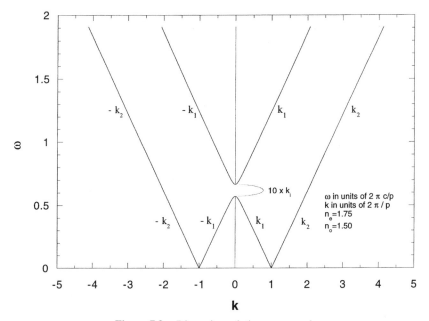

Figure 7.3. Dispersion relation ω versus k.

where p is the pitch of the twist in CLC. In the spectral regime defined by Eqs. (7.3-9) and (7.3-10) the wavenumbers $-k_1, k_1$, are pure imaginary. The imaginary part is plotted as dotted lines in Figure 7.3. Normal modes with purely imaginary propagation constants cannot propagate in the medium. This is the so-called Bragg regime when total reflection occurs. The spectral regime defined by Eqs. (7.3-9) and (7.3-10) is called the *bandgap* (or *stopband*). It is interesting to note that the stop band occurs only for one of the normal modes. As we examine the polarization states, we will discover that the stopband exists only for LHC polarized beam in a right-handed CLC medium. In other words, a beam of LHC polarized light will be reflected when the frequency falls in the spectral regime defined by Eqs. (7.3-9) and (7.3-10).

According to Figure 7.3, there are four wavenumbers for each frequency: $-k_2, -k_1, k_1, k_2$. It is important to note that two of them are propagating along the $+z$ axis, and the other two are propagating along the $-z$ axis. In the spectral regime below the stopband, we note that the mode with $0 < k_1$ has a negative group velocity, whereas the mode with $-k_1 < 0$ has a positive group velocity. As we know, the Bloch wave's wavenumber is only defined to within a multiple of $2q$. So the mode with $0 < k_1$ can be viewed as having a wavenumber of $k_1 - 2q < 0$. Similarly, the mode with $-k_1 < 0$ can be viewed as having a wavenumber of $0 < 2q - k_1$. Using these new wavenumbers, the phase velocity and the group velocity are in the same direction. In investigating the polarization state of propagation in the medium, it is important to select the appropriate wavenumbers for evaluating the phase retardation.

7.3.2. Polarization States of Normal Modes

We now examine the polarization states of the normal modes. The constants a and b in Eq. (7.3-5) for the normal modes are given by, according to Eq. (7.3-4)

$$\frac{a}{b} = \frac{\alpha k_0^2}{(k-q)^2 - \beta k_0^2} = \frac{(k+q)^2 - \beta k_0^2}{\alpha k_0^2} \tag{7.3-11}$$

To investigate the polarization states, we examine the x and y components of the normal modes. Using Eq. (7.3-5), we obtain

$$E_x = \tfrac{1}{2}(ae^{iqz} + be^{-iqz})e^{-ikz} \tag{7.3-12}$$

$$E_y = \frac{1}{2i}(ae^{iqz} - be^{-iqz})e^{-ikz} \tag{7.3-13}$$

For each wavenumber k, these equations represent an elliptically polarized wave. It can be shown that the major and minor axes of the ellipse are parallel to the local principal axes. Physically, as the modes propagate in CLC, the polarization ellipses of the modes follow exactly the twist of the local principal axes. In the principal coordinate, the polarization state can be written

$$E_1 = \tfrac{1}{2}(a + b) \tag{7.3-14}$$

$$E_2 = \frac{-i}{2}(a - b) \tag{7.3-15}$$

The ratio of the axes of the polarization ellipse (ellipticity) can be written

$$e = \frac{a - b}{a + b} \tag{7.3-16}$$

It can be shown that the two normal modes are elliptically polarized with opposite sense of handedness. The modes are not exactly orthogonal, in the sense that

$$e_1 e_2^* \neq -1 \tag{7.3-17}$$

where e_1, e_2 are the ellipticity of the modes. As both a and b depend on the frequency, the shape of the polarization ellipses (or ellipticity) of the normal modes depends on the frequency. Fig. 7.4 shows the polarization ellipses of the normal modes of propagation in a right handed CLC medium at various frequencies. Fig. 7.5 shows the ellipticity of the normal modes as a function of the wavelength / pitch ratio. We note that the ellipticity of the normal mode with wavenumber k_2 varies from 0 (linearly polarized) to 1 (circularly polarized) with a right-handedness. Except with a discontinuity at the bandgap, the ellipticity of the normal mode with wavenumber k_1 varies from $-$ infinity (linearly polarized) to -1 (circularly polarized) with a left-handedness.

We discuss the polarization ellipses in the following three spectral regimes:

1. *Mauguin Regime* ($\lambda \ll 0.5\,p\Delta n$). In the short wavelength spectral regime ($\lambda \ll 0.5\,p\Delta n$), the polarization ellipses of the normal modes have either near-zero or near-infinite ellipticity. The normal modes are almost linearly polarized with the long axis of the ellipses parallel or perpendicular to the local director **n**. Figure 7.4a gives an example for $\lambda = 0.001\,p$ with the ellipticity of the normal modes $= -10.1318, 0.09934$.

2. *Circular Regime* ($0.5\,np\Delta n \ll \lambda \ll p$). In this spectral regime ($0.5\,np\Delta n \ll \lambda \ll p$), the polarization ellipses of the normal modes have near-unity ellipticity. The normal modes are almost circularly polarized with opposite sense of handedness. In this spectral regime, the normal mode with wavenumber k_2 is RHC polarized, whereas the normal mode

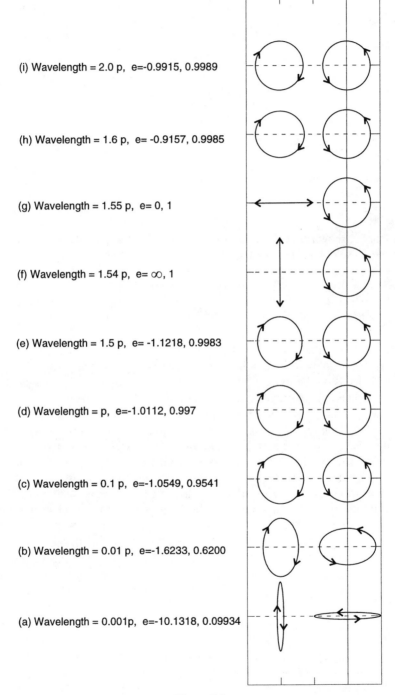

(i) Wavelength = 2.0 p, e=-0.9915, 0.9989

(h) Wavelength = 1.6 p, e= -0.9157, 0.9985

(g) Wavelength = 1.55 p, e= 0, 1

(f) Wavelength = 1.54 p, e= ∞, 1

(e) Wavelength = 1.5 p, e= -1.1218, 0.9983

(d) Wavelength = p, e=-1.0112, 0.997

(c) Wavelength = 0.1 p, e=-1.0549, 0.9541

(b) Wavelength = 0.01 p, e=-1.6233, 0.6200

(a) Wavelength = 0.001p, e=-10.1318, 0.09934

Figure 7.4

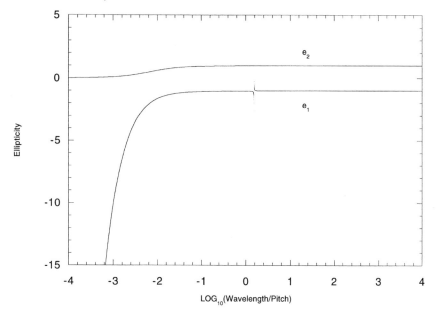

Figure 7.5. The ellipticities of the polarization ellipses of the normal modes in a right-handed CLC as functions of the wavelength/pitch ratio ($n_e = 1.55$, $n_o = 1.54$).

with k_1 is LHC polarized. With $k_1 < k_2$, the CLC medium exhibits a circular birefringence. Figure 7.4c shows an example for $\lambda = 0.1p$ with the ellipticity of the normal modes $= -1.0549$ and 0.9541. The specific optical rotatory power will be discussed in Section 7.6.

3. *Bragg Regime* ($n_o p < \lambda < n_e p$). In this spectral regime ($n_o p < \lambda < n_e p$), the wavenumber k_1 is purely imaginary. Thus the normal mode corresponds to k_1 is an evanescent wave that cannot propagate in the medium. The normal mode with k_2 is almost circularly polarized with a right-handed rotation. The polarization states of the normal mode with k_1 at the bandedges are linearly polarized (see Fig. 7.4 f, g).

4. *Circular Regime* ($n_e p \ll \lambda$). In the long wavelength spectral regime ($n_e p \ll \lambda$), both of the normal modes are almost circularly polarized. Again, the difference in the wavenumber of propagation leads to a circular

◄——

Figure 7.4. The polarization ellipses of the normal modes of propagation in a right handed CLC as functions of the wavelength/pitch ratio. (In this example, $n_e = 1.55$, $n_o = 1.54$): (a) wavelength $= 0.001p, e = -10.1318$, 0.09934; (b) wavelength $= 0.01p, e = -1.6233$, 0.6200; (c) wavelength $= 0.1p, e = -1.0549$, 0.9541; (d) wavelength $= p, e = -1.0112$, 0.997; (e) wave-length $= 1.5p, e = -1.1218$, 0.9983; (f) wavelength $= 1.54p, e = \infty$, 1; (g) wavelength $= 1.55p$, $e = 0$, 1; (h) wavelength $= 1.6p, e = -0.9157$, 0.9985; (i) wavelength $= 2.0p, e = -0.9915$, 0.9989.

birefringence. Figure 7.4i gives an example for $\lambda = 2.0p$ with ellipticity of the normal modes $= -0.9915$ and 0.9989. The specific rotatory power will be discussed in Section 7.6.

7.3.3. Power Orthogonality

According to Eq. (7.3-17), the two polarization ellipses do not have the same axial ratio. In other words, the two polarization states are not exactly orthogonal. It is important to note that a power orthogonal relation exists between the normal modes. This is written

$$\text{Re}\{\mathbf{E}_1 \times \mathbf{H}_2^* + \mathbf{E}_2 \times \mathbf{H}_1^*\} = 0 \qquad (7.3\text{-}18)$$

where $\mathbf{H}_1$, $\mathbf{H}_2$ are the corresponding magnetic field of the normal modes (see Problem 7.3). With the power orthogonality, the total power flow in the medium can be decomposed into the sum of the individual power flow of the normal modes.

7.4. BRAGG REGIME $(n_o p < \lambda < n_e p)$—COUPLED MODE ANALYSIS

In most practical cases, a given beam of polarized light is incident into a finite CLC medium (say, between $z = 0$ and $z = L$). The incident beam of light may not be exactly one of the normal modes of propagation. Thus, more than one normal modes can be excited in the CLC medium. Using the normal modes and the boundary conditions, we are able to obtain the transmitted wave and the reflected wave.

In the coupled mode approach, the twisting of the principal axes can be treated as a perturbation. In other words, we can rewrite the dielectric tensor as

$$\varepsilon(z) = \varepsilon_0 \begin{pmatrix} \beta & 0 & 0 \\ 0 & \beta & 0 \\ 0 & 0 & n_o^2 \end{pmatrix} + \varepsilon_0\alpha \begin{pmatrix} \cos 2qz & \sin 2qz & 0 \\ \sin 2qz & -\cos 2qz & 0 \\ 0 & 0 & 0 \end{pmatrix} \qquad (7.4\text{-}1)$$

where the first term is the averaged dielectric constant and the second term is a periodic perturbation due to the uniform twisting of the director. For propagation along the helical axis (z axis), the averaged dielectric constant supports plane waves of arbitrary polarization states. Thus, in the absence of the perturbation, any incident beam of light is a mode of propagation of the unperturbed structure described by the averaged dielectric constant. By using the coupled mode analysis, we are able to investigate the propagation of the incident beam of light in the presence of the periodic perturbation. In particular, we are interested in the reflection of the optical wave. For propagation along the z axis, the electric field vector is confined in the xy plane. We can ignore the third component of the dielectric tensor. Therefore Eq. (7.4-1) can be written (for z-direction propagation only)

$$\varepsilon(z) = \varepsilon_0 \begin{pmatrix} \beta & 0 \\ 0 & \beta \end{pmatrix} + \varepsilon_0 \alpha \begin{pmatrix} \cos 2qz & \sin 2qz \\ \sin 2qz & -\cos 2qz \end{pmatrix} \tag{7.4-2}$$

The dielectric perturbation (second term) can be written

$$\Delta\varepsilon(z) = \varepsilon_0 \frac{\alpha}{2} \begin{pmatrix} 1 & -i \\ -i & -1 \end{pmatrix} e^{i2qz} + \varepsilon_0 \frac{\alpha}{2} \begin{pmatrix} 1 & i \\ i & -1 \end{pmatrix} e^{-i2qz} \tag{7.4-3}$$

We now consider the following incident wave and reflected wave:

$$\text{Incident wave :} \qquad \mathbf{E}_i e^{i(\omega t - kz)} \tag{7.4-4}$$

$$\text{Reflected wave :} \qquad \mathbf{E}_o e^{i(\omega t + kz)} \tag{7.4-5}$$

where $\mathbf{E}_i$ and $\mathbf{E}_o$ are constant vectors in the xy plane and the wavenumber k is given by

$$k = \frac{\omega}{c} \sqrt{\frac{n_e^2 + n_o^2}{2}} \tag{7.4-6}$$

As the incident wave propagates in the medium, the presence of the dielectric perturbation $\Delta\varepsilon(z)$ gives rise to an additional dielectric polarization:

$$\Delta\mathbf{P} = \Delta\varepsilon(z)\mathbf{E}_i e^{i(\omega t - kz)} \tag{7.4-7}$$

This polarization represents a volume distribution of radiating dipoles. The polarization may radiate efficiently into the reflected wave $\mathbf{E}_o e^{i(\omega t + kz)}$, provided the following conditions are met:

$$\text{(a)} \quad 2k = 2q \tag{7.4-8}$$

$$\text{(b)} \quad \mathbf{E}_o^* \cdot \Delta\varepsilon(z)\mathbf{E}_i \neq 0 \tag{7.4-9}$$

Condition (a) is known as the *Bragg condition*, and (b) is known as the *dynamical coupling condition.* To illustrate the coupling, we consider the following cases of incidence into a right handed CLC medium:

Case 1: Linearly Polarized Incident Wave. For a linearly polarized incident wave, the Jones vector representation for $\mathbf{E}_i$ can be written

$$\mathbf{E}_i = \begin{pmatrix} 1 \\ 0 \end{pmatrix} e^{-ikz} \quad \text{or} \quad \mathbf{E}_i = \begin{pmatrix} 0 \\ 1 \end{pmatrix} e^{-ikz} \tag{7.4-10}$$

Using Eq. (7.4-7), we obtain a perturbation polarization that is proportional to

$$\Delta\mathbf{P} \propto \begin{pmatrix} 1 \\ -i \end{pmatrix} e^{ikz} \qquad (\text{LHC}, -z) \tag{7.4-11}$$

which can radiate into a left handed circularly (LHC) polarized reflected wave.

Case 2: Right-handed Circularly (RHC) Polarized Incident Wave. For a RHC polarized incident wave, the Jones vector representation for $\mathbf{E}_i$ can be written

$$\mathbf{E}_i = \begin{pmatrix} 1 \\ -i \end{pmatrix} e^{-ikz} \qquad (\text{RHC}, +z) \tag{7.4-12}$$

Using Eq. (7.4-7), we obtain a zero perturbation polarization

$$\Delta\mathbf{P} = 0 \tag{7.4-13}$$

which means that the incident wave is transmitted through the medium without coupling to any reflected wave.

Case 3: Left-Handed Circularly (LHC) Polarized Incident Wave. For a LHC polarized incident wave, the Jones vector representation for $\mathbf{E}_i$ can be written

$$\mathbf{E}_i = \begin{pmatrix} 1 \\ i \end{pmatrix} e^{-ikz} \qquad (\text{LHC}, +z) \tag{7.4-14}$$

Using Eq. (7.4-7), we obtain a perturbation polarization:

$$\Delta\mathbf{P} \propto \begin{pmatrix} 1 \\ -i \end{pmatrix} e^{ikz} \qquad (\text{LHC}, -z) \tag{7.4-15}$$

which can radiate into a LHC polarized reflected wave.

In the discussion of these three cases, we find that a beam of LHC polarized light will be reflected from a right-handed CLC medium provided the Bragg condition is satisfied. And the reflected beam is left handed circularly polarized. This is distinctly different from the reflection from a metal mirror. It is known that the handedness of a beam of circularly polarized light is reversed on reflection from a metal mirror. In addition, a beam of RHC polarized light can propagate through a right-handed CLC medium for all wavelengths in the transparent spectral regime.

We now investigate the extent of the reflection of the LHC polarized light by using coupled mode analysis. The total electric field is the sum of the electric field of the incident beam and the reflected beam:

$$\mathbf{E} = A(z)\begin{pmatrix} 1 \\ i \end{pmatrix} e^{i(\omega t - kz)} + B(z)\begin{pmatrix} 1 \\ -i \end{pmatrix} e^{i(\omega t + kz)} \tag{7.4-16}$$

where $A(z)$ and $B(z)$ are the amplitudes of the beams. The dependence on z is assumed to account for the coupling. Substituting Eq. (7.4-16) for $\mathbf{E}$ into the wave equation (7.3-1) and using Eqs. (7.4-2,3) for the dielectric tensor, we obtain a set of coupled equations for the amplitudes A and B:

$$\frac{d}{dz}A = -i\kappa B e^{i\Delta kz}, \qquad \frac{d}{dz}B = i\kappa A e^{-i\Delta kz} \tag{7.4-17}$$

with

$$\Delta k = 2k - 2q \tag{7.4-18}$$

$$\kappa = \frac{\pi\alpha\sqrt{2}}{\lambda\sqrt{n_e^2 + n_o^2}} \tag{7.4-19}$$

We note that the coupling constant κ is proportional to the dielectric anisotropy α. In arriving at the coupled equations (7.4-17), we assumed that both $A(z)$ and $B(z)$ are slowly varying amplitudes. The coupled equations (7.4-17) are identical to those of the periodic media [1]. The coupled equations can now be solved for the amplitudes.

To eliminate $B(z)$ in Eq. (7.4-17), we multiply both sides of the first equation by $\exp(-i\,\Delta kz)$ and then differentiate both sides with respect to z. After a few steps of substitution and algebra, we obtain

$$\frac{d^2}{dz^2}A - i\,\Delta k A - \kappa^2 A = 0 \tag{7.4-20}$$

This is a second order differential equation. The general solution can be written

$$A(z) = [C_1 \cosh sz + C_2 \sinh sz] e^{i(\Delta k/2)z} \tag{7.4-21}$$

where C_1, C_2 are arbitrary constants and s is given by

$$s^2 = \kappa^2 - \left(\frac{\Delta k}{2}\right)^2 = \kappa^2 - (k - q)^2 \qquad (7.4\text{-}22)$$

Once $A(z)$ is obtained, the reflected amplitude $B(z)$ can be obtained from the first equation in Eq. (7.4-17) and can be written as

$$B(z) = e^{-i\Delta kz}\frac{i}{\kappa}\frac{d}{dz}A(z) \qquad (7.4\text{-}23)$$

7.4.1. Reflectance of CLCs

We now investigate the reflectance of a beam of LHC polarized light from a CLC medium. Let the beam of light be incident at $z=0$. Thus the boundary condition is

$$B(L) = 0 \qquad (7.4\text{-}24)$$

The reflection coefficient is given by

$$r = \frac{B(0)}{A(0)} \qquad (7.4\text{-}25)$$

where $A(0)$ is the amplitude of the incident beam. Using Eqs. (7.4-21) and (7.4-23), the constants C_1, C_2 can be expressed in terms of the incident amplitude $A(0)$, and the amplitude $A(z)$, $B(z)$ can be written

$$A(z) = \frac{s\cosh s(L-z) + i\frac{\Delta k}{2}\sinh s(L-z)}{s\cosh sL + i\frac{\Delta k}{2}\sinh sL}A(0)e^{i(\Delta k/2)z} \qquad (7.4\text{-}26)$$

$$B(z) = \frac{-i\kappa\sinh s(L-z)}{s\cosh sL + i\frac{\Delta k}{2}\sinh sL}A(0)e^{-i(\Delta k/2)z} \qquad (7.4\text{-}27)$$

Using Eqs. (7.4-25)–(7.4-27), the reflection coefficient can be written

$$r = \frac{-i\kappa\sinh sL}{s\cosh sL + i\frac{\Delta k}{2}\sinh sL} \qquad (7.4\text{-}28)$$

By taking the absolute square of r, we obtain the following formula for the reflectance:

$$R = |r|^2 = \frac{\kappa^2 \sinh^2 sL}{s^2 \cosh^2 sL + \left(\dfrac{\Delta k}{2}\right)^2 \sinh^2 sL} \qquad (7.4\text{-}29)$$

Physically, the reflectance is a measure of the fraction of power transferred to the reflected beam $B(z)$ in the CLC medium between $z = 0$ and $z = L$. We note that the reflectance is strongly dependent on the momentum mismatch (wavenumber mismatch) Δk. Generally speaking, the reflectance decreases as Δk increases, according to Eq. (7.4-29). At the center of the stopband

$$\Delta k = 2k - 2q = 0 \qquad (7.4\text{-}30)$$

power transfer (reflectance) is maximum, and the reflectance can be written

$$R = \tanh^2 \kappa L \qquad (7.4\text{-}31)$$

We note this reflectance approaches unity when κL is large.

The calculated reflectance of a right-handed CLC medium for a beam of LHC polarized light is plotted in Figure 7.6 as a function of $\Delta k L$. We note that the reflectance spectrum is an even function that consists of a main peak with a sharp cutoff and a series of sidelobes. The bandwidth of the main peak is given approximately as

$$\Delta k = 4|\kappa| \qquad (7.4\text{-}32)$$

because for $|\Delta k| < 2|\kappa|$, the parameter s is real and the amplitude $A(z)$ decays exponentially as z increases. Thus, the region $\Delta k = 4|\kappa|$ is often called the "stopband." For CLC medium with $\Delta n = (n_e, -n_o) \ll n_e, n_o$ the bandgap $\Delta \lambda$ can be written

$$\frac{\Delta \lambda}{\lambda} = \frac{\Delta n}{n} \qquad (7.4\text{-}33)$$

where n is the average refractive index of the medium. This expression is in

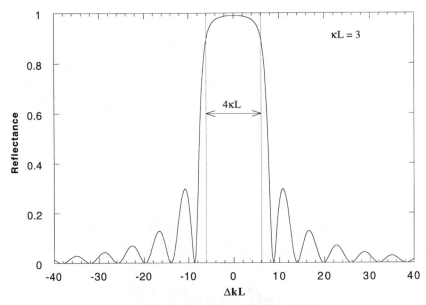

Figure 7.6. Reflectance as a function of ΔkL.

agreement with the exact result, according to Eq. (7.3-6)

$$\Delta\lambda = p\Delta n \qquad (7.4\text{-}34)$$

where p is the pitch of the CLC medium. For $\Delta n \ll n_e, n_o$, the Bragg condition occurs at

$$\lambda = np \qquad (7.4\text{-}35)$$

Using Eq. (7.4-35) for λ, we find that both Eq. (7.4-33) and (7.4-34) are in agreement in giving the same bandgap $\Delta\lambda$.

Example 7.1. Consider the incidence of a beam of light with wavelength $\lambda = 0.5\,\mu\text{m}$ into a thin layer of CLC medium with $n_e = 1.55$, $n_o = 1.54$, $p = 0.33\,\mu\text{m}$, and $L = 50\,\mu\text{m}$. Using Eq. (7.4-19), which can also be written approximately

$$\kappa = \frac{\pi\Delta n}{\lambda} \qquad (7.4\text{-}36)$$

provided $\Delta n = n_e - n_o \ll n_e, n_o$, we obtain $\kappa = 628\,\text{cm}^{-1}$. This leads to a peak reflectance of $R_{\text{max}} = 0.9925$. The stop bandgap is given by $\Delta\lambda = p\Delta n = 0.0033\,\mu\text{m}$. ∎

7.5. MAUGUIN REGIME ($\lambda \ll 0.5\,p\Delta n$)

In the short wavelength spectral regime ($\lambda \ll 0.5\,p\Delta n$), the Mauguin parameter u is much greater than unity:

$$1 \ll u = \frac{(n_e - n_o)p}{2\lambda} = \frac{\Delta n k_0}{2q} \qquad (7.5\text{-}1)$$

where we recall that k_0 is $2\pi/\lambda$ and p is the pitch of the CLC medium. The polarization ellipticity of the normal modes can be written, according to Eqs. (7.3-11) and (7.3-16)

$$e = \frac{\alpha k_0^2 - (k - q)^2 + \beta k_0^2}{\alpha k_0^2 + (k - q)^2 - \beta k_0^2} \qquad (7.5\text{-}2)$$

Using

$$\alpha = n\,\Delta n \qquad (7.5\text{-}3)$$

$$\beta = n^2 \qquad (7.5\text{-}4)$$

where n is the average refractive index, and neglecting the term $2q^2/nk_0$ in Eqs. (7.3-6,7), we obtain the following ellipticities for the normal modes

$$e_1 = -u - \sqrt{1 + u^2} \qquad (k = k_1) \qquad (7.5\text{-}5)$$

$$e_2 = -u + \sqrt{1 + u^2} \qquad (k = k_2) \qquad (7.5\text{-}6)$$

These expressions are in agreement with our earlier results Eqs. (4.3-31) and (4.3-32). In the Mauguin regime, $1 \ll u$, the ellipticity is either zero or infinity, according to Eqs. (7.5-5) and (7.5-6). This corresponds to two mutually orthogonal and linearly polarized modes. The phase retardation per unit propagation distance in this regime can be written

$$\Delta k = k_2 - k_1 = 2q\sqrt{1 + u^2} \qquad (7.5\text{-}7)$$

This is in agreement with our earlier result Eq. (4.3-35) using the conventional Jones matrix method. If we ignore the absolute phase shift and consider only the phase retardation between the normal modes, the limiting form of the normal modes in the Mauguin regime can be written

$$\mathbf{E}_1 = \begin{pmatrix} 1 \\ i(u + \sqrt{1+u^2}) \end{pmatrix} e^{-ink_0z} e^{iqz\sqrt{1+u^2}} \qquad (k = k_1) \qquad (7.5\text{-}8)$$

$$\mathbf{E}_2 = \begin{pmatrix} 1 \\ i(u - \sqrt{1+u^2}) \end{pmatrix} e^{-ink_0z} e^{-iqz\sqrt{1+u^2}} \qquad (k = k_2) \qquad (7.5\text{-}9)$$

where the Jones vectors are in the local principal coordinate system. We note that $\mathbf{E}_2$ is almost parallel to the local director, whereas $\mathbf{E}_1$ is almost perpendicular to the local director. Again, we note that these expressions are in exact agreement with our earlier results in Eqs. (4.3-31) and (4.3-32) using the conventional Jones matrix method. As the normal modes propagate in the CLC medium, the principal axes of the polarization ellipse will follow the twist of the local director. This is the phenomenon of waveguiding (or adiabatic following). We also note that in the Mauguin regime, the normal modes are mutually orthogonal:

$$e_1^* e_2 = -1 \qquad (7.5\text{-}10)$$

From on the preceding discussion, we find that the limiting forms of the exact solutions are identical to those obtained by using the conventional Jones calculus.

7.6. CIRCULAR REGIME

The normal modes are circularly polarized in most spectral regimes, except in the Bragg regime and the Mauguin regime. This can be seen in Figure 7.5, which plots the ellipticities of the normal modes as functions of the wavelength/pitch ratio. In this regime of circular birefringence, an important parameter to investigate is the optical rotatory power. The discussion of the optical rotatory power is divided into the following two spectral regimes.

7.6.1. Short Wavelength Circular Regime ($0.5\,np\Delta n \ll \lambda \ll p$)

In this regime, the polarization ellipses of the normal modes have near-unity ellipticity. The normal modes are almost circularly polarized with opposite sense of handedness. In this spectral regime, the normal mode with wavenumber k_2 is RHC polarized, whereas the normal mode with k_1 is LHC polarized. With $k_1 < k_2$, the CLC medium exhibits circular birefringence. The specific optical rotatory power is given by

$$\rho = \tfrac{1}{2}(k_1 - k_2) \qquad (7.6\text{-}1)$$

Using Eqs. (7.3-6,7), we obtain

$$\rho = -\frac{\pi(\Delta n)^2 p}{4\lambda^2} \qquad (7.6\text{-}2)$$

where p is the pitch of the CLC medium and $\Delta n = n_e - n_o$ is the birefringence of the liquid crystal medium in its nematic phase. In arriving at Eq. (7.6-2), we have ignored the term $2q$, which is the wavenumber of the periodic medium. We recall that the wavenumbers of the Bloch waves are defined within a multiple of $2q$. We have also used the condition in this regime

$$\frac{\alpha k_0}{2} \ll q \ll k_0 \qquad (7.6\text{-}3)$$

which is equivalent to $0.5\, np\Delta n \ll \lambda \ll p$. Note that the optical rotatory power is negative in the short wavelength regime, indicating a left-handed rotation of the plane of polarization in a right-handed CLC medium. The specific rotatory power is proportional to $(\Delta n)^2 p$ and inversely proportional to λ^2.

7.6.2. Long Wavelength Circular Regime ($n_e p \ll \lambda$)

In this spectral regime, both normal modes are also almost circularly polarized. Again, the difference in the wavenumber of propagation leads to circularly birefringence. In this regime, the normal mode with a positive k_1 has a negative group velocity. To evaluate the optical rotatory power, we must consider the modes with positive group velocities. Thus, the specific rotatory power is given by

$$\rho = \tfrac{1}{2}(-k_1 - k_2) \qquad (7.6\text{-}4)$$

Using Eqs. (7.3-6,7), we obtain

$$\rho = \frac{\pi n^2 (\Delta n)^2 p^3}{4\lambda^4} \qquad (7.6\text{-}5)$$

where p is the pitch of the CLC medium, n is the average refractive index, and $\Delta n = n_e - n_o$ is the birefringence of the LC medium in its nematic phase. In arriving at Eq. (7.6-5), we have again ignored the term $2q$, which is the wavenumber of the periodic medium. We have also used the following condition in this regime:

$$n k_0 \ll q \qquad (7.6\text{-}6)$$

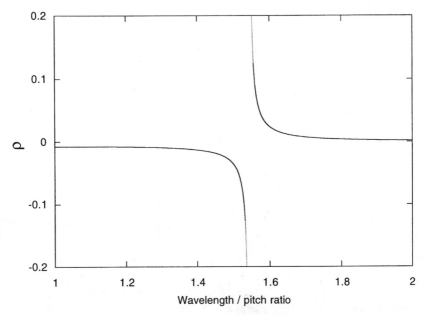

Figure 7.7. Optical rotatory power as a function of the wavelength/pitch ratio.

which is equivalent to $np \ll \lambda$. Note that the optical rotatory power is positive in the long wavelength regime, indicating a right-handed rotation of the plane of polarization in a right-handed CLC medium. The specific rotatory power is proportional to $(\Delta n)^2 p^3$ and inversely proportional to λ^4. Fig. 7.7 plots the optical rotatory power as a function of the wavelength/pitch ratio. We note that on the long wavelength side ($np < \lambda$), the optical rotatory power is positive and decreases as a function of the wavelength. On the short wavelength side of the bandgap, the optical rotatory power is negative. The rotatory power near the bandgap is very large and strongly dispersive, resembling a resonance.

REFERENCES

1. See, for example, P. Yeh, *Optical Waves in Layered Media*, Wiley, 1988.

SUGGESTED READINGS

P. G. de Gennes and J. Prost, *The Physics of Liquid Crystals*, Oxford Univ. Press, 1993.
S. Chandrasekhar, *Liquid Crystals*, Cambridge Univ. Press, 1992.

PROBLEMS

7.1. Derive the dielectric tensor Eq. (7.2-7) using coordinate rotation

$$\varepsilon(z) = \varepsilon_0 \begin{pmatrix} \beta + \alpha \cos 2qz & \alpha \sin 2qz & 0 \\ \alpha \sin 2qz & \beta - \alpha \cos 2qz & 0 \\ 0 & 0 & n_o^2 \end{pmatrix}$$

where

$$\beta = \tfrac{1}{2}(n_e^2 + n_o^2) \qquad \text{and} \qquad \alpha = \tfrac{1}{2}(n_e^2 - n_o^2)$$

For CLC media with small birefringence ($n_e - n_o \ll n_e$ or n_o), these parameters can be written $\alpha = n\,\Delta n$ and $\beta = n^2$, where n is the average refractive index, and Δn is the birefringence $\Delta n = n_e - n_o$.

7.2. (a) Derive Eqs. (7.3-6) and (7.3-7):

$$k^2 = k_1^2 = \beta k_0^2 + q^2 - \sqrt{4q^2 \beta k_0^2 + \alpha^2 k_0^4}$$

$$k^2 = k_2^2 = \beta k_0^2 + q^2 + \sqrt{4q^2 \beta k_0^2 + \alpha^2 k_0^4}$$

where k_0 is the wavenumber in vacuum:

$$k_0 = \frac{\omega}{c} = \frac{2\pi}{\lambda}$$

(b) Show that k_1^2 is negative for

$$\frac{q}{n_e} < \frac{\omega}{c} < \frac{q}{n_o}$$

or equivalently in terms of the wavelength, $n_o p < \lambda < n_e p$. In this spectral region, the wavenumbers $-k_1$, k_1 are pure imaginary. Note that the stopband exists only for LHC polarized beam in a right-handed CLC medium. In other words, a beam of LHC polarized light will be reflected when the frequency falls in the spectral regime defined above.

7.3. (a) Derive Eq. (7.3-11):

$$\frac{a}{b} = \frac{\alpha k_0^2}{(k-q)^2 - \beta k_0^2} = \frac{(k+q)^2 - \beta k_0^2}{\alpha k_0^2}$$

For each wavenumber k, these equations represent an elliptically polarized wave.

(b) Show that the major and minor axes of the ellipse are parallel to the local principal axes. Physically, as the modes propagate in CLC, the polarization ellipses of the modes follow exactly the twist of the local principal axes.

(c) Show that the two normal modes are elliptically polarized with opposite sense of handedness. Show that the modes are not exactly orthogonal, in the sense that $e_1 e_2^* \neq -1$ where e_1, e_2 are the ellipticity of the modes.

(d) Show that

$$Re\{\mathbf{E}_1 \times \mathbf{H}_2^* + \mathbf{E}_2 \times \mathbf{H}_1^*\} = 0$$

where $\mathbf{H}_1$, $\mathbf{H}_2$ are the corresponding magnetic fields of the normal modes.

(e) Consider a general solution

$$\mathbf{E} = c_1 \mathbf{E}_1 + c_2 \mathbf{E}_2$$

Find the corresponding magnetic field $\mathbf{H}$:

$$\mathbf{H} = c_1 \mathbf{H}_1 + c_2 \mathbf{H}_2$$

and show that the total power flow in the medium can be decomposed into the sum of the individual mode power flows.

7.4. (a) Derive Eq. (7.4-1):

$$\varepsilon(z) = \varepsilon_0 \begin{pmatrix} \beta & 0 & 0 \\ 0 & \beta & 0 \\ 0 & 0 & n_o^2 \end{pmatrix} + \varepsilon_0 \alpha \begin{pmatrix} \cos 2qz & \sin 2qz & 0 \\ \sin 2qz & -\cos 2qz & 0 \\ 0 & 0 & 0 \end{pmatrix}$$

(b) Derive Eq. (7.4-17):

$$\frac{d}{dz}A = -i\kappa B e^{i\Delta kz}, \qquad \frac{d}{dz}B = i\kappa A e^{-i\Delta kz}$$

and the general solution Eq. (7.4-21):

$$A(z) = [C_1 \cosh sz + C_2 \sinh sz]e^{i(\Delta k/2)z}$$

(c) Derive Eq. (7.4-28):

$$r = \frac{-i\kappa \sinh sL}{s \cosh sL + i\dfrac{\Delta k}{2} \sinh sL}$$

(d) Show that the bandgap can be written

$$\frac{\Delta\lambda}{\lambda} = \frac{\Delta n}{n}$$

or equivalently $\Delta\lambda = p\, \Delta n$

7.5. (a) Show that the normal modes in the Mauguin regime ($1 \ll u$) can be written

$$\mathbf{E}_1 = \left(\frac{1}{i(u + \sqrt{1+u^2})} \right) e^{-ink_0 z} e^{iqz\sqrt{1+u^2}} \qquad (k = k_1)$$

$$\mathbf{E}_2 = \left(\frac{1}{i(u - \sqrt{1+u^2})} \right) e^{-ink_0 z} e^{-iqz\sqrt{1+u^2}} \qquad (k = k_2)$$

where the Jones vectors are in the local principal coordinate system.

(b) Show that $\mathbf{E}_1$ is almost perpendicular to the local director, whereas $\mathbf{E}_2$ is almost parallel to the local director. As the normal modes propagate in the CLC medium, the principal axes of the polarization ellipse will follow the twist of the local director. This is the phenomenon of waveguiding (or adiabatic following).

(c) Show that in the Mauguin regime, the normal modes are mutually orthogonal: $e_1^* e_2 = -1$.

(d) Show that the limiting forms (Mauguin regime) of the exact solutions are identical to those obtained by using Jones calculus.

7.6. (a) Derive Eq. (7.6-2):

$$\rho = -\frac{\pi(\Delta n)^2 p}{4\lambda^2}$$

where p is the pitch of the CLC medium and $\Delta n = n_e - n_o$ is the birefringence of the LC medium in its nematic phase.

(b) Derive Eq. (7.6-5):

$$\rho = \frac{\pi n^2 (\Delta n)^2 p^3}{4\lambda^4}$$

where p is the pitch of the CLC medium, n is the average refractive index, and $\Delta n = n_e - n_o$ is the birefringence of the LC medium in its nematic phase.

8

Extended Jones Matrix Method

The Jones calculus, invented in 1941 by R. C. Jones [1] for studying the transmission characteristics of birefringent networks, and described in Chapter 4, is a powerful technique in which the state of polarization is represented by a two-component column vector and each optical element (e.g., wave plate, polarizer, liquid crystal cell) is represented by a 2×2 matrix. This matrix method, however, is limited to normally incident and paraxial rays only. To illustrate this situation, we point out, as an example, that the two polarization states of the incident beam that excite only the ordinary mode and extraordinary mode of a birefringent crystal plate, respectively, are, in general, not mutually orthogonal for off-axis light. This results from the Fresnel refraction and reflection of light at the plate surfaces, which are neglected in the Jones calculus. In addition, the conventional Jones matrix method does not offer an explanation for the leakage of off-axis light through a pair of crossed ideal polarizers. In Chapter 4, we introduced the conventional Jones matrix method and demonstrated its application. Referring to Figure 4.15, we note that the conventional Jones matrix method cannot accurately predict the intensity transmission pattern of a c plate between crossed polarizers. The intensity transmission pattern predicted by the extended Jones matrix method was also shown in Figure 4.15b, which agrees with experimental results. Many modern optical systems (e.g., liquid crystal displays) call for a birefringent network with a wide field of view. To accurately calculate the transmission characteristics of these systems for off-axis light, the effect of refraction and reflection at the plate interfaces cannot be ignored. The extended Jones matrix method, first introduced in 1982, is a powerful technique for treating the transmission of off-axis light in a general birefringent network. In this chapter, we will describe the extended Jones matrix method, demonstrate its applications, and compare them with the conventional Jones matrix method.

The transmission of light through birefringent networks has been treated using various methods. Exact solutions can be obtained by the 4×4 matrix method [2,3], which will be discussed in Section 8.3. The 4×4 matrix method takes into account the effect of refraction and multiple reflections between plate interfaces. If the effect of multiple reflections is neglected, a much easier 2×2 matrix (known as the *extended Jones matrix*) method [4–11] can be employed. In Section 8.1, we neglect the multiple reflections and derive the extended Jones matrix method that is much easier to manipulate algebraically and yet accounts for the effects of the Fresnel refraction and the single reflection at the interfaces. The assumption of no multiple reflection is legitimate for most practical LCD networks. When a spectral averaging is employed in the 4×4 matrix method, the results obtained using the two different methods are almost identical. The Fabry–Perot effect due to multiple reflections at the interfaces, which is strongly frequency dependent, is often unobservable by the spectral averaging in the detection process. For practical purposes, the 2×2 matrix method is adequate and easier to use. In what follows, we describe the extended Jones matrix derived originally by the authors. We then briefly describe a different formulation for the extended Jones matrix method in Section 8.2. The 4×4 matrix method will be discussed in Section 8.3.

8.1. MATHEMATICAL FORMULATION AND APPLICATIONS

In this section, we introduce the extended Jones matrix (2×2 matrix) method and discuss its applications. We first discuss the reflection and refraction at the interface between an isotropic medium and a uniaxial medium. On this basis, we derive the extended Jones matrix formulation. To describe the concept, we first examine the situation in which the c axis is parallel to the plate interfaces. As an example, the transmission of light through a pair of crossed ideal polarizers is calculated. Then we discuss the general case of an arbitrary c-axis orientation. As an example, the extended Jones matrix method is used to treat twisted nematic liquid crystal (TN-LC) cells.

8.1.1. Reflection and Refraction at the Interface

We have so far neglected the reflection of electromagnetic radiation from the surfaces of LC cells. This is legitimate when the surfaces are treated with antireflection coatings. To understand the transmission of light through a birefringent network, we begin by studying the reflection and refraction of electromagnetic radiation at an interface between an isotropic medium (e.g., air) and a uniaxial medium (e.g., a polarizer or a LC layer). Unlike the Fresnel reflection and refraction of a dielectric interface between two isotropic media, s and p waves are no longer independent (as a reminder, s wave is polarized perpendicular to the plane of incidence and p wave is polarized in the plane of incidence). They are coupled because of the optical anisotropy of the uniaxial medium. In other words, an incidence wave polarized along the s direction will generate a reflected wave that is a mixture of s and p waves. In addition, an o wave incident from the inside of the uniaxial medium will generate a reflected wave that is a mixture of o and e waves. The cross coupling disappears only at special angles of incidence and/or with particular polarization states.

Referring to Figure 8.1, we consider the incidence of a plane wave from the lower half-space ($z < 0$). The coordinates are chosen such that the (x, y) plane contains the interface and the z direction is perpendicular to the interface.

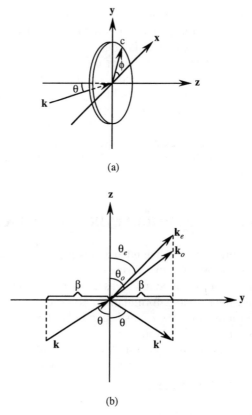

Figure 8.1. (a) *xyz* coordinate systems and the *c* axis of the crystal plate. The incident wave vector **k** lies in the *yz* plane and the *c* axis of the uniaxial medium lies in the *xy* plane; (b) reflection and refraction at the interface between an isotropic medium and a uniaxial medium ($\beta = \sin \theta$). The *z* axis is perpendicular to the plate surface.

Let **k**,**k**′ be the wavevectors of the incident and reflected waves, respectively, and $\mathbf{k}_o$,$\mathbf{k}_e$ be the wavevectors of the ordinary and extraordinary refracted waves, respectively. The electric fields of incident, reflected, and refracted waves are

Incident: $\mathbf{E} = (A_s\mathbf{s} + A_p\mathbf{p}) \exp\left[i(\omega t - \mathbf{k}\cdot\mathbf{r})\right]$ (8.1-1)

Reflected: $\mathbf{E} = (B_s\mathbf{s} + B_p\mathbf{p}') \exp\left[i\omega t - \mathbf{k}'\cdot\mathbf{r})\right]$ (8.1-2)

Refracted: $\mathbf{E} = (C_o o e^{-i\mathbf{k}_o\cdot\mathbf{r}} + C_e e e^{-i\mathbf{k}_e\cdot\mathbf{r}}) \exp(i\omega t)$ (8.1-3)

where **s** is a unit vector perpendicular to the plane of incidence and is given by $\mathbf{s} = \mathbf{x}$ in the chosen coordinate (see Fig. 8.1) and **p** and **p**′ are unit vectors parallel to the plane of incidence and are given by

$$\mathbf{p} = \frac{\mathbf{k} \times \mathbf{s}}{|\mathbf{k}|} \tag{8.1-4}$$

$$\mathbf{p}' = \frac{-\mathbf{k}' \times \mathbf{s}}{|\mathbf{k}'|} \tag{8.1-5}$$

The terms $\mathbf{o}$ and $\mathbf{e}$ represent unit vectors parallel to the electric field vector of the ordinary mode and extraordinary mode, respectively, in the uniaxially anisotropic medium. The wavevectors can be written as

$$\mathbf{k} = \beta \mathbf{y} + k_z \mathbf{z} \tag{8.1-6}$$

$$\mathbf{k}' = \beta \mathbf{y} - k_z \mathbf{z} \tag{8.1-7}$$

$$\mathbf{k}_o = \beta \mathbf{y} + k_{oz} \mathbf{z} \tag{8.1-8}$$

$$\mathbf{k}_e = \beta \mathbf{y} + k_{ez} \mathbf{z} \tag{8.1-9}$$

where β, which remains the same on both sides of the interface, is the tangential component of all the wavevectors. These components of wavevectors are related by Eq. (3.2-36). Once β is given, the z-components k_{ez} and k_{oz} can be obtained by solving Eq. (3.2-36). The unit vectors $\mathbf{o}$ and $\mathbf{e}$ can be obtained by normalizing Eqs. (3.3-6) and (3.3-7), respectively. It is important to note that Eqs. (3.2-36), (3.3-6), and (3.3-7) are expressed in terms of the components in the principal coordinate system. For media with small birefringence, Eqs. (3.3-8) and (3.3-9) can be used for these two unit vectors.

In Eqs. (8.1-1)–(8.1-3), A_s, A_p, B_s, B_p, C_o, and C_e are constants, where A_s and A_p are amplitudes of the incident s and p waves, respectively; B_s and B_p are those of the reflected waves; and C_o, and C_e are the amplitudes of the transmitted o and e waves, respectively.

The magnetic fields of the incident, reflected, and refracted waves can be derived from Eqs. (8.1-1)–(8.1-3) and by Maxwell's equation:

$$\mathbf{H} = \frac{i}{\omega \mu} \nabla \times \mathbf{E} \tag{8.1-10}$$

and are given, respectively, by

Incident: $\mathbf{H} = \dfrac{1}{\omega \mu} \mathbf{k} \times (A_s \mathbf{s} + A_p \mathbf{p}) \exp\left[i(\omega t - \mathbf{k} \cdot \mathbf{r})\right]$ $\qquad$ (8.1-11)

Reflected: $\mathbf{H} = \dfrac{1}{\omega \mu} \mathbf{k}' \times (B_s \mathbf{s} + B_p \mathbf{p}') \exp\left[i(\omega t - \mathbf{k}' \cdot \mathbf{r})\right]$ $\qquad$ (8.1-12)

Refracted: $\mathbf{H} = \dfrac{1}{\omega \mu} (C_o \mathbf{k}_o \times \mathbf{o}\, e^{-i\mathbf{k}_o \cdot \mathbf{r}} + C_e \mathbf{k}_e \times \mathbf{e}\, e^{-i\mathbf{k}_e \cdot \mathbf{r}}) \exp\left(i\omega t\right)$ $\quad$ (8.1-13)

The tangential component of $\mathbf{E}$ and $\mathbf{H}$ must be continuous at the boundary $z = 0$. In terms of the fields (8.1-1)–(8.1-3) and (8.1-11)–(8.1-13), these boundary conditions at $z = 0$ are

$$A_s + B_s = \mathbf{x} \cdot \mathbf{o} C_o + \mathbf{x} \cdot \mathbf{e} C_e \qquad (8.1\text{-}14)$$

$$\mathbf{y} \cdot \mathbf{p} A_p + \mathbf{y} \cdot \mathbf{p}' B_p = \mathbf{y} \cdot \mathbf{o} C_o + \mathbf{y} \cdot \mathbf{e} C_e \qquad (8.1\text{-}15)$$

$$\mathbf{x} \cdot (\mathbf{k} \times \mathbf{p}) A_p + \mathbf{x} \cdot (\mathbf{k}' \times \mathbf{p}') B_p = \mathbf{x} \cdot (\mathbf{k}_o \times \mathbf{o}) C_o + \mathbf{x} \cdot (\mathbf{k}_e \times \mathbf{e}) C_e \qquad (8.1\text{-}16)$$

$$\mathbf{y} \cdot (\mathbf{k} \times \mathbf{s}) A_s + \mathbf{y} \cdot (\mathbf{k}' \times \mathbf{s}) B_s = \mathbf{y} \cdot (\mathbf{k}_o \times \mathbf{o}) C_o + \mathbf{y} \cdot (\mathbf{k}_e \times \mathbf{e}) C_e \qquad (8.1\text{-}17)$$

These four equations can be used to solve for the four unknowns B_s, B_p, C_o, and C_e in terms of the amplitudes A_s and A_p of the incident wave. Using the expressions for $\mathbf{k}$, $\mathbf{k}'$, $\mathbf{p}$, and $\mathbf{p}'$ from Eqs. (8.1-4)–(8.1-7), these equations can be written as

$$k(A_s + B_s) = \mathbf{x} \cdot \mathbf{o} k C_o + \mathbf{x} \cdot \mathbf{e} k C_e \qquad (8.1\text{-}18)$$

$$k_z(A_p + B_p) = \mathbf{y} \cdot \mathbf{o} k C_o + \mathbf{y} \cdot \mathbf{e} k C_e \qquad (8.1\text{-}19)$$

$$-k(A_p - B_p) = \mathbf{x} \cdot (\mathbf{k}_o \times \mathbf{o}) C_o + \mathbf{x} \cdot (\mathbf{k}_e \times \mathbf{e}) C_e \qquad (8.1\text{-}20)$$

$$k_z(A_s - B_s) = \mathbf{y} \cdot (\mathbf{k}_o \times \mathbf{o}) C_o + \mathbf{y} \cdot (\mathbf{k}_e \times \mathbf{e}) C_e \qquad (8.1\text{-}21)$$

where $k_z = k \cos \theta$.

We now eliminate B_s and B_p and obtain

$$A C_o + B C_e = 2 k_z A_s \qquad (8.1\text{-}22)$$

$$C C_o + D C_e = 2 k_z A_p \qquad (8.1\text{-}23)$$

where A, B, C, and D are constants given by

$$A = \mathbf{o} \cdot (\mathbf{y} \times \mathbf{k}) + \mathbf{o} \cdot (\mathbf{y} \times \mathbf{k}_o) \qquad (8.1\text{-}24)$$

$$B = \mathbf{e} \cdot (\mathbf{y} \times \mathbf{k}) + \mathbf{e} \cdot (\mathbf{y} \times \mathbf{k}_e) \qquad (8.1\text{-}25)$$

$$C = \mathbf{o} \cdot \mathbf{y} k - \frac{(\mathbf{y} \times \mathbf{k}) \cdot (\mathbf{k}_o \times \mathbf{o})}{k} \qquad (8.1\text{-}26)$$

$$D = \mathbf{e} \cdot \mathbf{y} k - \frac{(\mathbf{y} \times \mathbf{k}) \cdot (\mathbf{k}_e \times \mathbf{e})}{k} \qquad (8.1\text{-}27)$$

In arriving at these expressions, we have used $\mathbf{x}k_z = (\mathbf{y} \times \mathbf{k})$. Equations (8.1-22) and (8.1-23) can now be solved for C_o and C_e in terms of A_s and A_p. This leads to the following linear relations:

$$C_o = A_s t_{so} + A_p t_{po} \qquad (8.1\text{-}28)$$

$$C_e = A_s t_{se} + A_p t_{pe} \qquad (8.1\text{-}29)$$

where t_{so}, t_{po}, t_{se}, and t_{pe} are the Fresnel transmission coefficients given by

$$t_{so} = \frac{2k_z D}{AD - BC} \qquad (8.1\text{-}30)$$

$$t_{po} = \frac{-2k_z B}{AD - BC} \qquad (8.1\text{-}31)$$

$$t_{se} = \frac{-2k_z C}{AD - BC} \qquad (8.1\text{-}32)$$

$$t_{pe} = \frac{2k_z A}{AD - BC} \qquad (8.1\text{-}33)$$

Note that we now have four transmission coefficients. Here t_{so} is the transmission coefficient for the case of an s-polarized incident wave and a transmitted o wave. The other coefficients have their similar physical meaning according to their subscripts.

The other two unknowns, B_s and B_p, which are the amplitudes of the reflected waves, can now be obtained by substituting Eqs. (8.1-28) and (8.1-29) into Eqs. (8.1-18)–(8.1-21). Similar linear relations are obtained:

$$B_s = A_s r_{ss} + A_p r_{ps} \qquad (8.1\text{-}34)$$

$$B_p = A_s r_{sp} + A_p r_{pp} \qquad (8.1\text{-}35)$$

where r_{ss}, r_{sp}, r_{ps}, and r_{pp} are the reflection coefficients given by

$$r_{ss} = \frac{A'D - B'C}{AD - BC} \qquad (8.1\text{-}36)$$

$$r_{ps} = \frac{AB' - A'B}{AD - BC} \qquad (8.1\text{-}37)$$

$$r_{sp} = -\frac{CD' - C'D}{AD - BC} \qquad (8.1\text{-}38)$$

$$r_{pp} = \frac{AD' - BC'}{AD - BC} \qquad (8.1\text{-}39)$$

with A', B', C', and D' given by

$$A' = \mathbf{o} \cdot (\mathbf{y} \times \mathbf{k}) - \mathbf{o} \cdot (\mathbf{y} \times \mathbf{k}_o) = 2\mathbf{x} \cdot \mathbf{o}k_z - A \qquad (8.1\text{-}40)$$

$$B' = \mathbf{e} \cdot (\mathbf{y} \times \mathbf{k}) - \mathbf{e} \cdot (\mathbf{y} \times \mathbf{k}_e) = 2\mathbf{x} \cdot \mathbf{e}k_z - B \qquad (8.1\text{-}41)$$

$$C' = \mathbf{o} \cdot \mathbf{y}k + \frac{(\mathbf{y} \times \mathbf{k}) \cdot (\mathbf{k}_o \times \mathbf{o})}{k} = 2\mathbf{o} \cdot \mathbf{y}k - C \qquad (8.1\text{-}42)$$

$$D' = \mathbf{e} \cdot \mathbf{y}k + \frac{(\mathbf{y} \times \mathbf{k}) \cdot (\mathbf{k}_e \times \mathbf{e})}{k} = 2\mathbf{e} \cdot \mathbf{y}k - D \qquad (8.1\text{-}43)$$

Among these four reflection coefficients, r_{ss} and r_{pp} are the direct-reflection coefficients, whereas r_{sp} and r_{ps} can be viewed as the cross-reflection coefficients. The cross-reflection coefficients r_{sp} and r_{ps} vanish when the anisotropy disappears (i.e., when $n_e = n_o$). The proof is left as an exercise for the students (see Problem 8.4).

We are now ready to introduce some important concepts. According to Eqs. (8.1-28) and (8.1-29), both ordinary and extraordinary waves are, in general, excited by the incidence of a polarized wave. It is interesting to note that there exist two input polarization states of the incident wave that will excite only normal modes (either ordinary or extraordinary). According to Eqs. (8.1-28) and (8.1-29), these two polarization states are given by

$$O\text{-wave excitation:} \qquad \left(\frac{A_p}{A_s} \right)_{C_e = 0} = -\frac{t_{se}}{t_{pe}} \qquad (8.1\text{-}44)$$

$$E\text{-wave excitation:} \qquad \left(\frac{A_p}{A_s} \right)_{C_o = 0} = -\frac{t_{so}}{t_{po}} \qquad (8.1\text{-}45)$$

The expressions for the four reflection coefficients reduce to simple forms at normal incidence. When the incidence angle θ is zero, all the wavevectors are parallel to the z axis, and the unit vectors $\mathbf{o}$ and $\mathbf{e}$ are given by

$$\mathbf{o} = -\mathbf{x} \sin\phi + \mathbf{y} \cos\phi \qquad (8.1\text{-}46)$$

$$\mathbf{e} = \mathbf{x} \cos\phi + \mathbf{y} \sin\phi \qquad (8.1\text{-}47)$$

Referring to Figure 8.1, we note that, in this case, $\mathbf{e}$ is parallel to the c axis and $\mathbf{o}$ is parallel to $\mathbf{z} \times \mathbf{c}$. In addition, we note that $\mathbf{s} = \mathbf{x}$, $\mathbf{p} = \mathbf{y}$ at normal incidence.

Substituting Eqs. (8.1-46) and (8.1-47) into Eqs. (8.1-22)–(8.1-27) and Eqs. (8.1-40)–(8.1-43), and according to Eqs. (8.1-36)–(8.1-39), we obtain

$$r_{ss} = \frac{(n^2 - n_o n_e) - n(n_e - n_o)\cos 2\phi}{(n + n_o)(n + n_e)} \tag{8.1-48}$$

$$r_{ps} = -\frac{(n_e - n_o)n \sin 2\phi}{(n + n_o)(n + n_e)} \tag{8.1-49}$$

$$r_{sp} = -\frac{(n_e - n_o)n \sin 2\phi}{(n + n_o)(n + n_e)} \tag{8.1-50}$$

$$r_{pp} = \frac{(n^2 - n_o n_e) - n(n_e - n_o)\cos 2\phi}{(n + n_o)(n + n_e)} \tag{8.1-51}$$

where n is the index of refraction of the incident medium.

Note that the cross-reflection coefficients vanish when the c axis is either perpendicular ($\phi = 0$) or parallel ($\phi = \pi/2$) to the plane of incidence. At these special angles, the s and p waves are not coupled.

The transmission coefficients for normal incidence are obtained in a similar fashion and are given by

$$t_{so} = \frac{-2n}{n + n_o} \sin \phi \tag{8.1-52}$$

$$t_{po} = \frac{2n}{n + n_o} \cos \phi \tag{8.1-53}$$

$$t_{se} = \frac{2n}{n + n_e} \cos \phi \tag{8.1-54}$$

$$t_{pe} = \frac{2n}{n + n_e} \sin \phi \tag{8.1-55}$$

According to Eqs. (8.1-46), (8.1-47), and (8.1-4), these expressions of transmission coefficients at normal incidence can also be written as

$$t_{so} = \frac{2n}{n + n_o} \mathbf{s} \cdot \mathbf{o} \tag{8.1-56}$$

$$t_{po} = \frac{2n}{n + n_o} \mathbf{p} \cdot \mathbf{o} \tag{8.1-57}$$

$$t_{se} = \frac{2n}{n + n_e} \mathbf{s} \cdot \mathbf{e} \tag{8.1-58}$$

$$t_{pe} = \frac{2n}{n + n_e} \mathbf{p} \cdot \mathbf{e} \tag{8.1-59}$$

We note that the transmission coefficients are mostly dependent on the scalar product between the electric field of the normal modes in the incident medium and the uniaxial medium (e.g., polarizer or LC).

8.1.2. Matrix Formulation

Referring to Figure 8.2, we now consider a uniaxial plate of finite thickness d. When the ordinary and extraordinary waves arrive at the output face of the plate, both the reflected and transmitted waves are generated. If we ignore the multiple reflections that are due to the two parallel faces of the plate, the amplitudes of the transmitted waves can be written as

$$A'_s = C_o t_{os} e^{-ik_{oz}d} + C_e t_{es} e^{-ik_{ez}d} \qquad (8.1\text{-}60)$$

$$A'_p = C_o t_{op} e^{-ik_{oz}d} + C_e t_{ep} e^{-ik_{ez}d} \qquad (8.1\text{-}61)$$

where t_{os}, t_{es}, t_{op}, and t_{ep} are another set of transmission coefficients that can also be derived from the continuity conditions.

For the incidence condition shown in Figure 8.1 where the c axis is parallel to the surface of the plate, the z component of the wave vectors (k_{ez}, and k_{oz}) are given by

$$k_{ez} = n_e \frac{\omega}{c} \sqrt{1 - \frac{(\sin\theta \sin\phi)^2}{n_o^2} - \frac{(\sin\theta \cos\phi)^2}{n_e^2}}$$

$$k_{oz} = n_o \frac{\omega}{c} \sqrt{1 - \frac{\sin^2\theta}{n_o^2}}$$

where we recall that the incident wave is propagating in the yz-plane with an angle of incidence of θ, and ϕ is the angle between the c axis and the x axis. For the case of a general incidence with an arbitrary orientation of the c axis, the

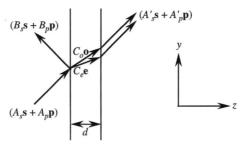

Figure 8.2.   Transmission of light through a uniaxial plate.

z components of the wavevectors are given by Equations (8.1-124) and (8.1-125).

The transmission of light through such a plate can be managed by using Eqs. (8.1-28), (8.1-29), (8.1-60), and (8.1-61), which are now rewritten in the 2×2 matrix form:

$$\begin{pmatrix} A'_s \\ A'_p \end{pmatrix} = \begin{pmatrix} t_{es} & t_{os} \\ t_{ep} & t_{op} \end{pmatrix} \begin{pmatrix} e^{-ik_{ez}d} & 0 \\ 0 & e^{-ik_{oz}d} \end{pmatrix} \begin{pmatrix} t_{se} & t_{pe} \\ t_{so} & t_{po} \end{pmatrix} \begin{pmatrix} A_s \\ A_p \end{pmatrix} \qquad (8.1\text{-}62)$$

This matrix equation relates the transmitted wave amplitudes A'_s and A'_p in terms of the amplitudes of the incident waves A_s and A_p. Note that this equation is valid provided the multiple reflection in the plate can be neglected. Equation (8.1-62) can be conveniently written as

$$\begin{pmatrix} A'_s \\ A'_p \end{pmatrix} = D_o P D_i \begin{pmatrix} A_s \\ A_p \end{pmatrix} \qquad (8.1\text{-}63)$$

where P is the propagation matrix and D_i and D_o are the input and output dynamical matrices, respectively. Then P, D_i, and D_o are given, respectively, by

$$P = \begin{pmatrix} e^{-ik_{ez}d} & 0 \\ 0 & e^{-ik_{oz}d} \end{pmatrix} \qquad (8.1\text{-}64)$$

$$D_i = \begin{pmatrix} t_{se} & t_{pe} \\ t_{so} & t_{po} \end{pmatrix}, \qquad D_o = \begin{pmatrix} t_{es} & t_{os} \\ t_{ep} & t_{op} \end{pmatrix} \qquad (8.1\text{-}65)$$

We note that both Eqs. (8.1-62) and (8.1-63) resemble the Jones matrix formulation. In fact, these equations are the generalization of the Jones matrix method. Therefore, this method is called the *extended Jones matrix method*. We must remember that D_i and D_o are no longer rotation matrices. In the special case of normal incidence with $n \approx n_o \approx n_e$, (**s**,**p**) and (**e**,**o**) become two sets of coplanar coordinate axes. These two matrices D_i and D_o reduce to rotation matrices, according to Eqs. (8.1-56)–(8.1-59).

The matrix formalism developed above can be used for calculating the transmission characteristics of a series of birefringent elements (i.e., LC cells, wave plates and polarizers) for off-axis light. To do this, we write down the Jones vector of the incident beam and then write down the 2×2 matrices of the various elements. The Jones vector of the emerging beam is obtained by carrying out the matrix multiplication in sequence:

$$\begin{pmatrix} A'_s \\ A'_p \end{pmatrix} = \begin{pmatrix} M_{11} & M_{12} \\ M_{21} & M_{22} \end{pmatrix} \begin{pmatrix} A_s \\ A_p \end{pmatrix} \qquad (8.1\text{-}66)$$

where the M_{ij} are the matrix elements of the overall transfer matrix obtained by the multiplication of all the 2×2 matrices in sequence. The fraction of energy transmitted is

$$T = \frac{|A_s'|^2 + |A_p'|^2}{|A_s|^2 + |A_p|^2} \qquad (8.1\text{-}67)$$

which depends on the matrix elements M_{ij} as well as on the polarization state of the incident beam (i.e., A_s, A_p). In practice, one often deals with an incident beam of unpolarized light. In this case, the Jones vector of the incident beam can be written as

$$\begin{pmatrix} A_s \\ A_p \end{pmatrix} = \frac{E_0}{\sqrt{2}} \begin{pmatrix} e^{i\alpha_1} \\ e^{i\alpha_2} \end{pmatrix} \qquad (8.1\text{-}68)$$

where α_1 and α_2 are random variables in a sense that the time-averaged quantities $[\cos(\alpha_1 - \alpha_2)]$ and $[\sin(\alpha_1 - \alpha_2)]$ vanish simultaneously. The fraction of energy transmitted in this case is thus, according to Eqs. (8.1-66)–(8.1-68), given by

$$T = \tfrac{1}{2}(|M_{11}|^2 + |M_{12}|^2 + |M_{21}|^2 + |M_{22}|^2) \qquad (8.1\text{-}69)$$

8.1.3. Small Birefringence Approximation

As we have seen, the general expression for the reflection and transmission coefficients are complicated, especially for off-axis light with the c axis of the crystal oriented at an arbitrary angle ϕ. It is desirable to have approximate expressions for these eight coefficients.

When the birefringence is small (i.e., $|n_e - n_o| \ll n_o, n_e$), the eight coefficients can be greatly simplified by an approximation. In this approximation, we derive the expressions for the reflection and transmission coefficients, disregarding the anisotropy of the elements. This is legitimate because the wavevectors $\mathbf{k}_o$ and $\mathbf{k}_e$ are almost equal (i.e., $\mathbf{k}_o \approx \mathbf{k}_e$) according to the discussion in Section 3.4, provided $|n_e - n_o| \ll n_o, n_e$. Also, the refraction angles θ_o and θ_e are almost equal (i.e., $\theta_e \approx \theta_o$), and the polarization vectors $\mathbf{o}$ and $\mathbf{e}$ can be given approximately by

$$\mathbf{o} = \frac{\mathbf{k}_o \times \mathbf{c}}{|\mathbf{k}_o \times \mathbf{c}|} \qquad (8.1\text{-}70)$$

and

$$\mathbf{e} = \frac{\mathbf{o} \times \mathbf{k}_o}{|\mathbf{o} \times \mathbf{k}_o|} \qquad (8.1\text{-}71)$$

respectively [note that Eq. (8.1-70) is exact]. Here we select a proper sign for these two unit vectors so that $(\mathbf{e}, \mathbf{o}, \mathbf{k}_o)$ form a right-handed set of orthogonal vectors. As a result of the small birefringence, an s wave retains its s polarization on refraction from the interfaces. The same thing happens to the p wave. Consequently, these eight coefficients can be given approximately by

$$r_{ss} = r_s \tag{8.1-72}$$

$$r_{sp} = r_{ps} = 0 \tag{8.1-73}$$

$$r_{pp} = r_p \tag{8.1-74}$$

and

$$t_{so} = \mathbf{s} \cdot \mathbf{o} t_s \tag{8.1-75}$$

$$t_{se} = \mathbf{s} \cdot \mathbf{e} t_s \tag{8.1-76}$$

$$t_{po} = \mathbf{p} \cdot \mathbf{o} t_p \tag{8.1-77}$$

$$t_{pe} = \mathbf{p} \cdot \mathbf{e} t_p \tag{8.1-78}$$

where $\mathbf{o}$ and $\mathbf{e}$ are unit vectors of the normal modes in the uniaxial medium and are given by Eqs. (8.1-70) and (8.1-71) and r_s, r_p, t_s, and t_p are the Fresnel reflection and transmission coefficients given by

$$r_s = \frac{n \cos \theta - n_o \cos \theta_o}{n \cos \theta + n_o \cos \theta_o} \tag{8.1-79}$$

$$r_p = \frac{n \cos \theta_o - n_o \cos \theta}{n_o \cos \theta + n \cos \theta_o} \tag{8.1-80}$$

and

$$t_s = \frac{2n \cos \theta}{n \cos \theta + n_o \cos \theta_o} \tag{8.1-81}$$

$$t_p = \frac{2n \cos \theta}{n \cos \theta_o + n_o \cos \theta} \tag{8.1-82}$$

where we recall that n is the index of refraction of the incident medium (usually glass or air), n_o is the index of refraction of the birefringent element (since $n_e \approx n_o$), θ is the incident angle, and θ_o is the refraction angle ($n_e \approx n_o$). Here t_s and t_p are the transmission coefficients for the s and p waves, respectively, on entering the birefringent element disregarding the anisotropy.

In a similar approach, the transmission coefficients on leaving the uniaxial plate are obtained as

$$t_{os} = \mathbf{o} \cdot \mathbf{s} t'_s \tag{8.1-83}$$

$$t_{es} = \mathbf{e} \cdot \mathbf{s} t'_s \tag{8.1-84}$$

$$t_{op} = \mathbf{o} \cdot \mathbf{p} t'_p \tag{8.1-85}$$

$$t_{ep} = \mathbf{e} \cdot \mathbf{p} t'_p \tag{8.1-86}$$

where t'_s and t'_p are the corresponding Fresnel transmission coefficients given by

$$t'_s = \frac{2n_o \cos \theta_o}{n_o \cos \theta_o + n \cos \theta} \tag{8.1-87}$$

$$t'_p = \frac{2n_o \cos \theta_o}{n_o \cos \theta + n \cos \theta_o} \tag{8.1-88}$$

These approximate expressions may be used to study the wide-field property of birefringent optical systems. In most liquid crystals, $|n_e - n_o| \sim 0.1$. Small birefringence approximation usually leads to satisfactory results in practical calculations.

Using Eqs. (8.1-75)–(8.1-78) and (8.1-83)–(8.1-86), the input and output dynamical matrices D_i and D_o [(8.1-65)] can be written as

$$D_i = \begin{pmatrix} \mathbf{s} \cdot \mathbf{e} t_s & \mathbf{p} \cdot \mathbf{e} t_p \\ \mathbf{s} \cdot \mathbf{o} t_s & \mathbf{p} \cdot \mathbf{o} t_p \end{pmatrix} \tag{8.1-89}$$

and

$$D_o = \begin{pmatrix} \mathbf{e} \cdot \mathbf{s} t'_s & \mathbf{o} \cdot \mathbf{s} t'_s \\ \mathbf{e} \cdot \mathbf{p} t'_p & \mathbf{o} \cdot \mathbf{p} t'_p \end{pmatrix} \tag{8.1-90}$$

respectively.

We assume that the refractive indices are well matched at the interface so that $(\mathbf{s}, \mathbf{p})$ and $(\mathbf{e}, \mathbf{o})$ are two sets of coplanar orthogonal unit vectors. Thus, we may define an angle ψ such that (see Fig. 8.3)

$$\mathbf{e} = \mathbf{s} \cos \psi + \mathbf{p} \sin \psi \tag{8.1-91}$$

$$\mathbf{o} = -\mathbf{s} \sin \psi + \mathbf{p} \cos \psi \tag{8.1-92}$$

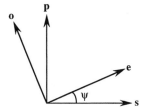

Figure 8.3. Rotation of coordinates.

In fact, the $(\mathbf{e},\mathbf{o})$ axes may be obtained by a right-handed rotation of the $(\mathbf{s},\mathbf{p})$ axes by an angle ψ about the wavevector $\mathbf{k}_0$ (see Fig. 8.3). If we further define two diagonal transmission matrices as

$$T_i = \begin{pmatrix} t_s & 0 \\ 0 & t_p \end{pmatrix} \tag{8.1-93}$$

$$T_o = \begin{pmatrix} t'_s & 0 \\ 0 & t'_p \end{pmatrix} \tag{8.1-94}$$

then the dynamical matrices D_i and D_o can be written, according to Eqs. (8.1-89)–(8.1-94), as

$$D_i = \begin{pmatrix} \cos\psi & \sin\psi \\ -\sin\psi & \cos\psi \end{pmatrix} \begin{pmatrix} t_s & 0 \\ 0 & t_p \end{pmatrix} \equiv R(\psi)T_i \tag{8.1-95a}$$

and

$$D_o = \begin{pmatrix} t'_s & 0 \\ 0 & t'_p \end{pmatrix} \begin{pmatrix} \cos\psi & -\sin\psi \\ \sin\psi & \cos\psi \end{pmatrix} \equiv T_o R(-\psi) \tag{8.1-95b}$$

respectively. The $R(\psi)$ is simply the transformation matrix for the rotation of the coordinate by an angle ψ.

Let ϕ be the angle between the c axis and the x axis (see Fig. 8.1) so that the unit vector $\mathbf{c}$ representing the c axis can be written as

$$\mathbf{c} = \mathbf{x}\cos\phi + \mathbf{y}\sin\phi \tag{8.1-96}$$

This angle ψ can be expressed in terms of ϕ and θ_o as, according to

Eqs. (8.1-70) and (8.1-71)

$$\sin \psi = \frac{\cos \theta_o \sin \phi}{(1 - \sin^2 \theta_o \sin^2 \phi)^{1/2}} \qquad (8.1\text{-}97)$$

$$\cos \psi = \frac{\cos \phi}{(1 - \sin^2 \theta_o \sin^2 \phi)^{1/2}} \qquad (8.1\text{-}98)$$

where we recall that θ_o is the refraction angle in the uniaxial medium ($n_o \sin \theta_o = n \sin \theta$). For normal incidence ($\theta = \theta_o = 0$), this angle is simply $\psi = \phi$.

By using these new definitions, we can now write matrix (8.1-63) as

$$\begin{pmatrix} A'_s \\ A'_p \end{pmatrix} = T_o R(-\psi) P R(\psi) T_i \begin{pmatrix} A_s \\ A_p \end{pmatrix} \qquad (8.1\text{-}99)$$

where each of the five matrices has its own physical meaning. Counting from the right-hand side, the first matrix T_i accounts for the Fresnel refraction (or transmission) of the incident beam at the input surface; the second matrix $R(\psi)$ accounts for the transformation that decomposes the light beam into a linear combination of the normal modes of propagation of the birefringent element (i.e., o wave and e wave). The third matrix P accounts for the propagation of these normal modes through the bulk of the plate; the fourth matrix $R(-\psi)$ transforms the light beam back into a linear combination of the s and p waves at the exit end of the plate; and the fifth matrix T_o accounts for the Fresnel refraction (or transmission) of the light beam at the output surface of the plate.

Example 8.1. Extended Jones Matrix Method for an a Plate Sandwiched between a Pair of Crossed Polarizers. Referring to Figure 4.12, we consider an a plate sandwiched between a pair of crossed polarizers. For simplicity, let us neglect the surface reflection at the air–polarizer interfaces. Assume that the two polarizers are ideal O-type polarizers, where the O wave is completely transmitted while the E wave is completely blocked. Let A_o and A_e be the amplitudes of the O and E waves in the entrance polarizer, respectively; and A'_o and A'_e, those in the exit polarizer, respectively. Similar to Eq. (8.1-62), we can obtain

$$\begin{pmatrix} A'_e \\ A'_o \end{pmatrix} = M \begin{pmatrix} A_e \\ A_o \end{pmatrix} = \begin{pmatrix} t'_{ee} & t'_{oe} \\ t'_{eo} & t'_{oo} \end{pmatrix} \begin{pmatrix} e^{-ik_{ez}d} & 0 \\ o & e^{-ik_{oz}d} \end{pmatrix} \begin{pmatrix} t_{ee} & t_{oe} \\ t_{eo} & t_{oo} \end{pmatrix} \begin{pmatrix} A_e \\ A_o \end{pmatrix}$$

where $t_{ij}(i, j = e, o)$ = transmission coefficient from i component in entrance polarizer to j component in a plate

k_{oz}, k_{ez} = z-components of ordinary and extraordinary waves inside a plate, respectively
d = thickness of a plate
$t'_{mn}(m, n = e, o)$ = transmission coefficient from the m component in a plate to n component in exit polarizer
M = extended Jones matrix.

Furthermore, we assume that the index of refraction of the polarizer is the same as that of the ordinary index of refraction of the a plate. Similar to what led to Eqs. (8.1-75)–(8.1-78) and (8.1-83)–(8.1-86), we arrive at the conclusion that under the small birefringence approximation each transmission coefficient is simply the inner product of the corresponding two polarization unit vectors. This statement will be proved later in this section [see the argument leading to Eqs. (8.1-126) and (8.1-127)]. For example, t_{oe} = o(entrance polarizer)·e(a plate).

The o and e vectors in each layer can be obtained from Eqs. (8.1-70) and (8.1-71) by using the corresponding c axis and the ordinary wave vector $\mathbf{k}_o$. Since we are assuming ideal O type polarizers, the total transmittance is given by

$$T = \frac{1}{2}\frac{|A'_o|^2}{|A_o|^2} = \frac{1}{2}|M_{11}|^2$$

where the factor $\frac{1}{2}$ comes from the assumption of unpolarized incident light. Using the parameters $n_o = 1.5$, $n_e = 1.6, d = 52.5\,\mu m$, and $\lambda = 0.5\,\mu m$, we obtain the transmittance according to the preceding equations. The resulting intensity transmission pattern is shown in Figure 8.4. Comparing Figures 8.4 with 4.13 (calculated with the ordinary Jones matrix method), we note that the

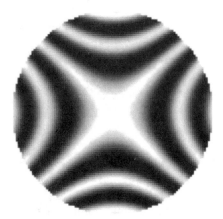

Figure 8.4. Intensity transmission pattern of an a plate between crossed polarizers. (The c axis is oriented at 45° relative to the polarizer transmission axes.)

intensity transmission patterns of an a plate between crossed polarizers calculated using the ordinary and extended Jones matrix methods are almost the same. However, this is not the case for a general birefringent network (e.g., a c plate between a pair of crossed polarizers). ∎

8.1.4. Comparison with the Conventional Jones Calculus

In the conventional Jones calculus, the input and output vectors are related by

$$\begin{pmatrix} A'_s \\ A'_p \end{pmatrix} = R(-\psi_o)PR(\psi_o)\begin{pmatrix} A_s \\ A_p \end{pmatrix} \tag{8.1-100}$$

where R is the rotation matrix, ψ_o is the angle between the c axis and the x axis ($\psi_o = \phi$), and P is the propagation matrix. In comparison with Eq. (8.1-99), we note that the newly developed 2×2 matrix method includes the effect of unequal transmittances for the s and p waves as well as the angular dependence of the azimuth angle ψ, which is different from ψ_o for off-axis light. Strictly speaking, the conventional Jones calculus is valid only at normal incidence ($\theta = \theta_0 = 0$) when the azimuth angle ψ becomes $\psi_0(\psi_0 = \phi)$ and the transmission coefficients become independent of the polarization states (i.e., $t_s = t_p$, $t'_s = t'_p$), provided $|n_e - n_o| \ll n_e, n_o$. In fact, if we put $\theta = \theta_o = 0$ in Eqs. (8.1-81), (8.1-82), (8.1-87), (8.1-88), (8.1-97), and (8.1-98), then substitute them into Eq. (8.1-99), we obtain

$$\begin{pmatrix} A'_s \\ A'_p \end{pmatrix} = t'tR(-\psi_o)PR(\psi_o)\begin{pmatrix} A_s \\ A_p \end{pmatrix} \tag{8.1-101}$$

where t and t' are the Fresnel transmission coefficients at normal incidence:

$$t = \frac{2n}{n + n_o} \tag{8.1-102}$$

$$t' = \frac{2n_o}{n + n_o} \tag{8.1-103}$$

Note that Eq. (8.1-101) is different from Eq. (8.1-100) by a factor of only $t't = 4nn_o/(n + n_o)^2$, which is simply the product of the Fresnel transmission coefficients associated with the two interfaces.

For off-axis light, the Fresnel transmission is polarization dependent even if $n_e = n_o$ (i.e., $t_s \neq t_p$). Generally speaking, the p wave has a higher transmission coefficient (i.e., $t_p > t_s$ and $t'_p > t'_s$). It is convenient to define a quantity σ that measures the fractional difference in these two coefficients:

$$\sigma = \frac{t_p - t_s}{t_p} = \frac{t_p' - t_s'}{t_p'} \qquad (8.1\text{-}104)$$

By using Eqs. (8.1-81), (8.1-82), (8.1-87), and (8.1-88), we can write Eq. (8.1-104) as

$$\sigma = \frac{(n_o - n)(\cos\theta_o - \cos\theta)}{n\cos\theta + n_o\cos\theta_o} \qquad (8.1\text{-}105)$$

This quantity σ is zero at normal incidence ($\theta = 0$) and increases monotonically as θ increases. It becomes $(n_o - n)^2/(n_o^2 + n^2)$ at Brewster's angle $\theta_B = \tan^{-1}(n_o/n)$, and reaches $(n_o - n)/n_o$ at grazing angle ($\theta = 90°$). For E7, this difference is about 1.8% at $\theta = 30°$, whereas σ is about 0.67% for ZLI-1646 at $\theta = 20°$. This difference is equivalent to a slight rotation of the polarization state (for linearly polarized light) and thus may contribute to the leakage of light in LCDs at large viewing angles. This leakage of light can be investigated analytically only by using the extended Jones matrix method.

Another interesting and important obliquity effect derived directly from the extended Jones matrix method is the fact that the azimuth angle ψ between the $(\mathbf{e}, \mathbf{o})$ axes and the $(\mathbf{s}, \mathbf{p})$ axes is different from $\psi_o (\psi_o = \phi)$ for off-axis light. According to Eqs. (8.1-97) and (8.1-98), ψ is a trigonometric function of ϕ. Thus the $\mathbf{e}$ axes (or $\mathbf{o}$ axes) of a pair of orthogonally oriented birefringent elements (i.e., in the sense that the c axes are orthogonal) may not be mutually perpendicular for off-axis light. In other words, $|\psi(\theta, \phi) - \psi(\theta, \phi + \pi/2)|$ is, in general, not $\pi/2$ for incident light with $\theta \neq 0$. For example, we consider a series of two quartz LC plates with their c axes oriented at $\phi = -45°, 45°$, respectively. For normally incident light, the azimuth angles of the $\mathbf{e}$ axis of the two plates are $\psi_o = -45°, 45°$, respectively. When the incident angle is $\theta = 30°$, taking $n_o = 1.55$, these azimuth angles are $\psi = -43.43°, 43.43°$, respectively. Note that the difference is 86.86°, which means that the two $\mathbf{e}$ axes are not orthogonal. This is the main reason why even a pair of crossed ideal sheet polarizers leaks obliquely incident light (see further discussion in Section 8.1.5).

The deviation of the azimuth angle ψ from ψ_o can be expressed in terms of ϕ and θ_o, according to Eqs. (8.1-97) and (8.1-98), as

$$\sin(\psi - \psi_o) = -\frac{\sin 2\phi \sin^2(\theta_o/2)}{(1 - \sin^2\theta_o \sin^2\phi)^{1/2}} \qquad (8.1\text{-}106)$$

Note that this deviation disappears at the special angles $\phi = 0, \pm\pi/2, \pm\pi$, and

$\theta_o = 0$. The magnitude of the azimuth deviation $|\psi - \psi_o|$ is a monotonically increasing function of θ_o (or θ).

8.1.5.   Crossed Polarizers

As an example, to illustrate the use of the extended Jones matrix method, we now consider the transmission of light through a pair of crossed ideal sheet polarizers. We assume that these sheet polarizers are characterized by their complex refractive indices with $n_e = n_o$, $\kappa_o = 0, 0 < \kappa_e \ll 1$, and thickness d such that $1 \ll 2\pi\kappa_e d/\lambda$. Thus the propagation matrix of these ideal polarizers is given by

$$P = \begin{pmatrix} 0 & 0 \\ 0 & 1 \end{pmatrix} \tag{8.1-107}$$

Since $n_e - n_o$ and $\kappa_o, \kappa_e \ll 1$, we may use the small birefringence approximation. Let ϕ and $\phi + \pi/2$ be the azimuth angles of the c axes of the polarizers, respectively, and θ be the angle of incidence. By using Eq. (8.1-99) and carrying out the multiplication, we obtain the 2×2 matrix that represents the polarizer with c axis oriented at ϕ:

$$M(\phi) = \begin{pmatrix} T_s \sin^2 \psi_1 & -T \sin \psi_1 \cos \psi_1 \\ -T \sin \psi_1 \cos \psi_1 & T_p \cos^2 \psi_1 \end{pmatrix}, \tag{8.1-108}$$

where

$$T_s = t'_s t_s, \qquad T_p = t'_p t_p, \qquad T = (T_s T_p)^{1/2} \tag{8.1-109}$$

and ψ_1 is given by

$$\sin \psi_1 = \frac{\cos \theta_o \sin \phi}{(1 - \sin^2 \theta_o \sin^2 \phi)^{1/2}} \tag{8.1-110}$$

$$\cos \psi_1 = \frac{\cos \phi}{(1 - \sin^2 \theta_o \sin^2 \phi)^{1/2}} \tag{8.1-111}$$

In a similar way, the 2×2 matrix associated with the polarizer with c axis oriented at $\phi + \pi/2$ is given by

$$M\left(\phi + \frac{\pi}{2}\right) = \begin{pmatrix} T_s \sin^2 \psi_2 & -T \sin \psi_2 \cos \psi_2 \\ -T \sin \psi_2 \cos \psi_2 & T_p \cos^2 \psi_2 \end{pmatrix}, \tag{8.1-112}$$

with ψ_2 given by

$$\sin \psi_2 = \frac{\cos \theta_o \cos \phi}{(1 - \sin^2 \theta_o \cos^2 \phi)^{1/2}} \tag{8.1-113}$$

$$\cos \psi_2 = \frac{- \sin \phi}{(1 - \sin^2 \theta_o \cos^2 \phi)^{1/2}} \tag{8.1-114}$$

To find the fraction of energy transmitted through the crossed polarizers, we multiply the two matrices $M(\phi)$ and $M(\phi + \pi/2)$ and use Eq. (8.1-69), assuming that the incident light is unpolarized. This leads to the following expression for the transmission:

$$\tau = \tfrac{1}{2}[s_1^2 s_2^2 (T_s^2 s_1 s_2 + T^2 c_1 c_2)^2 + T^2 s_1^2 c_2^2 (T_s s_1 s_2 + T_p c_1 c_2)^2 \\ + T^2 c_1^2 s_2^2 (T_s s_1 s_2 + T_p c_1 c_2)^2 + c_1^2 c_2^2 (T^2 s_1 s_2 + T_p^2 c_1 c_2)^2] \tag{8.1-115}$$

where $c_1 = \cos \psi_1, c_2 = \cos \psi_2, s_1 = \sin \psi_1$, and $s_2 = \sin \psi_2$. Note that τ is a function of θ and ϕ implicitly. Calculated transmission τ as a function of θ for various values of ϕ is shown in Figure 8.4. If we now ignore the difference between T_s and T_p and simply take $T_s = T_p = T$, the transmission τ [Eq. (8.1-115)] becomes

$$\tau = \tfrac{1}{2} T^4 \cos^2 (\psi_1 - \psi_2) \tag{8.1-116}$$

or, equivalently, by using Eqs. (8.1-110), (8.1-111), (8.1-113), and (8.1-114), this becomes

$$\tau = \frac{T^4 \sin^4 \theta_o \sin^2 \phi \cos^2 \phi}{2(1 - \sin^2 \theta_o \sin^2 \phi)(1 - \sin^2 \theta_o \cos^2 \phi)} \tag{8.1-117}$$

Expression (8.1-117) consists of two parts: (1) the factor T^4 accounts for the Fresnel transmission of light through the four interfaces of the two polarizers, and (2) the rest is a geometric contribution that results from the azimuth deviation discussed in Section 8.1.4. For any given angle of incidence θ, τ is maximum when $\phi = \pm \pi/4$ or $\pm 3\pi/4$. Since T is a decreasing function of θ, whereas the geometric factor $\cos^2 (\psi_1 - \psi_2)$ is an increasing function of θ, the cross-transmittance τ increases from zero as θ increases and reaches its maximum at a certain angle of incidence θ_{max} and then decreases to zero sharply as θ approaches $\pi/2$ (see Fig. 8.5) as $T(\pi/2) = 0$. The angle θ_{max} is usually

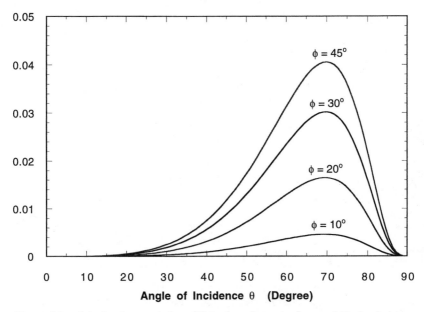

Figure 8.5. Calculated transmission of light through a pair of crossed ideal polarizers.

greater than Brewster's angle $\theta_B = \tan^{-1}(n_o/n)$ as T is almost uniform from $0 \leq \theta \leq \theta_B$ and drops sharply to zero for $\theta_B < \theta$.

The obliquity effects of a pair of crossed HN-22 polarizers have been studied experimentally [12]. The calculated results (8.1-115)–(8.1-117), based on the extended matrix method, agree with these experimental results.

8.1.6. Arbitrary c Axis Orientation

In the case of transmission through a uniaxial plate with an arbitrary c-axis orientation, the extended Jones matrix method [Eqs. (8.1-63)–(8.1-65)] as well as the small birefringence approximation are still applicable. Using the extended Jones matrix method, the z components of the ordinary and the extraordinary waves k_{oz} and k_{ez} can be derived from the expression of the normal surface. Here we recall that the z-axis is perpendicular to the plate and that the x and y components of the wavevectors are the same as those of the incident wavevector. Referring to Figure 8.6, we define θ_c as the angle between the c axis and the z axis and ϕ_c as the angle between the projection of the c axis on the (x, y) plane and the x axis. To calculate the z component of the extraordinary wave k_{ez}, we use the principal coordinate system of the uniaxial medium. The unit vector $\mathbf{c}$ can be written as

$$\mathbf{c} = \mathbf{x}\sin\theta_c\cos\phi_c + \mathbf{y}\sin\theta_c\sin\phi_c + \mathbf{z}\cos\theta_c \qquad (8.1\text{-}118)$$

where $\mathbf{x}$, $\mathbf{y}$, and $\mathbf{z}$ are unit vectors of the coordinate axes. The wavevector

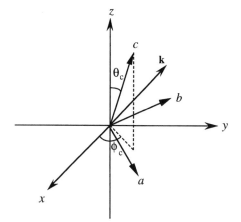

Figure 8.6. Orientation of the c axis of a uniaxial plate, where θ_c is the angle between the c axis and the z axis and ϕ_c is the angle between the projection of the c axis on the (x,y) plane and the x direction, respectively. The z axis is perpendicular to the plate surface (xy plane). The b axis is chosen to be perpendicular to the plane formed by the c axis and the z axis. (a,b,c) form a set of mutually orthogonal coordinate axes.

components of the extraordinary wave, in the principal coordinate system (see Fig. 8.6), can be written as

$$
\begin{aligned}
k_{ea} &= (\alpha \cos \phi_c + \beta \sin \phi_c) \cos \theta_c - k_{ez} \sin \theta_c \\
k_{eb} &= -\alpha \sin \phi_c + \beta \cos \phi_c \\
k_{ec} &= (\alpha \cos \phi_c + \beta \sin \phi_c) \sin \theta_c + k_{ez} \cos \theta_c
\end{aligned}
\tag{8.1-119}
$$

where α and β are the x and y components of the incident wavevector, respectively, and **a** and **b** directions are chosen such that **b** is perpendicular to **z** while both **a** and **b** are perpendicular to **c**. The normal surface for the extraordinary waves is given by

$$
\frac{k_{ea}^2 + k_{eb}^2}{n_e^2} + \frac{k_{ec}^2}{n_o^2} = \left(\frac{\omega}{c}\right)^2
\tag{8.1-120}
$$

where n_o and n_e are the ordinary and the extraordinary indices of refraction. Substituting Eqs. (8.1-119) into Eq.(8.1-120), we obtain

$$
u k_{ez}^2 - v k_{ez} + w = 0
\tag{8.1-121}
$$

where

$$
u = \frac{\sin^2 \theta_c}{n_e^2} + \frac{\cos^2 \theta_c}{n_o^2}
$$

$$
v = k_d \sin 2\theta_c \left(\frac{1}{n_e^2} - \frac{1}{n_o^2}\right)
\tag{8.1-122}
$$

$$
w = \frac{k_d^2 \cos^2 \theta_c + k_{eb}^2}{n_e^2} + \frac{k_d^2 \sin^2 \theta_c}{n_o^2} - \left(\frac{\omega}{c}\right)^2
$$

with

$$k_d = \alpha \cos \phi_c + \beta \sin \phi_c \qquad (8.1\text{-}123)$$

Solving Eq. (8.1-121), we obtain the z component of the extraordinary wave:

$$k_{ez} = \frac{v + \sqrt{v^2 - 4uw}}{2u} \qquad (8.1\text{-}124)$$

where we have taken the positive sign for the square root, since the light is transmitting in the $+z$ direction.

In the special case when the c axis is parallel to the surface of the crystal plate, we have $\theta_c = 90°$, the above equation reduces to

$$k_{ez} = n_e \sqrt{\left(\frac{\omega}{c}\right)^2 - \frac{(\alpha \cos \phi_c + \beta \sin \phi_c)^2}{n_o^2} - \frac{(-\alpha \sin \phi_c + \beta \cos \phi_c)^2}{n_e^2}}$$

where we recall that α and β are the x- and y-component of the incident wavevector, respectively.

The z component of the ordinary wave k_{oz} is independent of the orientation of the c axis and is given by

$$k_{oz} = \sqrt{\left(\frac{n_o \omega}{c}\right)^2 - \alpha^2 - \beta^2} \qquad (8.1\text{-}125)$$

Once the z components of the wavevector of the normal modes (k_{ez} and k_{oz}) are obtained, they can be used to calculate the propagation matrix by using Eq. (8.1-64) in the extended Jones matrix method. The unit vectors **e** and **o** in the birefringent medium can be obtained by using Eq. (8.1-70) and (8.1-71). Once these unit vectors are available, the dynamical matrices can be obtained by using Eq. (8.1-89) and (8.1-90). The overall Jones matrix is obtained by multiplying all the dynamical matrices and the propagation matrices in sequence.

In Chapter 4, Figure 4.15(*a,b*) showed the difference between the result obtained using the ordinary Jones matrix method and that using the extended Jones matrix method. We notice that the difference for a *c* plate between crossed polarizers is more dramatic than that for an *a* plate. For an arbitrary *c* axis orientation, the following example shows an application of the extended Jones matrix method.

Example 8.2: Extended Jones Matrix Method for an o-Plate. An *o* plate is a uniaxial plate with arbitrary *c* axis orientation. In other words, the *c* axis is

Figure 8.7. Conoscopic intensity pattern of an o plate between crossed polarizers.

oriented at an oblique angle relative to the surface. Consider an o plate with $\theta_c = 15°, \phi_c = 20°, n_o = 1.5, n_e = 1.6$, and $d = 50\,\mu m$, sandwiched between a pair of crossed polarizers with their transmission axes oriented at $\pm45°$. Using the approach as described in Section 8.1.3 for the a plate example and using Eqs. (8.1-124) and (8.1-125) to calculate the k_{ez} and k_{oz} values, we obtain the intensity transmission pattern as shown in Figure 8.7. In our calculation, the wavelength is chosen as $\lambda = 0.5\,\mu m$. Comparing Figure 8.7 with Figure 4.15b for a c plate, we note that as a result of the tilt of the c axis in the o plate, the rings and brushes are tilted and distorted. In Figure 8.7 the optical axis is shown pointing toward the upper right quadrant.   ■

The extended Jones matrix method with the small birefringence approximation can be summarized into the following steps:

Step 1. Given a general angle of incidence with a wavevector **k**, the plane of incidence is defined as the plane formed by the incident wavevector and the unit normal vector (**z**) of the surface of the birefringent plate. The unit vector **s** is perpendicular to the plane of incidence. The unit vector **p** is obtained by using Eq. (8.1-4).

Step 2. The wavevector $\mathbf{k}_o$ including the angle of refraction θ_o in the birefringent medium can be determined by using the conventional Snell's law. This wavevector $\mathbf{k}_o$ lies in the plane of incidence. The tangential components (α, β) of this wavevector $\mathbf{k}_o$ are the same as those of the incident wavevector **k**.

Step 3. The Fresnel transmission coefficients t_s, t_p, t'_s, t'_p can be obtained by using Eq. (8.1-81), (8.1-82), (8.1-87), and (8.1-88).

Step 4. The unit vectors **e** and **o** representing the polarization states of the normal modes in the birefringent medium can be obtained by using Eqs. (8.1-70) and (8.1-71).

Step 5. The dynamical matrices D_i and D_o can then be obtained by using Eqs. (8.1-89) and (8.1-90).

Step 6. The z components of the wavevector of the normal modes (k_{ez} and k_{oz}) can be obtained by using Eqs. (8.1-124) and (8.1-125).

Step 7. The propagation matrix **P** can then be obtained by substituting the z components of the wavevector of the normal modes (k_{ez} and k_{oz}) into Eq. (8.1-64).

Step 8. The Jones vector of the transmitted beam can then be obtained by using the matrix Eq. (8.1-63). The intensity transmission is obtained by using either Eq. (8.1-67) or Eq. (8.1-69).

8.1.7.   Application to Liquid Crystal Displays

Most LCD devices are made of a cell of twisted nematic liquid crystals (TN-LCs) sandwiched between transparent electrodes and polarizers. Although the LC medium is not homogeneous, it can be analyzed as a birefringent network. Referring to Figure 8.8, we divide a TN-LC into N layers. Each layer can be considered as a homogeneous and uniaxially birefringent medium. The orientation of the c axis may change from layer to layer. The ordinary and extraordinary indices of refraction, n_o and n_e, are constants for all layers. To derive the extended Jones matrix for such a LCD medium, we first examine the 2×2 matrix for a plane wave entering the $(n+1)$th layer from the nth layer. The 2×2 matrix can be derived by matching the boundary conditions directly. However, since we have already obtained the 2×2 dynamical matrix for a boundary between an isotropic medium and a uniaxial medium, we simply employ the previous results to derive the dynamical matrix between the nth and the $(n+1)$th layers.

A simple approach to derive the dynamical matrix between the nth and the $(n + 1)$th layers is to introduce an imaginary isotropic layer that is sandwiched between the nth and the $(n+1)$th layers. The imaginary isotropic layer has an index of refraction of n_o, with an infinitesimal thickness so that it does not introduce additional refraction or reflection. Denote the field amplitudes of the light before leaving the nth layer as $\begin{pmatrix} A'_e \\ A'_o \end{pmatrix}_n$ and those after entering the $(n + 1)$th layer as $\begin{pmatrix} A_e \\ A_o \end{pmatrix}_{n+1}$. They are related by

$$\begin{pmatrix} A_e \\ A_o \end{pmatrix}_{n+1} = \begin{pmatrix} t_{se} & t_{pe} \\ t_{so} & t_{po} \end{pmatrix}_{n+1} \begin{pmatrix} t_{es} & t_{os} \\ t_{ep} & t_{op} \end{pmatrix}_n \begin{pmatrix} A'_e \\ A'_o \end{pmatrix}_n \tag{8.1-126}$$

where the transmission coefficients are as given by Eqs. (8.1-75)–(8.1-78) and Eqs. (8.1-81)–(8.1-88). Note that since the index of refraction of the imaginary isotropic layer is $n = n_o$ and the Fresnel transmission coefficients under the small birefringence approximation are 1. Equation (8.1-126) can thus be written

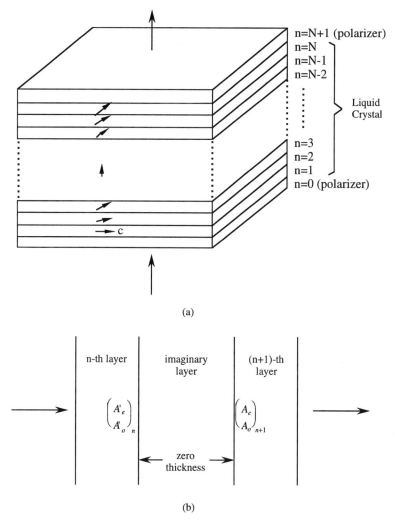

(a)

(b)

Figure 8.8. (*a*) A Liquid crystal display medium divided into N layers. Each layer can be considered as a uniaxial medium. The orientation of the c axis may change from layer to layer (e.g., a twisted nematic LCD). The ordinary and extraordinary indices of refraction n_o and n_e are constants for all layers. (*b*) Schematic drawing showing the imaginary layer used in deriving the interlayer dynamical matrix.

as

$$\begin{pmatrix} A_e \\ A_o \end{pmatrix}_{n+1} = \begin{pmatrix} \mathbf{s}\cdot\mathbf{e} & \mathbf{p}\cdot\mathbf{e} \\ \mathbf{s}\cdot\mathbf{p} & \mathbf{p}\cdot\mathbf{o} \end{pmatrix}_{n+1} \begin{pmatrix} \mathbf{e}\cdot\mathbf{s} & \mathbf{o}\cdot\mathbf{s} \\ \mathbf{e}\cdot\mathbf{p} & \mathbf{o}\cdot\mathbf{p} \end{pmatrix}_n \begin{pmatrix} A'_e \\ A'_o \end{pmatrix}_n$$

$$= \begin{pmatrix} \mathbf{e}_n\cdot\mathbf{e}_{n+1} & \mathbf{o}_n\cdot\mathbf{e}_{n+1} \\ \mathbf{e}_n\cdot\mathbf{o}_{n+1} & \mathbf{o}_n\cdot\mathbf{o}_{n+1} \end{pmatrix} \begin{pmatrix} A'_e \\ A'_o \end{pmatrix}_n$$

$$= \begin{pmatrix} t_{ee} & t_{oe} \\ t_{eo} & t_{oo} \end{pmatrix} \begin{pmatrix} A'_e \\ A'_o \end{pmatrix}_n \qquad (8.1\text{-}127)$$

where $\mathbf{e}_n$ and $\mathbf{o}_n$ represent the unit polarization vectors of the extraordinary and the ordinary waves in the nth layer, and $t_{ij}(i,j = e,o)$ is the transmission coefficient between the i component in the nth layer and the j component in the $(n+1)$th layer. Therefore, the dynamical matrix between the nth and the $(n+1)$th layers is given by

$$D_{n,n+1} = \begin{pmatrix} \mathbf{e}_n \cdot \mathbf{e}_{n+1} & \mathbf{o}_n \cdot \mathbf{e}_{n+1} \\ \mathbf{e}_n \cdot \mathbf{o}_{n+1} & \mathbf{o}_n \cdot \mathbf{o}_{n+1} \end{pmatrix} \tag{8.1-128}$$

It is important to note that the transmission coefficients between eigenmodes in the two layers are determined by the projections (inner product) of the corresponding polarization state vectors.

The overall Jones matrix for the LCD medium can thus be written as

$$\begin{pmatrix} A_s' \\ A_p' \end{pmatrix} = M \begin{pmatrix} A_s \\ A_p \end{pmatrix} \tag{8.1-129}$$

with

$$M = D_o P_{N+1} D_{N,N+1} P_N D_{N-1,N} \cdots D_{1,2} P_1 D_{0,1} P_0 D_i \tag{8.1-130}$$

where the output and input dynamical matrices D_o and D_i, the propagation matrix P_n, and the interlayer dynamical matrix $D_{n,n+1}(n = 1, 2, 3, \ldots, N)$ are as given by Eqs. (8.1-65), (8.1-64) and (8.1-128), respectively; the propagation matrices P_0 and P_{N+1} are those for the entrance and exit polarizers, respectively [Eqs. (8.1-107)]; and $D_{0,1}$ and $D_{N,N+1}$ are the dynamical matrices for the interfaces between the corresponding polarizer and its adjacent LC layer [Eq. (8.1-128), assuming the index of refraction of the polarizer is nearly the same as that of the LC]. For each layer, the $\mathbf{o}$ and $\mathbf{e}$ vectors can be calculated by using Eqs.(8.1-70) and (8.1-71), and the z component of the wavevector k_{oz} and k_{ez} can be obtained from Eqs. (8.1-125) and (8.1-124), respectively.

As an example, we analyze the leakage problem of the dark state of a normally white TN-LCD and suggest possible compensation methods (compensation methods will be discussed in detail in Chapter 9). Referring to Figure 8.9, we consider a TN-LC that is sandwiched between a pair of crossed polarizers (O-type polarizers [4], i.e., the transmission axis is perpendicular to their c axes at normal incidence). When there is no voltage applied across the TN-LC cell, the TN-LCD is at its ON state and the incident light can transmit through. To analyze the transmission property, we divide the TN-LC into 20 layers. Figure 8.10 shows the tilt and twist angles of the LC directors as functions of position inside the LC cell. We note that there is a 2° pretilt, which is assumed to be uniform throughout the cell. According to the definition described in

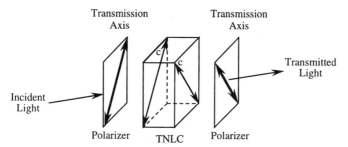

Figure 8.9. A TN-LCD with a TN-LC sandwiched between a pair of crossed polarizers.

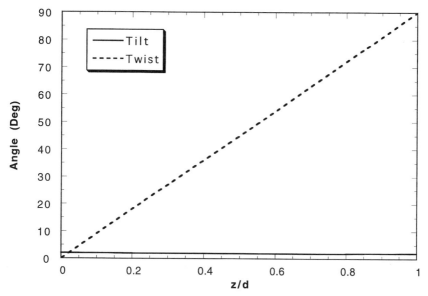

Figure 8.10. Distribution of the tilt and twist angles of the LC directors inside the LC cell (ON state, 0 V applied voltage).

Figure 8.6, we have $\theta_c = 88°$ and $\phi_c = 45°+$ twist angle. Using the extended Jones matrix method, we obtain the transmittance of an incident beam of unpolarized light. Figure 8.11a shows the intensity transmission pattern, and Figure 8.11b shows the corresponding contour plot. The contour lines are 0.1, 0.2, 0.3, and 0.4. In Figure 8.11, the incident angle varies from $\theta = 0–90°$ and $\phi = 0–360°$. In the numerical calculation the parameters are chosen as $n_o = 1.52$ for the two ideal O-type polarizers (the extraordinary component is supposed to be completely absorbed while the ordinary component is totally transmitted through); total LC thickness $d = 5.9$ μm, $n_o = 1.487$, and $n_e = 1.568$ for the TNLC; and $\lambda = 0.55$ μm for the incident light. In this calculation, we have included the Fresnel reflection at the air–polarizer interfaces and neglected the

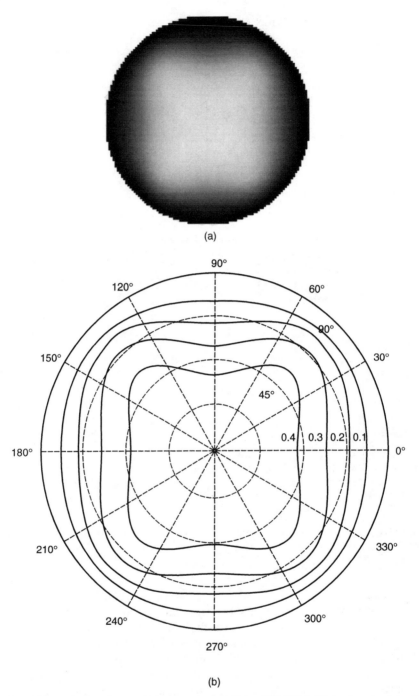

Figure 8.11. (*a*) Intensity transmission pattern of the NW TN-LCD in its ON state. (*b*) Transmission contour plot of the NW TN-LCD in its ON state with transmission = 0.4, 0.3, 0.2, and 0.1. The incident angle varies from $\theta = 0°$ to $90°$ and $\phi = 0°$ to $360°$. The parameters are chosen as $n_o = 1.52$ for the two ideal *O*-type polarizers; thickness $d = 5.9$ μm, $n_o = 1.487$, and $n_e = 1.568$ for the TN-LC; $\lambda = 0.55$ μm for the incident light.

reflection at the polarizer–LC interfaces. As we can see from Figure 8.11, the ON-state intensity distribution is reasonably uniform for all viewing angles.

When there is a drive voltage (e.g., 8 V) across the TN-LC cell, the TN-LCD is at its OFF state and the incident light is supposed to be blocked. However, at the OFF state, light beams with large incident angles are not completely blocked, as will be shown below. This leakage at OFF state severely affects the performance of the LCD. Fig. 8.12 shows the tilt and twist angles of the c axis (LC director) as functions of position at OFF state [5,13]. Using the extended Jones matrix method, we obtain the transmittance of an incident beam of unpolarized light. Figure 8.13a,b shows the intensity transmission pattern and the contour plot for the transmittance with arbitrary incident angles. Note that the maximum leakage can be as high as 30% (for comparison, the maximum transmittance at ON state is 50%).

To compensate for the leakage at OFF state, we consider a hypothetical case of a negatively birefringent ($\Delta n = n_e - n_o < 0$) and reversely twisted TN-LC as a compensator. The compensation method discussed below can be implemented with a stack of negatively birefringent uniaxial crystal plates (as will be shown below), which are more readily available. Referring to Figure 8.14, the compensator TN-LC has the c-axis orientations such that $\theta_c(z)|_{\text{compensator}} = \theta_c(d-z)|_{\text{original}}$ and $\phi_c(z)|_{\text{compensator}} = \phi_c(d-z)|_{\text{original}}$, where z is the distance measured from the front surface of each TN-LC and d is the common thickness of both the original and the compensator TN-LCs. In other words, the

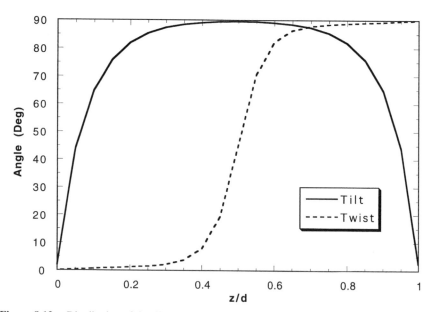

Figure 8.12. Distribution of the tilt and twist angles of the LC directors inside the LC cell (OFF state, 8 V applied voltage).

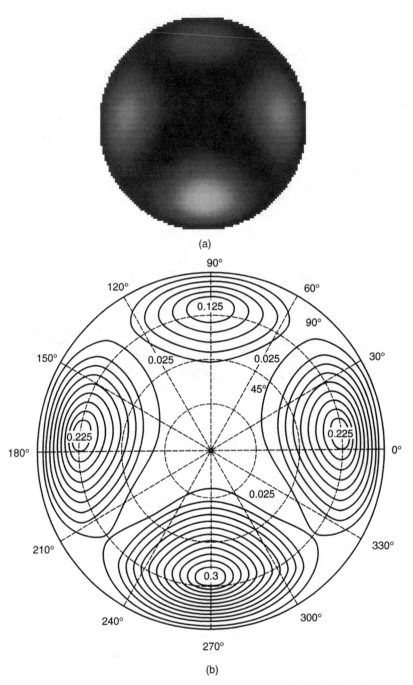

(a)

(b)

Figure 8.13. (a) Intensity transmission pattern $T(\theta,\phi)$ of the NW TN-LCD in its OFF state. (b) Transmission contour plot of the NW TN-LCD in its OFF state with transmission = 0.025, 0.05, 0.075, 0.1, 0.125, 0.15, 0.175, 0.2, 0.225, 0.25, 0.275, and 0.3. The incident angle varies from $\theta = 0°$ to $90°$ and $\phi = 0°$ to $360°$. The parameters are chosen as $n_o = 1.52$ for the two ideal O-type polarizers; thickness $d = 5.9$ μm, $n_o = 1.487$, and $n_e = 1.568$ for the TN-LC; $\lambda = 0.55$ μm for the incident light.

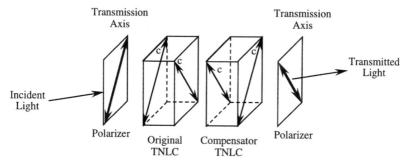

Figure 8.14. TN-LCD with a second, negatively birefringent TN-LC acting as a compensator. The compensator TN-LC twists in the opposite direction as compared with the original TN-LC.

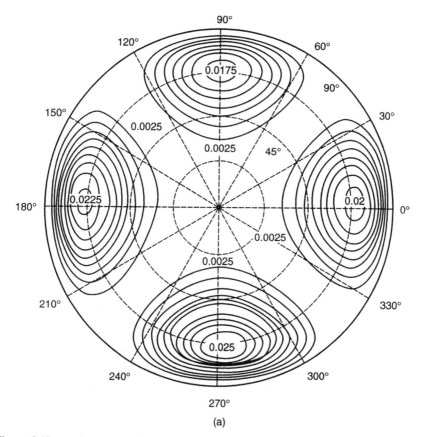

(a)

Figure 8.15. (*a*) Contour plot for the transmittance with arbitrary incident angles, with the compensator TN-LC. The incident angle varies from $\theta = 0°$ to $90°$ and $\phi = 0°$ to $360°$. The transmittances for the contour lines are 0.0025, 0.005, 0.0075, 0.01, 0.0125, 0.015, 0.0175, 0.02, 0.0225, and 0.025, respectively.

compensator TN-LC twists in the opposite direction as compared with the original TN-LC. The indices of refraction are given by $n_o|_{\text{compensator}} = n_e|_{\text{original}}$ and $n_e|_{\text{compensator}} = n_o|_{\text{original}}$. The Jones matrix for the original and the compensator TNLCs can be written as

$$M = \bar{M}_N \bar{M}_{N-1} \cdots \bar{M}_2 \bar{M}_1 M_N M_{N-1} \cdots M_2 M_1 \qquad (8.1\text{-}131)$$

where M_n and $\bar{M}_n$ represent the Jones matrices for the nth layer in the original and compensator TN-LCs, respectively, and $\bar{M}_{N-n+1} M_n \approx 1 (n = 1, 2, 3 \ldots, N)$ under the small birefringence approximation. Therefore, the phase retardation of the original TN-LC is canceled by that of the compensator TN-LC. This leads to $M \approx 1$ and the whole network is equivalent to a pair of crossed polarizers. Figure 8.15a shows the contour plot for the transmittance with arbitrary incident angles. The incident angle varies from $\theta = 0°$ to $90°$ and $\phi = 0°$ to $360°$. For

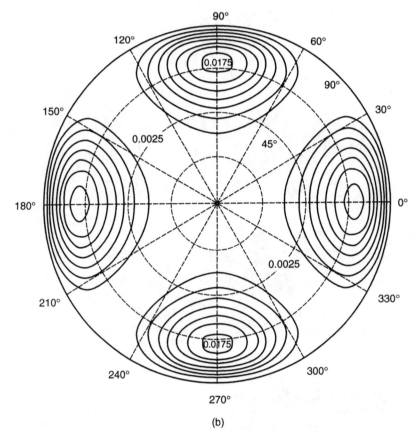

(b)

Figure 8.15. (b) Contour plot for the crossed polarizers. The transmittances for the contour lines are 0.0025, 0.005, 0.0075, 0.01, 0.0125, 0.015, and 0.0175, respectively.

comparison, Figure 8.15b shows the contour plot for the crossed polarizers. We note that with the presence of the compensator, the leakage is basically due to the crossed polarizers. As compared with conventional compensators such as a positively birefringent wave plate, it is easier to determine the thickness and birefringence for the compensator suggested here.

With this compensator, the ON state can be achieved by keeping the same drive voltage (e.g., 8 V) for the compensator and removing the applied voltage for the original TN-LC. Similar calculation shows that the ON state is not significantly affected by the compensator.

Since the applied voltage for the compensator is kept as a constant, the compensator TN-LC can be replaced by a stack of (e.g., 20) negatively birefringent uniaxial plates, which are readily available. This can be realized by using a negatively birefringent film of discotic compound with inclined axes [14]. Approximate compensation can be realized with fewer negatively birefringent uniaxial plates (e.g., three plates or even one plate), which are more readily available.

8.1.8. Generalized Jones Matrix Method

The results obtained above can be further generalized to cover any media including biaxial crystals and gyrotropic materials that exhibit optical rotation and Faraday rotation. In the case of small anisotropy, the 2×2 dynamical matrix at the boundary can be derived as follows.

Consider anisotropic materials with their principal axes oriented at arbitrary directions. The laboratory coordinate system is again chosen such that the z axis is normal to the interfaces. We will first derive the expressions for the eigen polarization states. Then we give the generalized Jones matrix formulations.

Since the medium is not isotropic, the propagation characteristics depend on the direction of propagation. The orientations of the crystal axes are described by the Euler angles θ_c, ϕ_c, and ψ_c with respect to a fixed xyz coordinate systems. The dielectric tensor in the xyz coordinate system is given by

$$\varepsilon = A \begin{pmatrix} \varepsilon_1 & 0 & 0 \\ 0 & \varepsilon_2 & 0 \\ 0 & 0 & \varepsilon_3 \end{pmatrix} A^{-1} \tag{8.1-132}$$

where ε_1, ε_2, and ε_3 are the principal dielectric constants and A is the coordinate rotation matrix given by

$A =$

$$\begin{pmatrix} \cos\psi_c \cos\phi_c - \cos\theta_c \sin\phi_c \sin\psi_c & -\sin\psi_c \cos\phi_c - \cos\theta_c \sin\phi_c \cos\psi_c & \sin\theta_c \sin\phi_c \\ \cos\psi_c \sin\phi_c + \cos\theta_c \cos\phi_c \sin\psi_c & -\sin\psi_c \sin\phi_c + \cos\theta_c \cos\phi_c \cos\psi_c & -\sin\theta_c \cos\phi_c \\ \sin\theta_c \sin\psi_c & \sin\theta_c \cos\psi_c & \cos\theta_c \end{pmatrix}. $$

$$\tag{8.1-133}$$

Since A is orthogonal, the dielectric tensor ε in the xyz coordinate must be symmetric, that is, $\varepsilon_{ij} = \varepsilon_{ji}$. The electric field can be assumed to have $\exp[i(\omega t - \alpha x - \beta y - \gamma z]$ dependence on each crystal layer, which is assumed to be homogeneous. Since the whole birefringent layered medium is homogeneous in the xy plane, α and β remain the same throughout the layered medium. Therefore, the two components (α, β) of the propagation vector are chosen as the dynamical variables characterizing the electromagnetic waves propagating in the layered media. Given α and β, the z component γ is determined directly from the wave equation in momentum space:

$$\mathbf{k} \times (\mathbf{k} \times \mathbf{E}) + \omega^2 \mu \varepsilon \mathbf{E} = 0 \qquad (8.1\text{-}134)$$

or equivalently

$$\begin{pmatrix} \omega^2 \mu \varepsilon_{xx} - \beta^2 - \gamma^2 & \omega^2 \mu \varepsilon_{xy} + \alpha\beta & \omega^2 \mu \varepsilon_{xz} + \alpha\gamma \\ \omega^2 \mu \varepsilon_{yx} + \alpha\beta & \omega^2 \mu \varepsilon_{yy} - \alpha^2 - \gamma^2 & \omega^2 \mu \varepsilon_{yz} + \beta\gamma \\ \omega^2 \mu \varepsilon_{zx} + \alpha\gamma & \omega^2 \mu \varepsilon_{zy} + \beta\gamma & \omega^2 \mu \varepsilon_{zz} - \alpha^2 - \beta^2 \end{pmatrix} \begin{pmatrix} E_x \\ E_y \\ E_z \end{pmatrix} = 0$$

$$(8.1\text{-}135)$$

To have nontrivial plane-wave solutions, the determinant of the matrix in Eq. (8.1-135) must vanish. This gives a quartic equation in γ that yields four roots $\gamma_\sigma, \sigma = 1, 2, 3, 4$. These roots may be either real or complex. For lossless dielectric media with real ε_{ij}, complex roots are always in conjugate pairs. These four roots can also be obtained graphically from Figure 8.16 if they are real. The plane of incidence is defined as the plane formed by $\alpha\mathbf{x} + \beta\mathbf{y}$ and $\mathbf{z}$. The intersection of this plane with the normal surface yields two closed curves that are symmetric with respect to the origin of the axes. Drawing a line from the tip of the vector $\alpha\mathbf{x} + \beta\mathbf{y}$ parallel to the $\mathbf{z}$ direction yields, in general, four points of intersection. The four wavevectors $\mathbf{k}_\sigma = \alpha\mathbf{x} + \beta\mathbf{y} + \gamma_\sigma\mathbf{z}$ all lie in the plane of incidence, which also remains the same throughout the layered medium because α and β are constants. However, the four group velocities associated with these partial waves are, in general, not lying in the plane of incidence. If all four

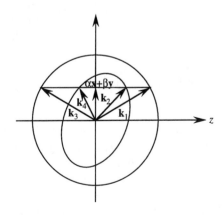

Figure 8.16. Graphic method to determine the propagation constants from the normal surface.

wavevectors $\mathbf{k}_\sigma$ are real, two of them have group velocities with positive z components, and the other two have group velocities with negative z components. The z component of the group velocity vanishes when γ_σ becomes complex. The polarization of these waves can be written

$$
\mathbf{p}_\sigma =
$$

$$
= N_\sigma \begin{pmatrix} (\omega^2 \mu \varepsilon_{yy} - \alpha^2 - \gamma_\sigma^2)(\omega^2 \mu \varepsilon_{zz} - \alpha^2 - \beta^2) - (\omega^2 \mu \varepsilon_{yz} + \beta\gamma_\sigma)^2 \\ (\omega^2 \mu \varepsilon_{yz} + \beta\gamma_\sigma)(\omega^2 \mu \varepsilon_{zx} + \alpha\gamma_\sigma) - (\omega^2 \mu \varepsilon_{xy} + \alpha\beta)(\omega^2 \mu \varepsilon_{zz} - \alpha^2 - \beta^2) \\ (\omega^2 \mu \varepsilon_{xy} + \alpha\beta)(\omega^2 \mu \varepsilon_{yz} + \beta\gamma_\sigma) - (\omega^2 \mu \varepsilon_{xz} + \alpha\gamma_\sigma)(\omega^2 \mu \varepsilon_{yy} - \alpha^2 - \gamma_\sigma^2) \end{pmatrix}
$$

$$(8.1\text{-}136)$$

where $\sigma = 1, 2, 3, 4$ and the N_σ values are the normalization constant such that $\mathbf{p}_\sigma \cdot \mathbf{p}_\sigma = 1$. The electric field of the plane electromagnetic waves can thus be written as

$$
\mathbf{E} = \sum_{\sigma=1}^{4} A_\sigma \mathbf{p}_\sigma \exp\left[i(\omega t - \alpha x - \beta y - \gamma_\sigma z)\right] \tag{8.1-137}
$$

where the A_σ terms are constants.

Partial waves with complex propagation vectors cannot exist in an infinite homogeneous birefringent medium. If the medium is semi-infinite, the exponentially damped partial waves are legitimate solutions near the interface, and the field envelope decays exponentially as a function of z, where z is the distance from the interface. These exponentially damped partial waves are called *evanescent waves*. The evanescent waves in birefringent media in general have complex γ values, that is $\gamma = \gamma_R + i\gamma_I$. In a uniaxially birefringent medium, the ordinary evanescent wave has a purely imaginary γ. If the three principal dielectric constants are all real, these partial waves with complex γ values can be shown to have their Poyting vectors parallel to the interface. In other words, the energy is flowing parallel to the interface, and the propagation is lossless, as it should be. A mathematical proof is given in Reference 2 for the special case of extraordinary evanescent waves in a uniaxially birefringent medium. For practical applications in LCDs, all partial waves are propagating with real γ values.

In the case of propagating modes (four real γ values), two of them will be positive and the other two will be negative. To discuss the transmission properties using 2×2 matrix method, we need only the two positive eigenvalues γ_1 and γ_2 (see Fig. 8.16). The propagation matrix within each medium of thickness d is simply given by

$$
P = \begin{pmatrix} e^{-i\gamma_1 d} & 0 \\ 0 & e^{-i\gamma_2 d} \end{pmatrix} \tag{8.1-138}
$$

Once the eigenmodes are obtained, the dynamical matrix at the interface between two media can be derived. Suppose that the electric fields inside the two media are given by

$$
\begin{aligned}
\text{Incident}: \quad & \mathbf{E} = (A_1 \mathbf{p}_{i1} e^{-i\mathbf{k}_{i1} \cdot \mathbf{r}} + A_2 \mathbf{p}_{i2} e^{-i\mathbf{k}_{i2} \cdot \mathbf{r}}) e^{i\omega t} \\
\text{Reflected}: \quad & \mathbf{E} = (B_1 \mathbf{p}_{r1} e^{-i\mathbf{k}_{r1} \cdot \mathbf{r}} + B_2 \mathbf{p}_{r2} e^{-i\mathbf{k}_{r2} \cdot \mathbf{r}}) e^{i\omega t} \\
\text{Refracted}: \quad & \mathbf{E} = (C_1 \mathbf{p}_{t1} e^{-i\mathbf{k}_{t1} \cdot \mathbf{r}} + A_2 \mathbf{p}_{t2} e^{-i\mathbf{k}_{t2} \cdot \mathbf{r}}) e^{i\omega t}
\end{aligned}
\tag{8.1-139}
$$

The dynamical matrix, which relates the refracted and the incident waves, is written

$$
\begin{pmatrix} C_1 \\ C_2 \end{pmatrix} = D_{12} \begin{pmatrix} A_1 \\ A_2 \end{pmatrix} = \begin{pmatrix} t_{11} & t_{21} \\ t_{12} & t_{22} \end{pmatrix} \begin{pmatrix} A_1 \\ A_2 \end{pmatrix}
\tag{8.1-140}
$$

where t_{11}, t_{12}, t_{21}, and t_{22} are transmission coefficients. To determine D_{12}, we assume that the anisotropy is small for both media so that the average indices of refraction can be written as two constants, n_1 and n_2, for medium 1 and medium 2, respectively. For the purpose of obtaining D_{12}, we insert two imaginary isotropic layers of zero thickness between the two media (see Fig. 8.17). The layer on the side of medium 1 (2) has an index of refraction n_1 (n_2). The Fresnel transmission coefficients between the two imaginary layers can be written

$$
t_s = \frac{2n_1 \cos\theta_1}{n_1 \cos\theta_1 + n_2 \cos\theta_2}, \qquad t_p = \frac{2n_1 \cos\theta_1}{n_1 \cos\theta_2 + n_2 \cos\theta_1}
\tag{8.1-141}
$$

where θ_1 and θ_2 are incident and refraction angles (angles between the propagation direction and the surface normal), respectively. From the previous discussion, the dynamical matrix between two media of the same refractive index is determined by the projection of polarization states of the eigenmodes. Therefore, we obtain the following, according to the method described in

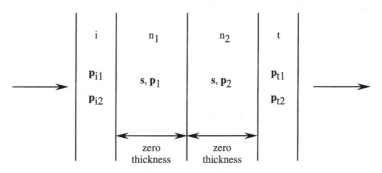

Figure 8.17.   Schematic drawing showing the two imaginary isotropic layers.

Section 8.1.2:

$$D_{12} = \begin{pmatrix} \mathbf{s} \cdot \mathbf{p}_{t1} & \mathbf{p}_2 \cdot \mathbf{p}_{t1} \\ \mathbf{s} \cdot \mathbf{p}_{t2} & \mathbf{p}_2 \cdot \mathbf{p}_{t2} \end{pmatrix} \begin{pmatrix} t_s & 0 \\ 0 & t_p \end{pmatrix} \begin{pmatrix} \mathbf{p}_{i1} \cdot \mathbf{s} & \mathbf{p}_{i2} \cdot \mathbf{s} \\ \mathbf{p}_{i1} \cdot \mathbf{p}_1 & \mathbf{p}_{i2} \cdot \mathbf{p}_1 \end{pmatrix}$$

$$= \begin{pmatrix} t_s(\mathbf{p}_{i1} \cdot \mathbf{s})(\mathbf{s} \cdot \mathbf{p}_{t1}) + t_p(\mathbf{p}_{i1} \cdot \mathbf{p}_1)(\mathbf{p}_2 \cdot \mathbf{p}_{t1}) & t_s(\mathbf{p}_{i2} \cdot \mathbf{s})(\mathbf{s} \cdot \mathbf{p}_{t1}) + t_p(\mathbf{p}_{i2} \cdot \mathbf{p}_1)(\mathbf{p}_2 \cdot \mathbf{p}_{t1}) \\ t_s(\mathbf{p}_{i1} \cdot \mathbf{s})(\mathbf{s} \cdot \mathbf{p}_{t2}) + t_p(\mathbf{p}_{i1} \cdot \mathbf{p}_1)(\mathbf{p}_2 \cdot \mathbf{p}_{t2}) & t_s(\mathbf{p}_{i2} \cdot \mathbf{s})(\mathbf{s} \cdot \mathbf{p}_{t2}) + t_p(\mathbf{p}_{i2} \cdot \mathbf{p}_1)(\mathbf{p}_2 \cdot \mathbf{p}_{t2}) \end{pmatrix}$$

$$(8.1\text{-}142)$$

where $\mathbf{p}_{i1}$, $\mathbf{p}_{i2}$, $\mathbf{p}_{t1}$, $\mathbf{p}_{t2}$, $\mathbf{s}$, $\mathbf{p}_1$, and $\mathbf{p}_2$ are unit vectors representing polarization states of the two incident eigenmodes, the two transmitted eigenmodes, s waves in the imaginary layers, and p waves in the imaginary layers 1 and 2, respectively. Note that when $n_1 = n_2$, $t_s = t_p = 1$, $\mathbf{p}_1 = \mathbf{p}_2$, and Eq. (8.1-142) reduces to

$$D_{12} = \begin{pmatrix} \mathbf{p}_{i1} \cdot \mathbf{p}_{t1} & \mathbf{p}_{i2} \cdot \mathbf{p}_{t1} \\ \mathbf{p}_{i1} \cdot \mathbf{p}_{t2} & \mathbf{p}_{i2} \cdot \mathbf{p}_{t2} \end{pmatrix} \qquad (8.1\text{-}143)$$

We note that again the dynamical matrix consists of elements that are inner products of the normalized polarization state eigenvectors (see Eq. (8.1-128). This is always valid in the small birefringence approximation. With the dynamical and propagation matrices, the overall Jones matrix for a system containing any dielectric media can be written in the form of Eqs. (8.1-129) and (8.1-130).

In Section 4.5.3, the intensity transmission patterns of a c-cut biaxial plate between crossed polarizers in Figure 4.16 were calculated using the generalized Jones matrix method.

In summary, we have described in this section the original extended Jones matrix method which is an extremely powerful technique for analyzing the transmission properties of LCDs at all angles of incidence. In this method, the matrix elements are expressed in the local principal coordinate system which consists of the unit vectors $(\mathbf{e}, \mathbf{o})$ representing the polarization states of the normal modes of propagation in each layer. These two vectors and the direction of propagation $\mathbf{k}_0$ form a mutually orthogonal triad $(\mathbf{k}_0, \mathbf{e}, \mathbf{o})$. In the small birefringence approximation, the matrix elements of the dynamical matrices are simply the inner products of the polarization unit vectors of the neighboring layers. This extended Jones matrix method is simple, physically intuitive, and involves only 2×2 matrices. It can be easily implemented numerically. This method has been employed to calculate the transmission properties of various LCDs. The numerical calculation requires the subdivision of the LC cell into a number of fine layers. A relatively simple computer program can then be written to calculate the transmission properties of the LCDs at all angles of incidence. Although the multiple reflections are neglected, the calculated transmissions are in excellent agreement with the measured results.

8.2. ANOTHER EXTENDED JONES MATRIX METHOD

As we recall, the extended Jones matrix method was originally introduced in 1982 [4]. A slightly different extended Jones matrix method was developed latter in 1990 [9]. In the latter approach, all formulations were given in the laboratory (x,y,z) coordinate system instead of the principal coordinate system. The dielectric tensor for a uniaxial LC medium in units of ε_0 is written in the xyz coordinate system as

$$\varepsilon = \varepsilon_0 \begin{pmatrix} \varepsilon_{xx} & \varepsilon_{xy} & \varepsilon_{xz} \\ \varepsilon_{yx} & \varepsilon_{yy} & \varepsilon_{yz} \\ \varepsilon_{zx} & \varepsilon_{zy} & \varepsilon_{zz} \end{pmatrix} \tag{8.2-1}$$

with [9]

$$\varepsilon_{xx} = n_o^2 + (n_e^2 - n_o^2)\sin^2\theta_c\cos^2\phi_c \tag{8.2-2}$$

$$\varepsilon_{xy} = \varepsilon_{yx} = (n_e^2 - n_o^2)\sin^2\theta_c\sin\phi_c\cos\phi_c \tag{8.2-3}$$

$$\varepsilon_{xz} = \varepsilon_{zx} = (n_e^2 - n_o^2)\sin\theta_c\cos\theta_c\cos\phi_c \tag{8.2-4}$$

$$\varepsilon_{yy} = n_o^2 + (n_e^2 - n_o^2)\sin^2\theta_c\sin^2\phi_c \tag{8.2-5}$$

$$\varepsilon_{yz} = \varepsilon_{zy} = (n_e^2 - n_o^2)\sin\theta_c\cos\theta_c\sin\phi_c \tag{8.2-6}$$

$$\varepsilon_{zz} = n_o^2 + (n_e^2 - n_o^2)\cos^2\theta_c \tag{8.2-7}$$

where n_o and n_e are the ordinary and extraordinary indices of refraction of the LC medium, θ_c is the angle between the LC director and the z axis, and ϕ_c is the angle between the projection of the LC director on the xy plane and the x axis (see Fig. 8.6).

Now consider a plane wave incident at an oblique angle on the surface of a TN-LCD as shown in Figure 8.9. Without losing generality, we can always choose an xyz coordinate system such that the wavevector **k** lies on the xz plane. The xy plane is parallel to the glass substrate surface, and the direction of the $+z$ axis is pointing from the entrance polarizer to the exit polarizer. The wavevector **k** can be written as

$$\mathbf{k} = k_0(\sin\theta_k, 0, \cos\theta_k) \tag{8.2-8}$$

where θ_k is the angle between **k** and the z axis, $k_0 = \omega/c = 2\pi/\lambda$, and λ is the wavelength of the incident light in free space. As illustrated in Figure 8.8, the entire LCD system is divided into $N + 2$ layers. The $n = 0$ layer is the entrance polarizer, and the $n = N + 1$ layer is the exit polarizer. These two polarizer layers are also characterized by a dielectric tensor whose components are given in Eqs. (8.2-2)–(8.2-7), where n_o and n_e are complex numbers, θ_c is the angle between the optic axis (or the absorption axis) of the polarizer layer and the z axis (usually $\theta_c = \pi/2$), and ϕ_c is the angle between the projection of the extraordinary axis of the polarizer layer on the xy plane and the x axis. The LC layer is divided into N sublayers, with each sublayer characterized by a corresponding dielectric tensor. From all this, one can show that there are four eigenwaves propagating in each layer; the two eigenwaves that propagate toward the $+z$ direction are transmitted waves, and the remaining two, which propagate toward the $-z$ direction, are reflected waves. However, in the case when the amplitudes of reflected waves are much smaller than those of the transmitted waves, the reflected waves can be neglected and only two transmitted waves are propagating in the medium. Under this assumption and using the boundary condition that tangential components of the electric field must be continuous at each layer interface, an extended Jones matrix representation for an LCD system at the oblique incidence can be obtained as

$$M = M_{N+1}M_N \cdots M_1 M_0 \tag{8.2-9}$$

where

$$M_n = (SPS^{-1})_n \qquad (n = 0, 1, 2, \ldots, N, N+1) \tag{8.2-10}$$

$$S = \begin{pmatrix} c_2 & 1 \\ 1 & c_1 \end{pmatrix} \tag{8.2-11}$$

and

$$P = \begin{pmatrix} e^{-ik_{ez}d} & 0 \\ o & e^{-ik_{oz}d} \end{pmatrix} \tag{8.2-12}$$

where d is the thickness of the corresponding layer, and [9]

$$\frac{k_{oz}}{k_0} = \sqrt{n_o^2 - \left(\frac{k_x}{k_0}\right)^2} \tag{8.2-13}$$

$$\frac{k_{ez}}{k_0} = -\frac{\varepsilon_{xz}}{\varepsilon_{zz}}\frac{k_x}{k_0} + \frac{n_o n_e}{\varepsilon_{zz}}\left[\varepsilon_{zz} - \left(1 - \frac{n_e^2 - n_o^2}{n_e^2}\sin^2\theta_c \sin^2\phi_c\right)\left(\frac{k_x}{k_0}\right)^2\right]^{1/2} \tag{8.2-14}$$

$$c_1 = \frac{[(k_x/k_0)^2 - \varepsilon_{zz}]\varepsilon_{yx} + [(k_x/k_0)(k_{oz}/k_0) + \varepsilon_{zx}]\varepsilon_{yz}}{[(k_{oz}/k_0)^2 + (k_x/k_0)^2 - \varepsilon_{yy}][(k_x/k_0)^2 - \varepsilon_{zz}] - \varepsilon_{yz}\varepsilon_{zy}} \tag{8.2-15}$$

$$c_2 = \frac{[(k_x/k_0)^2 - \varepsilon_{zz}]\varepsilon_{xy} + [(k_x/k_0)(k_{ez}/k_0) + \varepsilon_{xz}]\varepsilon_{zy}}{[(k_{ez}/k_0)^2 - \varepsilon_{xx}][(k_x/k_0)^2 - \varepsilon_{zz}] - [(k_x/k_0)(k_{ez}/k_0) + \varepsilon_{zx}][(k_x/k_0)(k_{ez}/k_0) + \varepsilon_{xz}]} \tag{8.2-16}$$

and $k_x = k_0 \sin \theta_k$. We note that, $k_0 = \omega/c$, θ_k is the incident angle in air. For a given incident $\mathbf{E}$ field (E_x, E_y), the transmitted $\mathbf{E}$ field (E'_x, E'_y) can be calculated by

$$\begin{pmatrix} E'_x \\ E'_y \end{pmatrix} = M \begin{pmatrix} E_x \\ E_y \end{pmatrix} \tag{8.2-17}$$

Thus, the total optical transmission is given as [9]

$$T_{op} = \frac{|E'_x|^2 + \cos^2 \theta_p |E'_y|^2}{|E_x|^2 + \cos^2 \theta_p |E_y|^2} \tag{8.2-18}$$

where

$$\theta_p = \sin^{-1}\left[\frac{\sin \theta_k}{\mathrm{Re}(n_p)}\right] \tag{8.2-19}$$

in which $\mathrm{Re}(n_p)$ stands for the average of the real parts of two indices of refraction (n_o and n_e) of the polarizer. Physically, θ_p is the approximate ray angle in the polarizer.

The transmission losses at the entrance and exit surfaces of an LCD cell are usually significant. Since the E_x and E_y components instead of E_s and E_p (or E_o and E_e) components are used here, we cannot use the results in Section 8.1 directly. We can take into account these transmission losses using the optical transmission modification procedure as follows. The modified optical transmission is given as [9]

$$T'_{op} = T_{op} T_{ent} T_{ext} \tag{8.2-20}$$

where T_{op} is as given by Eq. (8.2-18).

For $\theta_k = 0$ (Normal incidence): $\quad T_{ent} = T_{ext} = \dfrac{4\,\mathrm{Re}(n_p)}{[1 + \mathrm{Re}(n_p)]^2} \tag{8.2-21}$

For $\theta_k \neq 0$: $\quad\quad\quad\quad\quad\quad T_{ent} = T_p \cos^2 \alpha_{ent} + T_v \sin^2 \alpha_{ent} \tag{8.2-22}$

and

$$T_{ext} = T_p \cos^2 \alpha_{ext} + T_v \sin^2 \alpha_{ext} \tag{8.2-23}$$

where

$$T_p = \frac{\sin (2\theta_k) \sin (2\theta_p)}{\sin^2 (\theta_k + \theta_p) \cos^2 (\theta_k - \theta_p)} \qquad (8.2\text{-}24)$$

$$T_v = \frac{\sin (2\theta_k) \sin (2\theta_p)}{\sin^2 (\theta_k + \theta_p)} \qquad (8.2\text{-}25)$$

$$\cos \alpha_{\text{ent}} = \frac{E_x}{\sqrt{|E_x|^2 + \cos^2 \theta_k |E_y|^2}} \qquad (8.2\text{-}26)$$

$$\cos \alpha_{\text{ext}} = \frac{E'_x}{\sqrt{|E'_x|^2 + \cos^2 \theta_k |E'_y|^2}} \qquad (8.2\text{-}27)$$

We note that the method described in this section assumes no reflection at the interfaces. This is a reasonable assumption inside the liquid crystal medium which is divided into a number of layers for the numerical calculation. The Jones matrix elements are obtained by requiring a continuity of the tangential components of the electric field vector at the boundaries. This is similar to the scalar wave approximation in optics. It is important to note that ignoring the continuity of the tangential components of the magnetic field vector may lead to an error in the Jones matrix elements. This error may increase at large angles of incidence.

8.3. 4 × 4 MATRIX METHOD

The Jones matrix method introduced in Chapter 4 neglects the reflection of light completely. This is, of course, only an approximation. In practice, the reflection of light occurs at interfaces where dielectric discontinuity is present. The extended Jones matrix method introduced in Section 8.1 takes into account single reflection at each surface but neglects multiple reflections and their interference. To solve a birefringent network problem exactly, 4 × 4 matrices are needed in the matrix method.

8.3.1. Mathematical Formulation

The 4 × 4 matrix algebra, which provides a systematic way to analyze the propagation of monochromatic plane waves in birefringent layered medium, can now be introduced. Two mathematically equivalent approaches were introduced in 1972 and 1979. The approaches are general, so that the results can be used later for many special cases of propagation in anisotropic layered medium. The materials are assumed to be nonmagnetic so that $\mu = $ constant throughout the whole layered medium. The dielectric permittivity tensor ε in the xyz coordinates is given by

$$\varepsilon = \begin{cases} \varepsilon(0), & z < z_0 \\ \varepsilon(1), & z_0 < z < z_1 \\ \varepsilon(2), & z_1 < z < z_2 \\ \quad\vdots & \\ \varepsilon(N), & z_{N-1} < z < z_N \\ \varepsilon(s), & z_N < z \end{cases} \tag{8.3-1}$$

where N is the number of layers and 0 and s denote the incident and transmitted media, respectively. The electric field distribution within each homogeneous anisotropic layer can be expressed as a sum of four partial waves [as described in Eqs. (8.1-137)]. Generally speaking, two of the partial waves are propagating along the positive z direction, while the other two are propagating along the negative z direction (see Fig. 8.18). The complex amplitudes of these four partial waves constitute the components of a column vector. The electromagnetic field in the nth layer of the anisotropic layered medium can thus be represented by a column vector $A_\sigma(n), \sigma = 1, 2, 3, 4$.

As a result, the electric field distribution in the same nth layer can be written as

$$\mathbf{E} = \sum_{\sigma=1}^{4} A_\sigma(n)\mathbf{p}_\sigma(n)\exp\{i[\omega t - \alpha x - \beta y - \gamma_\sigma(n)(z - z_n)]\} \tag{8.3-2}$$

where $\mathbf{p}_\sigma$ ($\sigma = 1, 2, 3, 4$) is the eigen polarization vector, α, β, and γ_σ are the x, y, and z components of the corresponding wavevectors, respectively [see Eqs. (8.1-135)–(8.1-137)]. The column vectors are not independent of each other. They

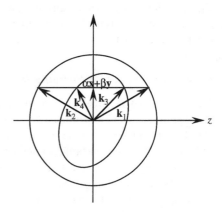

Figure 8.18. Graphic method to determine the propagation constants from the normal surface. Note the difference in numbering the wavevectors between this figure and Figure 8.16.

are related through the continuity conditions at the interfaces. In fact, only one vector (or any four components of four different vectors) can be arbitrarily chosen. The magnetic field distribution is obtained using Maxwell's equations and is given by

$$\mathbf{H} = \sum_{\sigma=1}^{4} A_\sigma(n)\mathbf{q}_\sigma(n)\exp\{i[\omega t - \alpha x - \beta y - \gamma_\sigma(n)(z - z_n)]\} \qquad (8.3\text{-}3)$$

with

$$\mathbf{q}_\sigma(n) = \frac{\mathbf{k}_\sigma(n)}{\omega\mu} \times \mathbf{p}_\sigma(n) \qquad (8.3\text{-}4)$$

$$\mathbf{k}_\sigma(n) = \alpha\mathbf{x} + \beta\mathbf{y} + \gamma_\sigma(n)\mathbf{z} \qquad (8.3\text{-}5)$$

Note that the $\mathbf{q}_\sigma(n)$ terms are not unit vectors.

Imposing the continuity of E_x, E_y, H_x, and H_y at the interface $z = z_{n-1}$, we obtain

$$\sum_{\sigma=1}^{4} A_\sigma(n-1)\mathbf{p}_\sigma(n-1)\cdot\mathbf{x} = \sum_{\sigma=1}^{4} A_\sigma(n)\mathbf{p}_\sigma(n)\cdot\mathbf{x}\exp\left[i\gamma_\sigma(n)t_n\right] \qquad (8.3\text{-}6)$$

$$\sum_{\sigma=1}^{4} A_\sigma(n-1)\mathbf{p}_\sigma(n-1)\cdot\mathbf{y} = \sum_{\sigma=1}^{4} A_\sigma(n)\mathbf{p}_\sigma(n)\cdot\mathbf{y}\exp\left[i\gamma_\sigma(n)t_n\right] \qquad (8.3\text{-}7)$$

$$\sum_{\sigma=1}^{4} A_\sigma(n-1)\mathbf{q}_\sigma(n-1)\cdot\mathbf{x} = \sum_{\sigma=1}^{4} A_\sigma(n)\mathbf{q}_\sigma(n)\cdot\mathbf{x}\exp\left[i\gamma_\sigma(n)t_n\right] \qquad (8.3\text{-}8)$$

$$\sum_{\sigma=1}^{4} A_\sigma(n-1)\mathbf{q}_\sigma(n-1)\cdot\mathbf{y} = \sum_{\sigma=1}^{4} A_\sigma(n)\mathbf{q}_\sigma(n)\cdot\mathbf{y}\exp\left[i\gamma_\sigma(n)t_n\right] \qquad (8.3\text{-}9)$$

where $t_n = z_n - z_{n-1}$ is the thickness of the nth layer, $n = 1, 2, \ldots, N$.

These four equations can be rewritten as a matrix equation:

$$\begin{pmatrix} A_1(n-1) \\ A_2(n-1) \\ A_3(n-1) \\ A_4(n-1) \end{pmatrix} = D^{-1}(n-1)D(n)P(n)\begin{pmatrix} A_1(n) \\ A_2(n) \\ A_3(n) \\ A_4(n) \end{pmatrix} \qquad (8.3\text{-}10)$$

where

$$
D(n) = \begin{pmatrix} \mathbf{x}\cdot\mathbf{p}_1(n) & \mathbf{x}\cdot\mathbf{p}_2(n) & \mathbf{x}\cdot\mathbf{p}_3(n) & \mathbf{x}\cdot\mathbf{p}_4(n) \\ \mathbf{y}\cdot\mathbf{q}_1(n) & \mathbf{y}\cdot\mathbf{q}_2(n) & \mathbf{y}\cdot\mathbf{q}_3(n) & \mathbf{y}\cdot\mathbf{q}_4(n) \\ \mathbf{y}\cdot\mathbf{p}_1(n) & \mathbf{y}\cdot\mathbf{p}_2(n) & \mathbf{y}\cdot\mathbf{p}_3(n) & \mathbf{y}\cdot\mathbf{p}_4(n) \\ \mathbf{x}\cdot\mathbf{q}_1(n) & \mathbf{x}\cdot\mathbf{q}_2(n) & \mathbf{x}\cdot\mathbf{q}_3(n) & \mathbf{x}\cdot\mathbf{q}_4(n) \end{pmatrix}, \tag{8.3-11}
$$

$$
P(n) = \begin{pmatrix} \exp[i\gamma_1(n)t_n] & 0 & 0 & 0 \\ 0 & \exp[i\gamma_2(n)t_n] & 0 & 0 \\ 0 & 0 & \exp[i\gamma_3(n)t_n] & 0 \\ 0 & 0 & 0 & \exp[i\gamma_4(n)t_n] \end{pmatrix}. \tag{8.3-12}
$$

The matrices $D(n)$ are called *dynamical matrices* because they depend only on the direction of polarization of those four partial waves. The dynamical matrices are defined in such a way that they are block diagonalized when the mode coupling disappears. This requires that A_1 and A_2 be the amplitudes of the plane waves of the same mode (polarization) such that the plane wave with amplitude A_1 propagates to the right, whereas the plane wave with amplitude A_2 propagates to the left. Likewise, A_3 and A_4 are the amplitudes of the plane waves of the same mode, propagating to the right or left, respectively. The matrices $P(n)$ are called *propagation matrices* and depend only on the phase shift of these four partial waves as they traverse through the bulk of the layer. Note that the exponential functions in P [see Eq. (8.3-12)] have positive exponents, while the propagation matrices in Section 8.1 [see, e.g, Eq. (8.1-64)] have negative exponents. This is a result of the unique ordering here. In Eq. (8.3-10), the Jones vector of the $(n-1)$th layer is on the left-hand side and that of the nth layer is on the right-hand side. In Section 8.1, we placed the incident Jones vector on the right-hand side of the corresponding equations. The choice in Section 8.1 emphasized on the input–output relationship. The choice in this section makes the equation follow the order in the graphical expression (see Eq. (8.3-15) and Figure 8.19). Both choices have been used in the literature. We define a transfer matrix as

$$
T_{n-1,n} = D^{-1}(n-1)D(n)P(n) \tag{8.3-13}
$$

Equation (8.3-10) can thus be written as

$$
\begin{pmatrix} A_1(n-1) \\ A_2(n-1) \\ A_3(n-1) \\ A_4(n-1) \end{pmatrix} = T_{n-1,n} \begin{pmatrix} A_1(n) \\ A_2(n) \\ A_3(n) \\ A_4(n) \end{pmatrix}. \tag{8.3-14}
$$

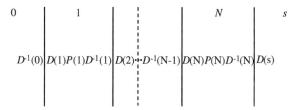

Figure 8.19.   Diagram representation of the matrix method.

The matrix equation that relates $A(0)$ and $A(s)$ is therefore given by

$$
\begin{pmatrix} A_1(0) \\ A_2(0) \\ A_3(0) \\ A_4(0) \end{pmatrix} = T_{0,1}T_{1,2}T_{2,3}\cdots T_{N-1,N}T_{N,s} \begin{pmatrix} A_1(s) \\ A_2(s) \\ A_3(s) \\ A_4(s) \end{pmatrix} \tag{8.3-15}
$$

where $s \equiv N + 1$ and $t_{N+1} \equiv 0$.

Equations (8.3-10) and (8.3-15) show how systematic the matrix method is for treating electromagnetic propagation in anisotropic layered media. If Eq. (8.3-15) is represented graphically in Fig. 8.19, two dynamical matrices can be seen to be associated with each interface and one propagation matrix is associated with the bulk of each layer. The overall transfer matrix is the product of all these matrices from left to right. This completes the theoretical formulation of the 4 × 4 matrix method.

The matrix method just described is an exact approach to the propagation of electromagnetic radiation in anisotropic layered media. Both birefringent phase retardation and thin film interference are considered. This differs from the conventional 2 × 2 Jones matrix method, which ignores the obliquity effect and neglects the reflection from each interface. It also differs from the extended 2 × 2 Jones matrix method, which neglects the interference from multiple reflections. Therefore, in calculating the transmission and reflection properties of some birefringent devices, the results obtained from these methods are expected to be different.

In the 4 × 4 matrix method, all reflected waves are included. The most important difference is the effect of interference due to multiple reflections between the interfaces of dielectric (refractive index) discontinuity. These interfaces include air-polarizer interface, polarizer-glass interface, glass-ITO interface, ITO-LC interface, and so forth. The interference effect is also known as the Fabry–Perot interference effect. As a result of the multiple reflections, fine interference structures exist in the transmission spectrum. This is consistent with the numerical results obtained by using the 4 × 4 matrix method. These fine structures are rapid variation of the transmission as a function of the wavelength. In most cases, the transmission oscillates from a maximum to a minimum within a wavelength range of $\Delta\lambda = \lambda^2/2D$, where D is the distance between the interfaces of dielectric (refractive index) discontinuity. With a typical thickness of $D = 1$ mm, the transmission oscillates within a wavelength range of only one Angstrom. Such a fast oscillation in the transmission spectrum is often unobservable in the display systems, as most human eyes and detectors are not sensitive to such a fine spectral detail. In other words, a spectral averaging is automatically performed in most detection systems, including the human eyes. It can be shown that the transmission spectra obtained by a spectral averaging of the results from the 4×4 matrix method are in agreement with those obtained by using the extended Jones matrix method.

In addition to the difference in numerical results, there are several interesting optical phenomena in periodic birefringent layered media that have been analyzed by using this 4×4 matrix algebra. These include the indirect optical bandgap, exchange Bragg reflection, and exchange Solc–Bragg transmission. Interested readers are referred to Reference 2.

8.3.2.   Reflection and Transmission

The matrix method just discussed is very useful in the calculation of the reflectance and transmittance of an anisotropic layered medium. Because of the anisotropy of the medium, mode coupling appears at the interfaces. Therefore, there are four complex amplitudes associated with the reflection and another four associated with the transmission. These eight complex amplitudes can be expressed in terms of the matrix elements of the overall transfer matrix. To illustrate this, one considers, without loss of generality, the case of an anisotropic layered medium sandwiched between two isotropic ambient and substrate media. Assume that the light is incident from the left side of the structure, and let A_s, A_p, B_s, B_p, and C_s, C_p be the incident, reflected, and transmitted electric field amplitudes, respectively. By using the 4×4 matrix method, a transfer matrix can be found for any given anisotropic layered structure such that

$$\begin{pmatrix} A_s \\ B_s \\ A_p \\ B_p \end{pmatrix} = \begin{pmatrix} M_{11} & M_{12} & M_{13} & M_{14} \\ M_{21} & M_{22} & M_{23} & M_{24} \\ M_{31} & M_{32} & M_{33} & M_{34} \\ M_{41} & M_{42} & M_{43} & M_{44} \end{pmatrix} \begin{pmatrix} C_s \\ 0 \\ C_p \\ 0 \end{pmatrix} \qquad (8.3\text{-}16)$$

The reflection and transmission coefficients are defined and expressed in terms of the matrix elements as follows:

$$r_{ss} = \left(\frac{B_s}{A_s}\right)_{A_p=0} = \frac{M_{21}M_{33} - M_{23}M_{31}}{M_{11}M_{33} - M_{13}M_{31}} \qquad (8.3\text{-}17)$$

$$r_{sp} = \left(\frac{B_p}{A_s}\right)_{A_p=0} = \frac{M_{41}M_{33} - M_{43}M_{31}}{M_{11}M_{33} - M_{13}M_{31}} \qquad (8.3\text{-}18)$$

$$r_{ps} = \left(\frac{B_s}{A_p}\right)_{A_s=0} = \frac{M_{11}M_{23} - M_{21}M_{13}}{M_{11}M_{33} - M_{13}M_{31}} \qquad (8.3\text{-}19)$$

$$r_{pp} = \left(\frac{B_p}{A_p}\right)_{A_s=0} = \frac{M_{11}M_{43} - M_{41}M_{13}}{M_{11}M_{33} - M_{13}M_{31}} \qquad (8.3\text{-}20)$$

$$t_{ss} = \left(\frac{C_s}{A_s}\right)_{A_p=0} = \frac{M_{33}}{M_{11}M_{33} - M_{13}M_{31}} \qquad (8.3\text{-}21)$$

$$t_{sp} = \left(\frac{C_p}{A_s}\right)_{A_p=0} = \frac{-M_{31}}{M_{11}M_{33} - M_{13}M_{31}} \qquad (8.3\text{-}22)$$

$$t_{ps} = \left(\frac{C_s}{A_p}\right)_{A_s=0} = \frac{-M_{13}}{M_{11}M_{33} - M_{13}M_{31}} \qquad (8.3\text{-}23)$$

$$t_{pp} = \left(\frac{C_p}{A_p}\right)_{A_s=0} = \frac{M_{11}}{M_{11}M_{33} - M_{13}M_{31}} \qquad (8.3\text{-}24)$$

These equations can also be written in matrix form as

$$\begin{pmatrix} B_s \\ B_p \end{pmatrix} = \begin{pmatrix} r_{ss} & r_{ps} \\ r_{sp} & r_{pp} \end{pmatrix} \begin{pmatrix} A_s \\ A_p \end{pmatrix} \qquad (8.3\text{-}25)$$

$$\begin{pmatrix} C_s \\ C_p \end{pmatrix} = \begin{pmatrix} t_{ss} & t_{ps} \\ t_{sp} & t_{pp} \end{pmatrix} \begin{pmatrix} A_s \\ A_p \end{pmatrix} \qquad (8.3\text{-}26)$$

These reflection and transmission formulas are extremely useful in the calculation of the spectral transmission characteristics of anisotropic layered structures, including LCDs. The matrix elements are obtained by carrying out the matrix multiplication in Eq. (8.3-21). The general explicit forms are normally not available. For fast results, a computer program is generally required. Even for the special case of periodic layered medium or TN-LCD, closed forms for the reflectance and transmittance are too complicated to derive. It is important to note that these eight complex amplitudes are spectrally correlated (see Ref. 2).

8.3.3. Berreman's 4 × 4 Matrix Method

Similar 4 × 4 matrix methods were developed earlier by Teitler and Henvis, and by Berreman. In 1970, Teitler and Henvis first introduced the 4 × 4 matrix technique [15], retaining two electric and two magnetic field variables throughout the computation. A little later, Berreman [16] developed an essentially equivalent technique to solve the problem of reflection and transmission of obliquely incident light by planar layers of cholesteric liquid crystals. A differential formulation was also developed by Berreman [3]. In Berreman's general approach [3], he started with a 6×6 matrix representation of Maxwell's equations

$$
\begin{pmatrix}
0 & 0 & 0 & 0 & -\dfrac{\partial}{\partial z} & \dfrac{\partial}{\partial y} \\[2mm]
0 & 0 & 0 & \dfrac{\partial}{\partial z} & 0 & -\dfrac{\partial}{\partial x} \\[2mm]
0 & 0 & 0 & -\dfrac{\partial}{\partial y} & \dfrac{\partial}{\partial x} & 0 \\[2mm]
0 & \dfrac{\partial}{\partial z} & -\dfrac{\partial}{\partial y} & 0 & 0 & 0 \\[2mm]
-\dfrac{\partial}{\partial z} & 0 & \dfrac{\partial}{\partial x} & 0 & 0 & 0 \\[2mm]
\dfrac{\partial}{\partial y} & -\dfrac{\partial}{\partial x} & 0 & 0 & 0 & 0
\end{pmatrix}
\begin{pmatrix} E_x \\ E_y \\ E_z \\ H_x \\ H_y \\ H_z \end{pmatrix}
= \frac{\partial}{\partial t}
\begin{pmatrix} D_x \\ D_y \\ D_z \\ B_x \\ B_y \\ B_z \end{pmatrix}
\qquad (8.3\text{-}27)
$$

which can include Faraday rotation and optical activity. From this, he derived expressions for 16 differential matrix elements so that a wide variety of specific problems can be attacked without repeating a large amount of tedious algebra. Since only four out of the six field vector components are independent variables, the E_z and H_z components can be eliminated. The 6×6 matrix formulation can be reduced to a 4×4 matrix equation

$$
\frac{i}{\omega} \frac{\partial}{\partial z}
\begin{pmatrix}
0 & 0 & 0 & 1 \\
0 & 0 & -1 & 0 \\
0 & -1 & 0 & 0 \\
1 & 0 & 0 & 0
\end{pmatrix}
\begin{pmatrix} E_x \\ E_y \\ H_x \\ H_y \end{pmatrix}
= S
\begin{pmatrix} E_x \\ E_y \\ H_x \\ H_y \end{pmatrix}
\qquad (8.3\text{-}28)
$$

where S is a 4×4 matrix. This equation can then be used to treat various problems, including reflection and transmission through layered media. Note that all four elements—E_x, E_y, H_x, and H_y in Eq. (8.3-28)—are continuous at the boundaries in a layered medium such as that shown in Figure 8.19, where the medium is homogenous in the xy directions. This continuity condition can be used to derive reflectance and transmittance.

Notice that the four-component vector in Section 8.3.1 is chosen differently from the one used in this section. For example, $\mathbf{E}$ in this section is the total electric field, including both the incident and the reflected components. The choice in Section 8.3.1 makes it easier to analyze transmission and reflection problems.

Berreman's 4×4 matrix method has the advantage of a general formulation. It can be used to analyze almost all birefringent network problems. In fact, the 4×4 matrix method introduced in Section 8.3.1, which is particularly powerful in dealing with layered media, can be derived from Berreman's formulation. In addition, Berreman's 4×4 matrix method is suitable for continuously varying media. Interested readers are referred to Reference 3.

REFERENCES

1. R. C. Jones, *J. Opt. Soc. Am.* **31**, 488 (1941).
2. See, for example, P. Yeh, "Electromagnetic propagation in birefringent layered media," *J. Opt. Soc. Am.* **69**, 742–756 (1979).
3. D. W. Berreman, "Optics in stratified and anisotropic media: 4×4-matrix formulation," *J. Opt. Soc. Am.* **62**, 502–510 (1972).
4. P. Yeh, "Extended Jones matrix method," *J. Opt. Soc. Am.* **72**, 507–513 (1982).
5. C. Gu and P. Yeh, "Extended Jones matrix method II," *J. Opt. Soc. Am. A* **10**, 966–973 (1993).
6. See, for example, P. Yeh, *Optical Waves in Layered Media*, Wiley, 1988.
7. A. R. MacGregor, "Method for computing homogeneous liquid-crystal conoscopic figures," *J. Opt. Soc. Am. A* **7**, 337–347 (1990).
8. A. Lien, "The general and simplified Jones matrix representations for the high pretilt twisted nematic cell," *J. Appl. Phys.* **67**, 2853–2856 (1990).
9. A. Lien, "Extended Jones matrix representation for the twisted nematic liquid-crystal display at oblique incidence," *Appl. Phys. Lett.* **57**, 2767–2769 (1990).
10. H. L. Ong, "Electro-optics of electrically controlled birefringence liquid-crystal displays by 2×2 propagation matrix and analytic expression at oblique angle," *Appl. Phys. Lett.* **59**, 155–157 (1991).
11. H. L. Ong, "Electro-optics of a twisted nematic liquid-crystal display by 2×2 propagation matrix at oblique angle," *Jpn. J. Appl. Phys.* Part 2 — Letters **30**, L1028–L1031 (1991).
12. L. Baxter, *J. Opt. Soc. Am.* **46**, 435–442 (1956).
13. D. Taber, Department of Optical Devices, Rockwell International Science Center, Thousand Oaks, CA 91360 (personal communication, 1992).
14. H. Mori, Y. Itoh, Y. Nishiura, T. Nakamura, and Y. Shinagawa, *Jpn. J. Appl. Phys.* **36**, 143–147 (1997); H. Mori, *Jpn. J. Appl. Phys.* **36**, 1068–1072 (1997).
15. S. Teitler and B. W. Henvis, *J. Opt. Soc. Am.* **60**, 830 (1970).
16. D. W. Berreman and T. J. Sheffer, *Phys. Rev. Lett.* **25**, 577 (1970); *Mol. Cryst. Liq. Cryst.* **11**, 395 (1970).

PROBLEMS

8.1. (a) Using extended Jones matrix method, plot $T(\theta,\phi)$ for various λ values for a NW (normally white) $90°$ TN-LC sandwiched between a pair of crossed polarizers with E-mode configuration. Use $n_o = 1.5, n_e = 1.6$, $L = 10 \, \mu m$, and $\lambda = 0.45, 0.55, 0.65 \, \mu m$.

    (b) Repeat (a) with O mode configuration.

    (c) Explain the difference in $T(\theta,\phi)$ between the E-mode and O-mode configurations.

8.2. (a) Using the extended Jones matrix method, plot $T(\theta,\ \phi)$ for various λ values for a NB (normally black) $90°$ TN-LC sandwiched between a pair of parallel polarizers with E-mode configuration. Use $n_o = 1.5, n_e = 1.6, L = 10\,\mu m$, and $\lambda = 0.45, 0.55, 0.65\,\mu m$. Compare the results with those of Problem 8.1.

    (b) Repeat (a) with O-mode configuration.

    (c) Explain the difference in $T(\theta,\phi)$ between the E mode and O mode configurations.

8.3. Derive equations (8.1-28)–(8.1-33).

8.4. *Cross-reflection coefficients*: Show that r_{sp} and r_{ps} vanish when $n_o = n_e$ for an arbitrary angle of incidence.

8.5. *Crossed polarizers*: Derive Eqs. (8.1-108) and (8.1-112).

8.6. Show that k_{ez} in Eq. (8.1-124) is positive.

8.7. Derive Eqs. (8.2-2)–(8.2-7) and (8.2-10)–(8.2-12).

8.8. Derive the expressions for the azimuth angle ψ in Eq. (8.1-97), (8.1-98) and (8.1-106). Sketch $\psi - \psi_0$ versus angle of incidence θ.

8.9. Show the equality in Eq. (8.1–104). Sketch σ verus angle of incidence θ.

8.10. Derive Eq. (8.1-121)–(8.1-123). Show that w is always negative for positive k_{ez}.

8.11. Show that Eq. (8.2-18) is equivalent to $\{|E'_s|^2 + |E'_p|^2\}/\{|E_s|^2 + |E_p|^2\}$.

9

Optical Compensators for LCDs

From the discussion in the previous chapters, we have a very good understanding of the properties of conventional twisted nematic LCDs (TN-LCDs), which have been and will continue to be very important display devices in conjunction with active matrix (AM) addressing technology. High quality (contrast, gray-scale stability) information display can be obtained only within a narrow range of viewing angles centered about the normal incidence by using these conventional TN-LCDs. In fact, the viewing angle characteristic is a fundamental property of virtually all modes and configurations of LCDs. The angular dependence of the viewing is due to the fact that both the phase retardations and optical path in most LC cells are functions of the viewing angles. The narrow viewing angle characteristics have been a significant problem in advanced applications requiring high quality displays, such as avionics displays and wide-screen displays. In the case of avionics displays, the LCDs must provide the same (or nearly the same) contrast and gray scale for viewing angles from both the pilot and the copilot. Such high-information-content and high quality displays require LCDs whose contrast and gray scale must be as invariant as possible with respect to viewing angles.

Various methods and LCD modes of operation have been proposed so far to improve the viewing angle characteristics. In the method of optical birefringence compensation, a thin film of birefringent material is inserted into the LCD at a proper location to neutralize the angular dependence. It was first suggested that a film of negative birefringence can be employed to improve the viewing characteristics of LCDs based on vertically aligned nematic cells (VA-LCDs) [1,2]. The same film of negative birefringence can also be employed in TN-LCDs to improve the viewing angle characteristics [3,4]. There are several methods of producing the films of negative birefringence. These include the use of negative form birefringence in multilayers of alternating thin films [5], negative form birefringence of coplanar alignment of polymers (e.g., spin-coated polyimide) [6], and negative birefringent films of discotic compound with inclined optical axes [7]. Using multiple domains (e.g., two or four) with different LC orientations in each pixel of TFT-LCDs, the angular dependence of the optical transmission characteristics can also be significantly reduced [8,9]. The viewing characteristics can also be improved by using various different LCD modes of operation. These include optically compensated birefringence (OCB) mode LCDs [10,11], in-plane switching (IPS) mode LCDs [12–13], the halftone gray-scale (HTGS) method [14]. The angular dependence of the optical transmission can be reduced or eliminated by using a beam of collimated light as the backlight. A high quality diffuser is needed in this case for wide-angle viewing [15].

In this chapter we discuss the principle of phase retardation compensation using birefringent thin films to achieve high contrast ratios and gray-level stability in LCDs. We will also briefly discuss various techniques of implementing and manufacturing these compensation films. Both films of uniaxial and biaxial birefringence will be considered. We first summarize the basic properties of TN-LCDs at large viewing angles.

9.1. VIEWING ANGLE CHARACTERISTICS OF LCDs

As mentioned earlier, the transmission properties of most LCDs depend on the angle of viewing for virtually all modes of operations based on a thin film of LC materials, including VA-LCDs, TN-LCDs, BA-LCDs, and STN-LCDs. Consider the simplest case of a normally black (NB) LCD based on a vertically aligned LC cell. In the field-OFF state (dark state), the cell is effectively a c plate with a positive birefringence sandwiched between a pair of crossed polarizers. For on-axis viewing, the light is propagating along the c axis with a zero phase retardation. The transmission is zero due to the crossed polarizers. As we know, leakage of light occurs at off-axis viewing as the result of a finite phase retardation. The leakage becomes severe at large viewing angles, leading to reduced contrast ratios. In the following paragraphs we consider the viewing angle characteristics of TN-LCDs.

Contrast and stability of gray levels are important attributes in determining the quality of a liquid crystal display (LCD). The primary factor limiting the contrast ratio achievable in a LCD is the amount of light that leaks through the display in the dark state. To illustrate the viewing angle characteristics of a typical conventional TN-LCD, we examine the transmission curves at various horizontal and vertical viewing angles. Referring to Figure 9.1, we consider a typical normally white (NW) TN-LCD with either an E-mode or O-mode operation. The LC cell is expanded to show the rubbing directions and the tilt of the LC director at midlayer. In an E-mode operation, the transmission axis of the polarizer is parallel to the director of the liquid crystal on the surface adjacent to the polarizer. In a NW display configuration, the $90°$ twisted nematic liquid crystal cell is placed between a pair of crossed polarizers. In the absence of an applied electric field, the direction of polarization (E-field vector) of an incoming beam of polarized light will follow the twist of the director (known as *waveguiding*) in traveling through the LC layer. The polarization state of the light will thus be aligned parallel to the transmission axis of the analyzer, leading to a white state (high transmission). When a sufficient voltage is applied to the transparent electrodes, a strong electric field along the z axis is established in the LC cell. The applied electric field causes the director of the LC material to tend

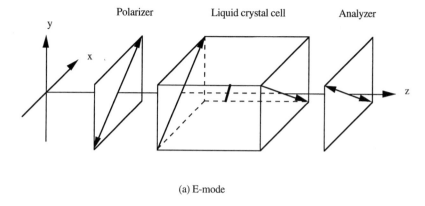

(a) E-mode

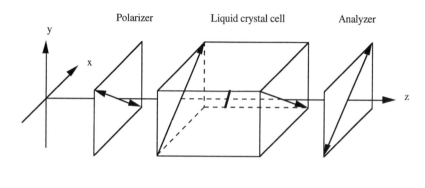

(b) O-mode

Figure 9.1. Schematic drawing of a NW TN-LCD. (*a*) *E*-mode operation with polarizer transmission axis parallel to the director at adjacent surface of LC cell. (*b*) *O*-mode operation with polarizer transmission axis perpendicular to the director at adjacent surface of LC cell. The arrows on the surfaces of the LC cell indicate the rubbing directions.

to align parallel to the field (z axis). With the LC material in this state, the cell exhibits no phase retardation to the incoming beam of polarized light. The light transmitted through the polarizer is thus extinguished by the analyzer, leading to a dark state (zero or low transmission).

To avoid confusion in the discussion, we choose a right-handed coordinate (xyz) where the z axis is pointed toward the general direction of the transmission of light. The x-axis is pointed in the horizontal direction, whereas the y axis is pointed toward the vertical direction. Thus, light propagates toward the viewer in the positive z direction. To describe the orientation of the polarizers and the director of the liquid crystal, we define the azimuth ϕ as the angle in the xy plane measured from the x axis. We define the tilt θ as the angle between the director and the xy plane, measured from the xy-plane. With this definition, $\phi = 0$ is parallel to the x axis, $\phi = 90°$ is parallel to the y axis, $\theta = 0$ is parallel to the xy

Table 9.1. Orientation of Various Components in Field-OFF State of NW TN-LCD with an E-Mode Operation

Component	Location	Orientation θ	Orientation φ
Polarizer	Rear	0	45°
Rubbing	Rear	0	45°
LC director	$z = 0$	$1°^a$	45°
LC director	Midlayer	$1°^a$	90°
LC director	$z = d$	$1°^a$	135°
Rubbing	Front	0	−45°
Analyzer	Front	0	−45°

^{a} A small pretilt angle usually exists at the rubbed surface; d is the thickness of the LC layer.

plane, and $\theta = 90°$ is parallel to the z axis. In the configuration shown in Figure 9.1, the LC director exhibits a right-handed twist of $90°$ in the absence of an applied field. The midlayer director is pointed toward $(+y, +z)$ direction under the influence of an intermediate field.

Using these definitions, the E-mode TN-LCD sketched in Figure 9.1a with no applied field $(V = 0)$ is described in Table 9.1.

As a result of the rubbing directions and the pretilt, the director of the liquid crystal will twist in a right-handed direction. Thus, in a nonenergized state (field-OFF state), the director is parallel (except for the pretilt) to the xy plane with the midlayer director pointed approximately in the vertical direction ($+y$ axis). When an intermediate voltage is applied, the LC director at midlayer ($z = d/2$) is pointed toward the upper viewing angle at around $\theta = 45°$ depending on the voltage applied. For the purpose of discussion, we reproduce the following figures for the twist and tilt angles as functions of position in the cell for various applied voltages.

Figures 9.2 and 9.3 show the distributions of the twist ϕ and tilt θ angles of the LC director as functions of position z under the application of various applied voltages chosen to yield several intermediate transmission states between the white and dark states. The midlayer tilt angle $\theta(z = d/2)$ is a measure of the electrodistortion (electroopatical distortion) of the cell. We note that the midlayer tilt angle is zero in the absence of the field and is an increasing function of the applied voltage. In the extreme case when the applied voltage is significantly higher than the threshold voltage (e.g., 5.49 V), the liquid crystal is approximately homeotropically aligned, with a midlayer tilt of $90°$. In other words, the LC director is approximately parallel to the z axis, resembling a positive c plate. At somewhat lower voltages, the LC director has a continuous distribution with its midlayer tilt angle ranging between $0°$ (zero voltage) and $90°$ (maximum voltage). The midlayer tilt angle is a monotonically increasing function of the applied voltage. At low voltages, the twist is approximately uniform throughout the cell. At high voltages (5.49 V), the twist of $90°$ occurs in a small region near the center of the cell.

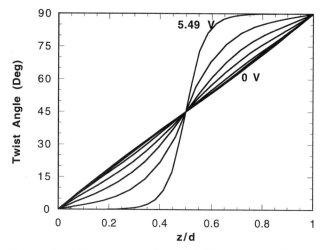

Figure 9.2. Twist angle of director versus z in a TN-LCD at various applied voltages. (In this figure, the twist is measured from the rubbing direction.)

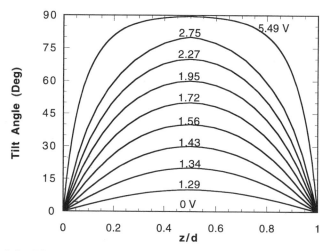

Figure 9.3. Tilt angle of director versus z in a TN-LCD at various applied voltages.

As a result of the different optical path, the birefringence, and distribution of LC director $\mathbf{n}(z)$, the transmission of the display is intrinsically dependent on the angle of incidence (θ, ϕ). Figures 9.4 and 9.5 show the transmission properties of the LCD as functions of horizontal and vertical viewing angles at various applied voltages.

Referring to Figures 9.4a and 9.5a, we first discuss the contrast ratio of the normally white TN-LCDs. Notice that the display exhibits an exact left–right symmetric horizontal viewing characteristics. The vertical viewing character- istics, however, are quite asymmetric because of the distribution of the tilt angle

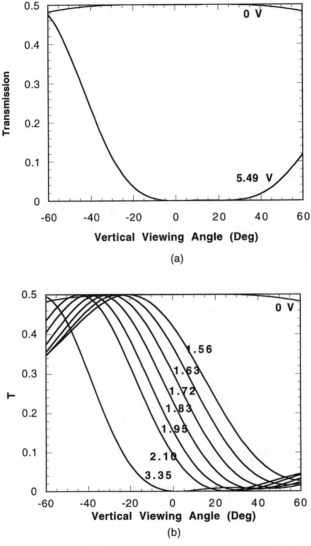

Figure 9.4. Vertical viewing characteristics $T(\theta, \phi = 90°)$ of a typical NW TN-LCD with an O-mode operation, at various applied voltages (in the O-mode operation, the transmission axis of the polarizer is perpendicular to the director adjacent to the polarizer): (*a*) binary operation; (*b*) gray-level operation.

of the LC director. In the nonselect state (0 V), the display exhibits a high transmission (50%) at normal incidence ($\theta = 0°$), with a slight decrease at large viewing angles. We recall that 50% of the light energy is absorbed by the polarizer. In the select state (dark state with 5.49 V), the display exhibits a zero transmission at normal incidence ($\theta = 0°$) as desired. In the horizontal viewing

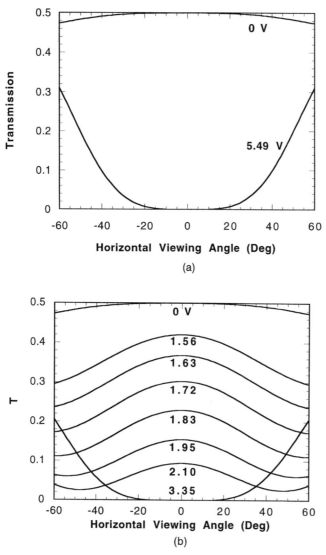

Figure 9.5. Horizontal viewing characteristics $T(\theta, \phi = 0°)$ of a typical NW TN-LCD with an
O-mode operation, at various applied voltages (in the O-mode operation, the transmission axis of
the polarizer is perpendicular to the director adjacent to the polarizer) (*a*) binary operation;
(*b*) gray-level operation.

angles, we note that the leakage of light (dark state with 5.49 V) becomes severe
for horizontal viewing angles greater than 20°, reaching over 30% at 60° in the
horizontal direction. In the vertical viewing angles, the leakage is extremely
severe in the lower viewing directions, reaching almost 50% at $\theta = -60°$. The
leakage of light (dark state with 5.49 V) in the upper vertical viewing angles is

not as severe. As a result of the leakage of light at these large angles of viewing, the contrast ratios decrease accordingly. The leakage of light at these large angles cannot be eliminated by increasing the applied voltage, even if the LC director is homeotropically aligned.

We now discuss the issue of gray levels. To ensure a high quality display with gray levels, it is important that the transmission follow the applied voltage in a monotonic fashion. In the case of NW operation, the transmission must decrease monotonically with the applied voltage. This can be easily obtained at normal incidence ($\theta = 0$). Referring to Figures 9.4b and 9.5b, we note that the transmission at normal incidence indeed decreases with the applied voltage. To ensure a high quality display of gray levels, the transmission curves at various voltages must be well separated. Examining Figures 9.4b and 9.5b, we find that some of the transmission curves cross at large viewing angles. For viewing angles beyond the crossing point, the gray levels are reversed, meaning that the transmission no longer follows the applied voltage monotonically. This leads to a degradation of the so-called gray-level stability. In the horizontal viewing angles, we find that the transmission curves at various voltages are well separated, except for the transmission curve at 3.35 V. In the vertical viewing angles, each transmission curve reaches a minimum near zero in the upper viewing directions. The transmission curves rebound beyond these minima. Thus, at these large viewing angles, the gray levels are actually reversed. This severely degrades the gray-level stability in TN-LCDs.

In what follows, we describe some of the techniques of using optical compensators to improve the viewing characteristics at large viewing angles. The numerical modeling is performed with the extended Jones matrix method (see Chapter 8 and Refs. 16, and 17, this chapter).

9.2. NEGATIVE c PLATE COMPENSATORS

As mentioned earlier, a thin film of negative birefringence can be employed as a phase retardation compensator for LCDs based on vertically aligned LC cells (VA-LCDs) [1,2]. In the field-OFF state, the phase retardation due to the positive birefringence of the LC cell can be compensated by a thin film of negative birefringence. This leads to a net phase retardation of near zero for virtually all angles of incidence. The same idea can be employed for phase retardation compensation in the field-ON state of TN-LCDs. As discussed earlier, the LC cell is approximately a positive c plate in the field-ON state when the LC molecules are approximately homeotropically aligned (vertically aligned). When a positive c plate is sandwiched between a pair of crossed polarizers, a dark state is achieved only for light propagation along the z axis (normal incidence). As we already know, the phase retardation is zero for propagation along the c axis. For a beam of light with an oblique incidence, the light is no longer propagating along the c axis of the LC cell. The LC cell thus exhibits its birefringence and a finite phase retardation that can change the polarization state of the light, leading to a

Polarizer   TN-LC cell   Negative c-plate   Analyzer

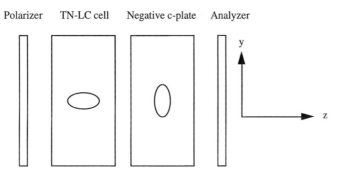

Figure 9.6. Basic concept of negative *c*-plate compensator (side view). In the field-ON state, the TN-LC cell is approximately a positive *c* plate with a prolate index ellipsoid (elongated sphere). Note that this is very similar to a vertically aligned cell. The negative *c* plate has an oblate index ellipsoid (flattened sphere).

leakage of light at the analyzer. The phase retardation increases with the angle of incidence θ (see Eq. (4.5-5)). This is the main cause of leakage of light at large viewing angles in the field-ON state.

A simple and convenient way of improving the transmission characteristics of TN-LCDs at large viewing angles is to include a negative *c* plate with the same birefringence thickness product (Δnd). The basic concept of such an optical compensator is illustrated in Figure 9.6.

The negative *c* plate must be placed between the polarizer and the analyzer. Because it is in a homeotropic alignment, the TN-LC with a positive biref-ringence ($n_o < n_e$) can be represented optically by a prolate ellipsoid (sphere elongated along the *z* axis). The birefringence property of the negative *c* plate is represented by an oblate ellipsoid (sphere flattened along the *z* axis). The presence of the negative *c* plate with its *c* axis aligned along the *z* axis does not create any additional phase retardation for normally incident light. For beams with an oblique incidence, the positive phase retardation due to the LC cell can be compensated by the negative phase retardation of the negative *c* plate. A similar compensation scheme is illustrated in Figure 9.7, in which the negative *c* plate in Figure 9.6 is divided equally into two plates. The two halves are then placed on both sides of the TN-LC cell. The compensation scheme described in Figure 9.7 preserves the left–right viewing symmetry.

To illustrate the effectiveness of the phase retardation compensation, we recalculate the transmission properties of the TN-LCD by including a negative *c* plate. Figures 9.8 and 9.9 show the transmission properties of the TN-LCD (with negative *c* plate compensators) as functions of horizontal and vertical viewing angles.

Referring to Figures 9.8 and 9.9, we focus our attention on the transmission curves of the field-ON state (5.49 V). We note that the inclusion of a negative *c* plate leads to an improvement in reducing the leakage of light in the dark state at large viewing angles. By comparing Figures 9.5*a* and 9.9*b*, we find that the leakage at a horizontal viewing angle of 60° is reduced from over 30% to about

Polarizer   TN-LC cell   Analyzer

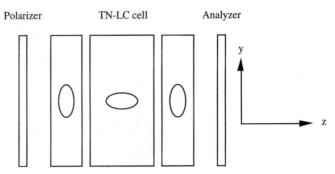

Figure 9.7. A different scheme of a compensator with two negative *c* plates. This scheme preserves the left-right viewing symmetry.

8%. This leads to a significant improvement in the contrast ratio. Further improvement can be obtained at higher applied voltages (e.g., >5.49 V). As discussed above, a negative *c* plate compensator is designed to correct for the angular dependence of the phase retardation introduced by the optical propagation through the central region of the TN-LC, which is almost homeotropically aligned. Such a compensator is effective to the extent that the optical property of this region dominates the field-ON state of the LC cell. This implies that the negative *c* plate works best when strong fields are employed for the energized state to achieve nearly homeotropic alignment. The use of a negative *c* plate in TN-LCDs has been experimentally demonstrated to significantly reduce the leakage of light in the dark state over an extended field of view, leading to an improved contrast ratio and a reduced color desaturation [5]. Referring to Figure 9.9*b*, we also note that the split *c* plate configuration as shown in Figure 9.7 indeed preserves the left–right viewing symmetry.

Although the negative *c* plate is capable of providing a significant improvement in reducing the leakage of light at the dark state, further reduction of the leakage is possible by using more complicated compensators. As we know, homeotropic alignment of the LC molecules occurs only in the central region of the cell, which behaves like a positive *c* plate. The regions, near each surface of the cell, behave like positive *a* plates, each with its optic axis aligned with the rubbing direction of the proximate substrate. Thus, a more precise cancellation of the phase retardation would require the inclusion of split negative *c* plates and split negative *a* plates. This is illustrated in Figure 9.10*a*.

The TN-LC cell is sandwiched between the two negative *a* plates, which are employed to cancel the phase retardation due to the LC regions near the surfaces. The optic axes of the *a* plates are aligned parallel to the rubbing directions of the adjacent surfaces. The *c* plates are aligned with their optic axes parallel to the *z* axis. We note that the leakage of light at field-ON state (5.49 V) is further decreased with the addition of the two *a* plates (see Fig. 9.10). By comparing Figures 9.5*a* and 9.10*c*, we find that the leakage at a horizontal viewing angle of 60° is further reduced from around 8% to 1%. This leads to a significant improvement in the contrast ratio.

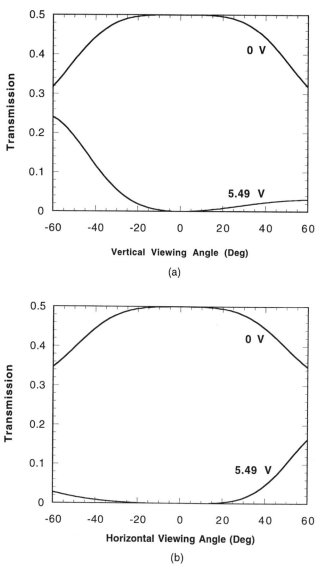

Figure 9.8. Viewing characteristics of a typical NW TN-LCD (with a negative *c* plate compensator) with an *O*-mode operation, at various applied voltages: (*a*) vertical viewing characteristics $T(\theta, \phi = 90°)$; (*b*) horizontal viewing characteristics $T(\theta, \phi = 0°)$.

Referring to Figures 9.2 and 9.3 for the tilt and twist angles of the LC director of TN-LCDs, respectively, we find that even with a strong applied field, the LC cell is a positive birefringent medium with a continuous variation of its director in the cell. Thus, strictly speaking, a complete cancellation of phase retardation would require negative birefringent plates with a similar continuous variation of

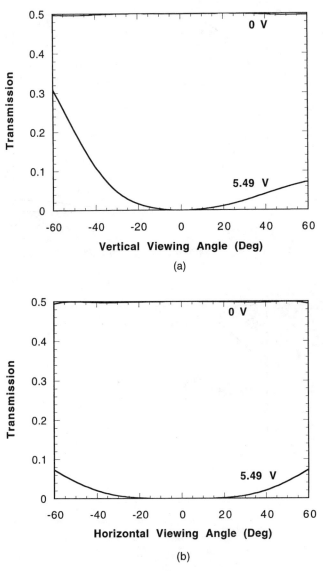

Figure 9.9. Viewing characteristics of a typical NW TN-LCD (with split negative c plates) with an O-mode operation, at various applied voltages: (a) vertical viewing characteristics $T(\theta, \phi = 90°)$; (b) horizontal viewing characteristics $T(\theta, \phi = 0°)$.

Figure 9.10. (a) Optical compensators using negative a plates and c plates. The index ellipsoids for the TN-LC cell are prolate, and those of the compensators ($-a, -c$) are oblate. The c axes of the negative a plates are parallel to the rubbing direction of the proximate substrate. (b, c) Viewing characteristics of a typical NW TN-LCD (with split negative c plates and split negative a plates) with an O-mode operation, at two different applied voltages: (b) vertical viewing characteristics $T(\theta, \phi = 90°)$; (c) horizontal viewing characteristics $T(\theta, \phi = 0°)$. The y'-axis is oriented at 45° from the x-axis.

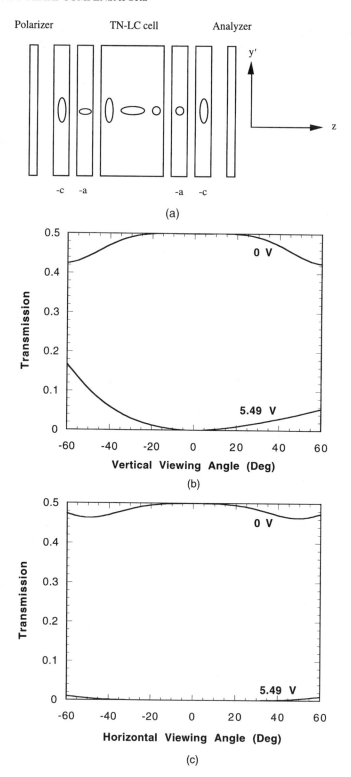

(a)

(b)

(c)

the tilt and twist angles of the c axis (as discussed in Chapter 8). Each thin slice of the TN-LC cell with a given tilt and twist can be compensated by a thin layer of the negative birefringent plate with the same tilt and twist angles. An multilayer stack of thin TN-LC layers can be compensated by a corresponding multilayer stack of negative birefringence layers in a proper sequence. Such a novel but complicated compensator has been proposed and demonstrated by using films of a discotic compound that exhibits negative birefringence [7]. Such a compensator effectively suppresses the leakage of light at the dark state of TN-LCDs from almost any angle of viewing. The inclusion of such a compensator in TN-LCDs leads to a very high contrast ratio at wide viewing angles. The compensation film also provides a reasonable improvement in the gray-scale stability at large viewing angles.

9.3. COMPENSATION FILM WITH POSITIVE BIREFRINGENCE (o PLATE)

While the use of the negative c plate is important for improving the contrast ratio and reducing the color desaturation, the issue of gray-scale stability remains unsolved. This can be easily seen in Figures 9.4b and 9.5b, where the transmission curves are not well separated at large viewing angles. A high quality display with gray-scale capability requires a gray-scale linearity over the field of view. In other words, the brightness levels between the select state (level 0 with minimum or zero brightness in NW operation) and the nonselect state (level 7 with maximum brightness in NW operation) must vary linearly with the assigned gray levels. The linearity (or near linearity) requires that the transmission curves be well separated without crossing in the range of viewing angles. To understand the gray-scale linearity problem of a conventional TN-LCD, we reexamine the transmission curves in Figures 9.4 and 9.5. Note that the brightness level near the normal incidence ($\theta = 0$) is a monotonic function of the applied voltage. Thus, linearly spaced (brightness) gray levels can be chosen with a set of properly assigned applied voltages. We also note that the linearly spaced brightness levels are reasonably well maintained at large horizontal viewing angles (see Fig. 9.5b), except the transmission curve of 3.35 V. The gray-scale linearity problem appears when the vertical viewing angle varies (see Fig. 9.4b). We note that each transmission curve decreases initially with the positive vertical viewing angle, and then rebounds after reaching a minimum (near zero). Note that the zero transmission angle (point of rebound) decreases with the applied voltage. The transmission curves cross after the rebounds. The crossing of the transmission curves removes the separation needed for linearity. This is the region of gray-scale reversal (inversion). In a conventional TN-LCD, the gray-scale linearity needed for high quality displays disappears in the region near the zero transmission angles. The gray-scale reversal can lead to a severe problem in the display of desired colors as the gray-scale is varied.

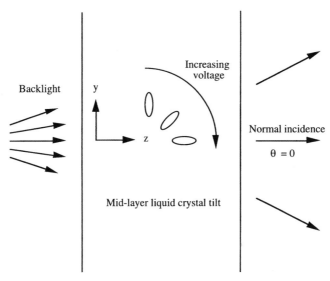

Figure 9.11.   Schematic drawing of the midlayer liquid crystal tilt (in *yz* plane) in the cell.

To solve the gray-scale linearity problem, we must first understand the variation of the transmission (brightness level) as a function of the applied voltage at various viewing angles. Referring to Figure 9.11, we examine the midlayer tilt angle of the LC director in the cell as the applied voltage increases. The midlayer tilt angle is a measure of the average orientation of the LC director in the cell. The tilt angle of the LC director is initially zero (except the pretilt) with no applied voltage. The tilt angle then increases with the applied voltage and reaches almost 90° at the highest applied voltage. We note that for viewing at normal incidence, the phase retardation is highest at the nonselect state (zero voltage) when the LC director is perpendicular to the viewing direction. Maximum transmission (brightness) is obtained in this state as a result of the waveguiding and the crossed polarizers. As the applied voltage is increased, the LC director is tilted toward the upper forward direction, leading to a smaller phase retardation and less waveguiding for the viewing at normal incidence. This leads to a correspondingly lower transmission. The phase retardation decreases with the midlayer tilt angle, which increases monotonically with the applied voltage. Thus, for viewing at normal incidence ($\theta = 0°$), the brightness level decreases monotonically with increasing voltage. This explains the gray-scale linearity for viewing at normal incidence.

Let us now consider the viewing from positive vertical angles (light propagating in the upper forward direction for viewers looking down at the TN-LCD). At some intermediate applied voltage level, the midlayer tilt angle is pointed toward the viewing direction. For this applied voltage and the viewing angle, the viewer is looking along the director of the liquid crystal at midlayer in conventional TN-LCDs. The transmission in this case is near zero, due to the vanishing phase retardation. For viewing at this angle, the viewer sees a brightness that initially decreases with the voltage, reaching a minimum (near zero) at an intermediate voltage when the midlayer director tilts toward this particular direction. The brightness then rebounds beyond this intermediate voltage, destroying the linearity and creating a gray-level reversal problem. For the negative vertical viewing angles, the midlayer LC director is always oriented at a large angle around 90° relative to the viewing direction. The average director at these orientations provide large phase retardations and waveguiding, which lead to higher transmissions in the negative viewing angles. As a result of the large phase retardation at these viewing angles, the viewer can see a monotonic decrease in the brightness as the voltage is increased.

On the basis of this discussion, we can eliminate the reversal of gray levels and improve the gray-scale stability by including a special birefringence thin film as a phase retardation compensator. The birefringent thin film must provide proper positive phase retardations for positive vertical viewing angles. In addition, the film must also provide a zero phase retardation for normal incidence. We first consider the possibility of using a thin film of positive uniaxial birefringent material. The most important issue, then, is the orientation of its c axis. As we know, a positive c plate will severely degrade the contrast ratios at large viewing angles. Thus, such a positive birefringent thin film must be oriented at an angle other than the z axis. To maintain the left–right symmetry of the viewing, it is natural to orient the c axis in the central vertical plane (yz plane).

Knowing the severe problem due to the viewing along the average director direction of the liquid crystal in the positive vertical angles, we can orient the c axis of the positive birefringent plate along a direction that is perpendicular to the midlayer LC director at the middle gray level. The midlayer tilt angle for most TN-LCDs is around 45° toward the upper forward direction. Thus, a natural choice for the orientation of the c axis of the positive birefringent compensator is at 45° with the c axis along the lower forward direction. This is known as an o plate, because of its substantially oblique orientation of the c axis relative to the display plane (xy plane). Such an orientation with the proper thickness would somewhat symmetrize the vertical viewing characteristics for the gray level when the midlayer tilt is about 45°. The o plate would also push the zero transmission angles toward larger angles beyond the viewing angles of interest. To maintain the same viewing characteristics at normal incidence, the compensator must be configured to ensure that no phase retardation is introduced for light traversing the LC cell at normal incidence. This can be accomplished by adding a positive a plate with its c axis perpendicular to the o plate. Further numerical simulation can be employed to fine-tune the orientations of the c axes of both the o plate and the a plate for optimized overall performance, including gray scales and contrast ratios. Figure 9.12a shows a schematic drawing of the concept of o plate compensator. The birefringent wave plates can also be split

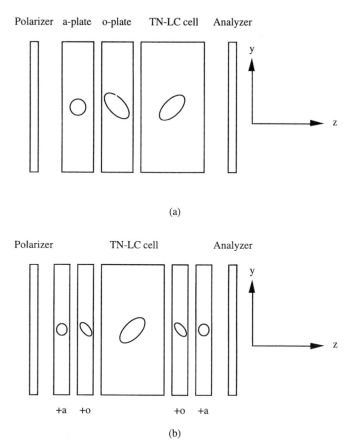

Figure 9.12. (*a*) Schematic drawing of the concept of *o*-plate compensator (side view). The *c* axis of the *a* plate is parallel to the *x* axis. (*b*) A different scheme of a compensator with split positive *o* plates and split positive *a* plates. This scheme preserves the left-right viewing symmetry. All ellipsoids shown in the figures are prolate in shape.

into two equal parts and then placed on both sides of the TN-LC cell as in Figure 9.12*b*. The latter scheme of birefringence compensation preserves the left–right viewing symmetry.

Figures 9.13 and 9.14 show the vertical and horizontal viewing characteristics at various gray levels of a typical TN-LCD with the inclusion of *o* plate and *a* plate compensators. Comparing these with the uncompensated TN-LCD (Figs. 9.4*b* and 9.5*b*), we find that the inclusion of *o* plates and *a* plates indeed improves the gray-scale linearity over a large range of viewing angles. We also note that the split configuration provides a better compensation, in addition to symmetric horizontal viewing. According to Figure 9.14, there are no gray-level crossings throughout the horizontal viewing angles from − 60° to 60°. In the vertical

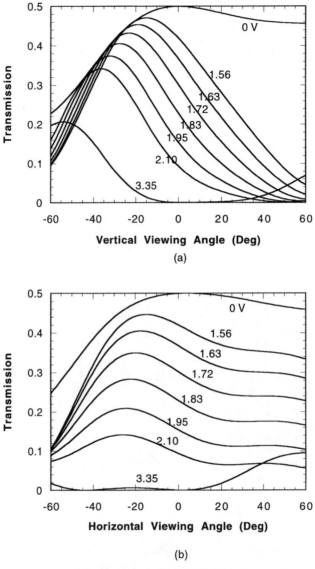

Figure 9.13. Viewing characteristics of a typical NW TN-LCD (with one *o* plate and one *c* plate) with an *O*-mode operation, at various applied voltages: (*a*) vertical viewing characteristics $T(\theta, \phi = 90°)$; (*b*) horizontal viewing characteristics $T(\theta, \phi = 0°)$.

viewing direction, the gray-level crossings are pushed from $+20°$ (uncompensated) to $+35°$ (compensated).

There are several different variations of the arrangement of the *o* plate and *a* plate, including the addition of c plates in TN-LCDs. Interested readers are referred to References 18 and 19.

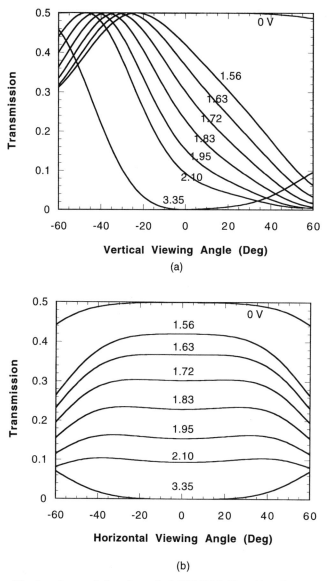

Figure 9.14. Viewing characteristics of a typical NW TN-LCD (with split *o* plates and split *c* plates) with an *O*-mode operation, at various applied voltages: (*a*) vertical viewing characteristics $T(\theta, \phi = 90°)$; (*b*) horizontal viewing characteristics $T(\theta, \phi = 0°)$.

9.4. BIAXIAL COMPENSATION FILM

We consider the possibility of using a biaxially birefringent thin film as a compensator to improve the gray-scale characteristics. For the purpose of discussion, let the principal refractive indices be $n_a < n_b < n_c$. Here, *a*,*b*,*c* are the principal axes of the biaxial medium. Because it is biaxial, the thin film has two optic axes. There is no phase retardation for the propagation of light along the optic axes. These two optic axes are in the ac plane on both sides of the c axis in the direction, with

$$\tan \theta = \frac{n_c}{n_a} \sqrt{\frac{(n_b^2 - n_a^2)}{(n_c^2 - n_b^2)}} \qquad (9.4\text{-}1)$$

where θ is measured from the c axis. We note that in the limit when $n_b = n_a$, these two optic axes merge into one which is parallel to the c axis.

We now consider the proper orientation of the principal axes of the thin biaxial film. To maintain the left–right viewing symmetry, we consider the orientation in which the b axis is parallel to the x axis, which is in the horizontal direction, and the two optic axes and the c axis of the film, which are in the vertical plane (yz plane) containing the z axis. Figure 9.15 is a schematic drawing of the biaxial compensator. For the purpose of maintaining the same viewing characteristics for normal incidence, we can orient the principal axes of the biaxial thin film such that one of the optic axes is pointed along the z axis. This ensures a zero phase retardation for light propagating along the z axis (normal incidence). The c axis is pointed in the lower forward (having no polarity, this is the same as upper backward) direction.

For viewing at normal incidence, the biaxial thin film exhibits no phase retardation because one of the optic axes is oriented parallel to the z axis. Thus, a single biaxial film has the combined effect of an o plate plus an a plate. For positive viewing angles, the light is propagating along the direction of the midlayer director, which offers no phase retardation. The presence of the biaxial film offers a birefringence of $(n_c - n_b)$ as the a axis is also in the same direction. This additional phase retardation can remove the zero transmission at these

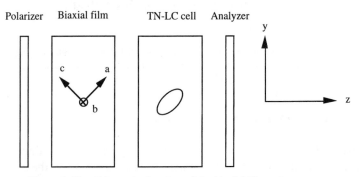

Figure 9.15. Schematic drawing of the biaxial film compensator.

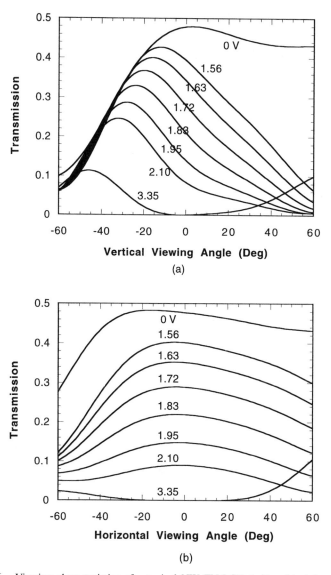

Figure 9.16. Viewing characteristics of a typical NW TN-LCD (with a biaxial compensator) with an *O*-mode operation, at various applied voltages: (*a*) vertical viewing characteristics $T(\theta, \phi = 90°)$; (*b*) horizontal viewing characteristics $T(\theta, \phi = 0°)$.

angles. Figures 9.16 and 9.17 show the viewing characteristics of a NW TN-LCD with a single biaxial compensator and with split biaxial compensators, respectively. We note that the viewing characteristics of a TN-LCD with biaxial compensators is similar to those in Figures 9.13 and 9.14 with *o*- and *a*-plate compensators.

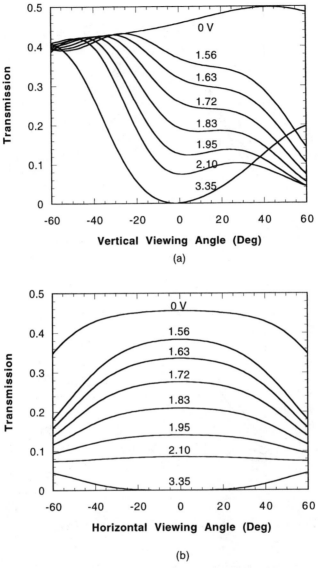

Figure 9.17. Viewing characteristics of a typical NW TN-LCD (with split biaxial compensators) with an O-mode operation, at various applied voltages: (a) vertical viewing characteristics $T(\theta, \phi = 90°)$; (b) horizontal viewing characteristics $T(\theta, \phi = 0°)$.

To further improve viewing characteristics, we need to reduce the leakage of light at large angles. This can be achieved by additional negative c and a plates. By adjusting various parameters for compensators, we can optimize the gray-level stability and viewing angles. An example of compensated viewing characteristics is shown in Figure 9.18, where we note that the compensators

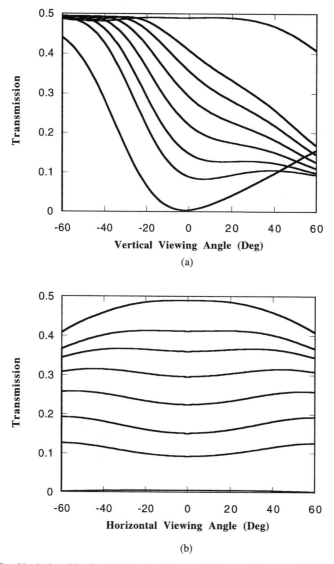

Figure 9.18. Vertical and horizontal viewing characteristics at various gray levels of a typical TN-LCD with the inclusion of a pair of biaxial plus uniaxial compensators.

have dramatically improved the gray-level stability of the LCD. In the horizontal direction, the curves are much flatter than before. In the vertical direction, there is no crossing between $-20°$ and $+40°$. However, the lowest gray level in the vertical viewing has a significant leakage of light. This leads to a lower contrast. Further optimization by using computer simulation is needed to achieve a better compensation.

9.5. MATERIALS FOR OPTICAL PHASE RETARDATION COMPENSATION

Phase retardation compensation for enhancing the performance of LCDs requires thin films of optically anisotropic materials with prescribed orientations of the principal axes. We now discuss various birefringent thin films that are useful for phase retardation compensation.

9.5.1   Plate Types

a Plates

a plates of positive birefringent materials may be fabricated by the use of uniaxially stretched polymer films, such as polyvinyl alcohol (PVA), or other suitably oriented organic birefringent materials. *a* plates of negative birefringent materials may be fabricated by using uniaxially aligned films of discotic compound.

c Plates

c plates of positive birefringent materials may be fabricated by the use of homeotropically aligned (vertically aligned) films of a rodlike LC compound. *c* plates of negative birefringent materials may be fabricated by the use of uniaxially compressed polymer films. In the uniaxially compressed films, the axes of the polymers are confined in the same plane with random azimuthal orientations. It has been shown that spin-coated polyimide films exhibit a negative birefringence as a result of the coplanar alignment of the rodlike molecules [6]. *c* plates of negative birefringent materials may also be fabricated by the use of homeotropically aligned films of discotic compound [7].

It is interesting to note that a negative *c* plate may also be fabricated by the use of alternating multilayer thin films of two different isotropic materials [5,20]. This is especially convenient for flat panel LCDs, where the thin films can be deposited on the glass plates. Optical birefringence due to microstructures in normally isotropic materials is known as *form birefringence*. In the following section, we describe the phenomenon of form birefringence in thin films of optically isotropic materials.

9.5.2.   Form Birefringence

Stratified media, consisting of a stack of alternating thin layers of isotropic optical materials having different refractive indices, are known to exhibit optical birefringence [20]. In the spectral regime where the layer thicknesses are much smaller than the wavelength, the whole stack behaves like a homogeneous and optically uniaxial medium with its *c* axis perpendicular to the layer interfaces. It is also known that thin films with microstructure have been shown, both theoretically and experimentally, to possess optical anisotropy [21–23]. Here again, the dimension of the microstructure is assumed to be much smaller than the wavelength of light.

Form Birefringence in Thin Layered Media

Referring to Figure 9.19, we consider the transmission of light through a periodic layered medium that consists of two different optical materials with refractive indices n_1 and n_2. In the spectral regime where the layer thicknesses a and b are much smaller than the wavelength of light ($a,b \ll \lambda$), the whole structure behaves like a homogeneous medium. Although each individual layer is optically isotropic, the whole structure is not optically isotropic. Optical waves with different polarization states are found to propagate with different phase shifts and transmissions. It can be shown that the whole stack behaves optically like a homogeneous and uniaxially birefringent medium, if the thicknesses a and b are sufficiently small compared to the wavelength. As a result of symmetry, the c axis is perpendicular to the layer interfaces. The effective indices of the equivalent birefringent medium is given by [20]

$$n_o^2 = \frac{a}{a+b} n_1^2 + \frac{b}{a+b} n_2^2 \qquad (9.5\text{-}1)$$

$$\frac{1}{n_e^2} = \frac{a}{a+b} \frac{1}{n_1^2} + \frac{b}{a+b} \frac{1}{n_2^2} \qquad (9.5\text{-}2)$$

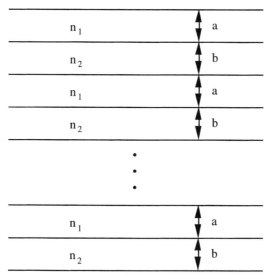

Figure 9.19. Schematic drawing of a stratified medium consisting of alternating thin layers of two different materials with refractive indices n_1, n_2 and thicknesses a,b.

It can be easily shown that $n_e < n_o$. In other words, the whole stack behaves optically like a negative birefringent medium.

The birefringence $\Delta n = n_e - n_o$ depends on the ratio $f = b/(a+b)$. It can be shown that n_o/n_e is maximum when $f = \frac{1}{2}$. At this ratio, the birefringence is given by

$$\Delta n = n_e - n_o = -\frac{(n_2 - n_1)^2}{\sqrt{2(n_1^2 + n_2^2)}} \tag{9.5-3}$$

Example 9.1. Consider a special case of alternating layers of silicon dioxide ($n_1 = 1.53$) and titanium oxide ($n_2 = 2.13$). Using Eqs. (9.5-1, 9.5-2), we obtain $n_e = 1.75$ and $n_o = 1.85$, with a birefringence of $\Delta n = -0.10$. ■

The form birefringence for a stack of alternating layers of equal thickness is independent of the actual layer thickness according to Eqs. (9.5-1) and (9.5-2), which are valid provided the layer thicknesses are sufficiently smaller than the wavelength of the interrogating radiation. For LCD compensators, the layers must be very thin compared to visible wavelength so that no reflection bands exist in this spectral regime. An average layer thickness of 20 nm is adequate for this purpose.

A compensator based on the form birefringence of layered media was designed and fabricated for a LC display [4,5]. The compensator consists of 20 pairs of alternating layers of SiO_2 and TiO_2 with a layer thickness of 20 nm. The form birefringence of $\Delta n = -0.10$ was confirmed experimentally by using ellipsometric measurements. With a total of 40 layers, the total thickness of the stack is 800 nm. This leads to a negative c plate with $\Delta n\,d = -80$ nm.

Form Birefringence in Composite Media

Referring to Figure 9.20, we consider a composite medium that consists of a uniform distribution of aligned ellipsoids of refractive index n_2 in a medium of refractive index n_1. It can be shown that [24] the effective principal refractive indices are given by

$$\frac{n_\alpha^2}{n_1^2} - 1 = \frac{f[(n_2^2/n_1^2) - 1]}{1 + (1-f)Q_\alpha[(n_2^2/n_1^2) - 1]} \qquad \alpha = a,b,c \tag{9.5-4}$$

where f is the filling factor defined as the fraction of volume occupied by the ellipsoids. The Q_α are the depolarization factors, which account for the local screening of electric fields due to the dielectric structure of the ellipsoids. The subscript α refers to the principal axes (a, b, or c) of the ellipsoids.

The depolarization factor Q_α depends on the principal axes a, b, and c. For general ellipsoids with three different semiaxes (a,b,c), the depolarization factor

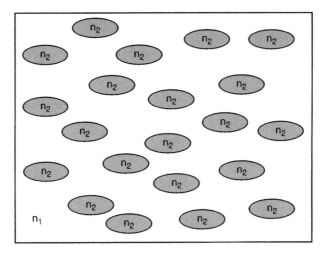

Figure 9.20. Schematic drawing of a composite medium consisting of a uniform distribution of aligned ellipsoids of refractive index n_2 in a medium of refractive index n_1.

can be written in terms of a, b, and c by using elliptic integrals [25]. For general ellipsoids with three different semiaxes (a,b,c), the depolarization factors are all different, leading to a composite medium with biaxial anisotropy. For ellipsoids of revolution (spheroids) with $a = b$, the depolarization factors depend on the ellipticity $m = c/a$. In this case the ellipsoids are cylindrically symmetric about the c axis. The uniaxial symmetry implies that $Q_a = Q_b$, leading to a composite medium with uniaxial anisotropy. Accordingly, depolarization factors related to the ordinary and extraordinary waves may be defined. These depolarization factors are designated Q_o and Q_e, respectively. Simple analytic expressions for the depolarization factors are as follows:

Case 1: $m < 1$ (Oblate Spheroids—Pancakelike)

$$Q_c = Q_e = \frac{1}{1 - m^2}\left\{1 - \frac{m}{\sqrt{1 - m^2}}\cos^{-1}m\right\}$$

$$Q_a = Q_b = Q_o = \frac{1}{2}(1 - Q_c)$$
(9.5-5)

Case 2: $1 < m$ (Prolate Spheroids—Cigarlike)

$$Q_c = Q_e = \frac{1}{m^2 - 1}\left\{\frac{m}{\sqrt{m^2 - 1}}\ln[m + \sqrt{m^2 - 1}] - 1\right\}$$

$$Q_a = Q_b = Q_o = \frac{1}{2}(1 - Q_c)$$
(9.5-6)

Case 3: $m = 1$ (Spheres)

$$Q_a = Q_b = Q_c = \tfrac{1}{3}$$
(9.5-7)

Case 4: $1 \ll m$ *(Parallel Cylinders)*

$$Q_c = Q_e = 0, \qquad Q_a = Q_b = Q_o = \tfrac{1}{2} \qquad (9.5\text{-}8)$$

Case 5: $m \ll 1$ *(Parallel Plates)*

$$Q_c = Q_e = 1, \qquad Q_a = Q_b = Q_o = 0 \qquad (9.5\text{-}9)$$

We note that cases 3–5 can be obtained as special cases of 1 and 2. Using Eqs. (9.5-4) and (9.5-9), we can obtain the effective refractive indices of the layered media n_e and n_o:

$$n_o^2 = (1-f)n_1^2 + fn_2^2$$
$$\frac{1}{n_e^2} = (1-f)\frac{1}{n_1^2} + f\frac{1}{n_2^2} \qquad (9.5\text{-}10)$$

which are exactly the same as those in Eqs. (9.5-1) and (9.5-2), where the filling factor f is $b/(a+b)$.

REFERENCES

1. J. F. Clerc, M. Aizawa, S. Yamaguchi, and J. Duchene, *Jpn. Display'89* 188 (1989); J. F. Clerc, *Digest SID'91* 758 (1991).
2. S. Yamaguchi, M. Aizawa, J. F. Clerc, T. Uchida, and J. Duchene, *Digest SID'89* 378 (1989); T. Yamamoto et al., *Digest SID'91* 762 (1991).
3. H. L. Ong, *Digest Jpn. Display'92* 247 (1992).
4. P. Yeh et al., "Compensator for liquid crystal display," U.S. Patent 5,196,953 (1993).
5. J. P. Eblen, Jr., W. J. Gunning, J. Beedy, D. Taber, L. Hale, P. Yeh, and M. Khoshnevisan, *Digest SID'94* 245 (1994); J. P. Eblen, Jr., W. J. Gunning, D. Taber, P. Yeh, M. Khoshnevisan, J. Beedy, and L. Hale, *Proc. SPIE* **2262**, 234–245 (1994).
6. S.-T. Wu, *J. Appl. Phys.* **76** (10), 5975–5980 (1994).
7. H. Mori, Y. Itoh, Y. Nishiura, T. Nakamura, and Y. Shinagawa, *Jpn. J. Appl. Phys.* **36**, 143–147 (1997); H. Mori, *Jpn. J. Appl. Phys.* **36**, 1068–1072 (1997).
8. K. H. Yang, *IDRC'91 Digest* 68 (1991).
9. K. Takatori, K. Sumiyoshi, Y. Hirai, and S. Kaneko, *Digest Jpn Display'92* 521 (1992).
10. P. J. Bos and J. A. Rahman, *Digest SID'93* 273 (1993).
11. Y. Yamaguchi, T. Miyashita, and T. Uchida, *Digest SID'93* 277 (1993).
12. M. Oh-e, M. Ohta, S. Aratani, and K. Kondo, *Digest Asia Display'95* 577 (1995).
13. M. Ohta, M. Oh-e, and K. Kondo, *Digest Asia Display'95* 68 (1995).
14. K. R. Sarma, R. I. McCartney, B. Heinze, S. Aoki, Y. Ugai, T. Sunata, and T. Inada, *Digest SID'91* 68 (1991).

15. See, for example, M. McFarland, S. Zimmerman, K. Beeson, J. Wilson, T. J. Credelle, K. Bingaman, P. Ferm, and J. T. Yardley, *Digest Asia Display '95* 739 (1995).
16. P. Yeh, *J. Opt. Soc. Am.* **72**, 507–513 (1982).
17. C. Gu and P. Yeh, *J. Opt. Soc. Am. A* **10**, 966–973 (1993).
18. B. K. Winker et al., "Optical compensator for improved gray scale," U.S. Patent 5,504,603 (1996); D. B. Taber, L. G. Hale, B. K. Winker, W. J. Gunning, III, M. C. Skarohlid, J. D. Sampica, and T. A. Seder, "Gray scale and contrast compensator for LCDs using obliquely oriented anisotropic network," SPIE, Orlando, 1998.
19. P. Yeh and C. Gu, "Birefringent optical compensators for TN-LCDs," *Proc. SPIE* **3421** (*Display Technologies II*), 224–235 (1998).
20. See, for example, P. Yeh, *Optical waves in layered media*, Wiley, 1988, pp. 135–138.
21. J. K. Moyle, W. J. Gunning, and W. H. Southwell, *Proc. SPIE* **821**, 157 (1987).
22. I. J. Hodgekinson and P. W. Wilson, *CRC Crit. Rev. Solid State Mater. Sci.* **15**(1) 27 (1988).
23. Q. Wu and I. J. Hodgekinson, "Materials for birefringence coatings," Supplement *Opt. Soc. Am.* **2**(2), S9 (May 1994).
24. V. Twersky, "Form and intrinsic birefringence," *J. Opt. Soc. Am.* **65**, 239–245 (1975).
25. E. C. Stoner, "The demagnetization factors for ellipsoids," *Phil. Mag.* **36**, 803 (1945).

PROBLEMS

9.1. Consider a pair of crossed *a* plates with their *c* axes mutually perpendicular to each other. Show that the total phase retardation for $\theta \ll 1$ can be written

$$\Gamma = -\frac{2\pi}{\lambda}\, d\, \sin^2\theta\, \frac{(n_e^2 - n_o^2)}{2n_o^2 n_e}\, (\cos^2\phi - \sin^2\phi)$$

where θ, ϕ are the angles of incidence in air. It is important to note that a pair of crossed *a* plates does not behave like a negative *c* plate for small angles of incidence.

9.2. Consider a homogeneous LC medium that consists of rodlike molecules with a statistical angular distribution of the director **n**. The statistical distribution can be due to rotational Brownian motion of the rodlike molecules. Let $\langle \mathbf{n} \rangle$ denote the statistical average of the director. In a spherical coordinate with its polar axis (*z* axis) parallel to $\langle \mathbf{n} \rangle$, the director can be written

$$\mathbf{n} = (\sin\theta\cos\phi,\ \sin\theta\sin\phi,\ \cos\theta)$$

where θ is the polar angle and ϕ is the azimuth angle. Let $p(\theta, \phi)$ be the angular distribution function of the director, then

$$\langle \mathbf{n} \rangle = \int \mathbf{n} p(\theta, \phi) \, d\Omega$$

where the integral is over all solid angles $d\Omega$. Let $\alpha_{\|}, \alpha_{\perp}$ be the principal polarizability of the molecules, so that the polarizability tensor in its principal coordinate can be written

$$\alpha = \begin{pmatrix} \alpha_{\perp} & 0 & 0 \\ 0 & \alpha_{\perp} & 0 \\ 0 & 0 & \alpha_{\|} \end{pmatrix}$$

For rodlike molecules, we assume $\alpha_{\perp} < \alpha_{\|}$.

(a) Using coordinate transformations, derive the following expression for the polarizability tensor for a general orientation of the molecule in the spherical coordinate:

$$\alpha = \begin{pmatrix} \alpha_{xx} & \alpha_{xy} & \alpha_{xz} \\ \alpha_{yx} & \alpha_{yy} & \alpha_{yz} \\ \alpha_{zx} & \alpha_{zy} & \alpha_{zz} \end{pmatrix}$$

with

$$\alpha_{xx} = \alpha_{\perp}(\cos^2\theta \cos^2\phi + \sin^2\phi) + \alpha_{\|}\sin^2\theta \cos^2\phi$$
$$\alpha_{yy} = \alpha_{\perp}(\cos^2\theta \sin^2\phi + \cos^2\phi) + \alpha_{\|}\sin^2\theta \sin^2\phi$$
$$\alpha_{zz} = \alpha_{\perp}\sin^2\theta + \alpha_{\|}\cos^2\theta$$
$$\alpha_{xy} = \alpha_{yx} = (\alpha_{\|} - \alpha_{\perp})\sin^2\theta \sin\phi\cos\phi$$
$$\alpha_{xz} = \alpha_{zx} = (\alpha_{\|} - \alpha_{\perp})\sin\theta\cos\theta\cos\phi$$
$$\alpha_{yz} = \alpha_{zy} = (\alpha_{\|} - \alpha_{\perp})\sin\theta\cos\theta\sin\phi$$

(b) Show that if the azimuth angle is random with a uniform distribution over $(0, 2\pi)$, the statistical average of the polarizability tensor becomes

$$\langle \alpha_{xx} \rangle = \alpha_{\perp}\frac{1 + \cos^2\theta}{2} + \alpha_{\|}\frac{\sin^2\theta}{2}$$
$$\langle \alpha_{yy} \rangle = \alpha_{\perp}\frac{1 + \cos^2\theta}{2} + \alpha_{\|}\frac{\sin^2\theta}{2}$$
$$\langle \alpha_{zz} \rangle = \alpha_{\perp}\sin^2\theta + \alpha_{\|}\cos^2\theta$$
$$\langle \alpha_{xy} \rangle = \langle \alpha_{yx} \rangle = 0$$
$$\langle \alpha_{xz} \rangle = \langle \alpha_{zx} \rangle = 0$$
$$\langle \alpha_{yz} \rangle = \langle \alpha_{zy} \rangle = 0$$

Note that the averaged polarizability tensor exhibits a uniaxial symmetry.

(c) Using

$$\langle \cos^2\theta \rangle = \frac{2S+1}{3} \quad \text{and} \quad \langle \sin^2\theta \rangle = 1 - \langle \cos^2\theta \rangle$$

show that the averaged polarizability tensor can be written

$$\alpha = \begin{pmatrix} \dfrac{2\alpha_\perp + \alpha_\parallel}{3} - S\dfrac{\alpha_\parallel - \alpha_\perp}{3} & 0 & 0 \\ 0 & \dfrac{2\alpha_\perp + \alpha_\parallel}{3} - S\dfrac{\alpha_\parallel - \alpha_\perp}{3} & 0 \\ 0 & 0 & \dfrac{2\alpha_\perp + \alpha_\parallel}{3} + 2S\dfrac{\alpha_\parallel - \alpha_\perp}{3} \end{pmatrix}$$

where S is the order parameter. Note that the averaged polarizability tensor exhibits a positive uniaxial symmetry, provided S is positive.

9.3. *In-plane alignment of rodlike molecules*:

Consider a LC-like medium consisting of rodlike molecules with an in-plane alignment in the xy plane. The director $\mathbf{n}$ can be written $\mathbf{n} = (\cos\phi, \sin\phi, 0)$, with ϕ uniformly distributed over $(0,2\pi)$.

(a) Show that the order parameter S is $S = -\frac{1}{2}$.

(b) Find the averaged polarizability tensor and show that the averaged polarizability tensor exhibits a negative uniaxial symmetry. Uniplanar compression of materials with rodlike molecules can be employed to obtain films with a negative birefringence.

9.4. *In-plane bidirectional alignment*:

Consider a LC-like medium consisting of rodlike molecules with an in-plane bidirectional alignment in the xy plane. The director $\mathbf{n}$ can be written $\mathbf{n} = (\cos\phi, \sin\phi, 0)$, with ϕ distributed over for distinct angles $0, \pi/2, \pi, 3\pi/2$ with equal probability.

(a) Show that the order parameter S is $S = -\frac{1}{2}$.

(b) Find the averaged polarizability tensor and show that it exhibits a negative uniaxial symmetry. Note that bidirectional alignment can also be employed to obtain films with a negative birefringence.

9.5. *Form birefringence of periodic birefringent layered media*:

Consider a layered medium consisting of alternating layers of isotropic material and birefringent material. Let the thickness and refractive index of the isotropic layers be a, n_i, and those of the birefringent layers be b, n_e, n_o. We assume that the c axis of the birefringent layers is uniformly aligned parallel to the layer surfaces along the y axis (e.g., unidirectionally stretched PVA films). Let z be the axis perpendicular to the layers. Assume that $a,b \ll \lambda$ and $n_o < n_e$.

(a) Show that the principal axes are the x, y and z axes defined above, and that the effective principal refractive indices are

$$n_{xx}^2 = \frac{a}{a+b}n_i^2 + \frac{b}{a+b}n_o^2$$

$$n_{yy}^2 = \frac{a}{a+b}n_i^2 + \frac{b}{a+b}n_e^2$$

$$n_{zz}^{-2} = \frac{a}{a+b}n_i^{-2} + \frac{b}{a+b}n_o^{-2}$$

(b) Show that the form birefringence is biaxial with $n_{zz} < n_{xx} < n_{yy}$ regardless of the magnitude of n_i.

(c) Show that the optic axes are in the yz plane.

9.6. *Form birefringence of periodic birefringent layered media*:

Consider a layered medium which consists of alternating layers of uniaxially birefringent material. Let the thickness be a and b and the refractive indices of layers be n_e, n_o. We assume that the c axes of the neighboring layers are mutually orthogonal and aligned parallel to the layer surfaces along the x and y axes. These layers can be, for example, unidirectionally stretched PVA films. Let z be the axis perpendicular to the layers. Assume that $a, b \ll \lambda$ and $n_o < n_e$.

(a) Show that the principal axes are the x, y, and z axes defined above, and that the effective principal refractive indices are

$$n_{xx}^2 = \frac{a}{a+b}n_o^2 + \frac{b}{a+b}n_e^2$$

$$n_{yy}^2 = \frac{a}{a+b}n_e^2 + \frac{b}{a+b}n_o^2$$

$$n_{zz}^2 = n_o^2$$

The birefringence is, in general, biaxial.

(b) Show that the form birefringence becomes uniaxial when $a = b$, with the principal refractive indices:

$$n_{xx}^2 = \tfrac{1}{2}(n_o^2 + n_e^2)$$

$$n_{yy}^2 = \tfrac{1}{2}(n_o^2 + n_e^2)$$

$$n_{zz}^2 = n_o^2$$

Note that the form birefringence is negative uniaxial.

(c) Consider the case when the c axes of the neighboring layers are not orthogonal. Let θ be the angle between the c axes. Find the principal axes and the principal refractive indices.

9.7. *Phase retardation of o plates*:

Consider a plate of uniaxial crystal with its c axis pointed at a polar angle α measured from its normal, the z axis in the yz plane. The unit vector $\mathbf{c}$

along the c axis can be written $\mathbf{c} = (0, \sin \alpha, \cos \alpha)$. Now consider the incidence of a beam of light along the angle (θ, ϕ) such that the incident wavevector can be written

$$\mathbf{k} = k(\sin \theta \cos \phi, \sin \theta \sin \phi, \cos \theta) = (k_x, k_y, k_z)$$

(a) Show that the wavevectors of the O wave and E wave in the crystal can be written

$$\mathbf{k}_o = (k_x, k_y, k_{oz}) \qquad \text{and} \qquad \mathbf{k}_e = (k_x, k_y, k_{ez})$$

with

$$k_{oz} = \sqrt{n_o^2 k^2 - k_x^2 - k_y^2}$$

$$k_{ez} = n_{ez} \Bigg[k^2 - \frac{k_x^2}{n_e^2} - \frac{k_y^2}{n_{ey}^2} + k_y^2 n_{ez}^2 \sin^2 \alpha \cos^2 \alpha \left(\frac{1}{n_o^2} - \frac{1}{n_e^2} \right)^2$$

$$- n_{ez}^2 k_y \sin \alpha \cos \alpha \left(\frac{1}{n_o^2} - \frac{1}{n_e^2} \right) \Bigg]$$

where

$$\frac{1}{n_{ez}^2} = \frac{\cos^2 \alpha}{n_o^2} + \frac{\sin^2 \alpha}{n_e^2} \qquad \text{and} \qquad \frac{1}{n_{ey}^2} = \frac{\cos^2 \alpha}{n_e^2} + \frac{\sin^2 \alpha}{n_o^2}$$

(b) Let the phase retardation be written

$$\Gamma(\theta, \phi, \alpha) = (k_{ez} - k_{oz})d$$

Show that Γ exhibits a left–right symmetry. In other words

$$\Gamma(\theta, \phi, \alpha) = \Gamma(\theta, \pi - \phi, \alpha)$$

Show that, in general, Γ exhibits no up–down symmetry. In other words

$$\Gamma(\theta, \phi, \alpha) \neq \Gamma(\theta, -\phi, \alpha)$$

(c) Consider a series of two o plates with the c axes given by

$$\mathbf{c}_1 = (0, \sin \alpha, \cos \alpha) \qquad \text{and} \qquad \mathbf{c}_2 = (0, \sin \alpha, -\cos \alpha)$$

Both c axes are in the yz plane with polar angles α and $\pi - \alpha$. Define a phase retardation

$$\Gamma_2(\theta, \phi, \alpha) = \Gamma(\theta, \phi, \alpha) + \Gamma(\theta, \phi, \pi - \alpha)$$

Show that the total phase retardation $\Gamma_2(\theta, \phi, \alpha)$ exhibits both left–right and up–down symmetry. In other words

$$\Gamma_2(\theta, \phi, \alpha) = \Gamma_2(\theta, \pi - \phi, \alpha) = \Gamma_2(\theta, -\phi, \alpha)$$

(d) Plot $\Gamma_2(\theta, \phi, \alpha)$, $\Gamma(\theta, \phi, \alpha)$, and $\Gamma(\theta, \phi, \pi - \alpha)$ as functions of θ.

Appendix A

Elastic and Electromagnetic Energy Density

In the discussion follows, we consider a nematic LC cell in the xy plane. The z axis is chosen to be perpendicular to the cell, so that the cell is located between $z = 0$ and $z = d$. Initially, the LC molecules are aligned parallel to the xy plane. An electric field is then applied in the LC cell. We consider the redistribution of the orientation of the LC molecules (director).

First, we consider the elastic energy density due to the distortion produced by the presence of an electric field. It is important to remember that the directors $\mathbf{n}$ of the LC molecules are anchored at the inner surface of the glass plates with the directors $\mathbf{n}$ parallel to the rubbed directions, regardless of the presence of the electric field. Thus, the applied electric field affects only the director distribution $\mathbf{n}(z)$ of the LC molecules away from the boundaries. This leads to a distribution of the director $\mathbf{n}(z)$ in the cell. The distribution is a result of the balance between the elastic energy and the electrostatic energy.

ELASTIC ENERGY DENSITY

We consider first the elastic energy density due to the redistribution of the directors. As described in Chapter 1, the elastic energy density is given by

$$U_{EL} = \tfrac{1}{2}k_1(\nabla \cdot \mathbf{n})^2 + \tfrac{1}{2}k_2(\mathbf{n} \cdot \nabla \times \mathbf{n})^2 + \tfrac{1}{2}k_3(\mathbf{n} \times \nabla \times \mathbf{n})^2 \qquad \text{(A-1)}$$

where $\mathbf{n}$ is a unit vector representing the director distribution in the cell, and k_1, k_2, k_3 are the three principal elastic constants. To illustrate the electrooptical distortion, we consider three important examples of the electrooptical distortion.

1. *Tilt Mode.* Consider a nematic cell with an initial director distribution $\mathbf{n}(z)$ parallel to the y axis. An electric field is applied along the z axis. As a result, the director $\mathbf{n}(z)$ will be tilted toward the z axis, subject to the boundary condition that $\mathbf{n}(0)$ and $\mathbf{n}(d)$ remain parallel to the y axis. Letting $\theta(z)$ be the tilt angle at position z, the director can be written

$$\mathbf{n} = (0, \cos\theta, \sin\theta) \qquad \text{(A-2)}$$

where the tilt angle $\theta(z)$ is measured from the y axis. With this definition, $\theta(z)$ is zero at $z = 0$ and $z = d$.

To evaluate the elastic energy density, we need to find the divergence and the curl of the director $\mathbf{n}(z)$. Using Eq. (A-1), the divergence and the curl of the director $\mathbf{n}$ can be written

$$\nabla \cdot \mathbf{n} = \cos \theta \frac{d\theta}{dz} \tag{A-3}$$

$$\nabla \times \mathbf{n} = \left(\sin \theta \frac{d\theta}{dz}, 0, 0 \right) \tag{A-4}$$

Using Eq. (A-1), we obtain the elastic energy density

$$U_{EL} = \left[\frac{1}{2} k_1 \cos^2 \theta + \frac{1}{2} k_3 \sin^2 \theta \right] \left(\frac{d\theta}{dz} \right)^2 \tag{A-5}$$

2. *Twist Mode.* Consider a nematic LC cell with the director $\mathbf{n}$ parallel to the x axis. A uniform electric field is applied along the y axis. As a result, the directors are twisted toward the direction of the electric field. This leads to a distribution of the director as a function of z. Letting $\phi(z)$ be the twist angle at position z, the director can be written

$$\mathbf{n} = (\cos \phi, \sin \phi, 0) \tag{A-6}$$

The divergence and the curl of the director $\mathbf{n}$ can be written

$$\nabla \cdot \mathbf{n} = 0 \tag{A-7}$$

$$\nabla \times \mathbf{n} = (-\cos \phi, -\sin\phi, 0) \frac{d\phi}{dz} \tag{A-8}$$

Using Eq. (A-1), we obtain

$$U_{EL} = \frac{1}{2} k_2 \left(\frac{d\phi}{dz} \right)^2 \tag{A-9}$$

3. *Twist and Tilt Mode.* Consider a twisted nematic LC cell with the director $\mathbf{n}$ parallel to the xy plane. The director $\mathbf{n}$ is uniformly twisted as a function of z. In other words, the initial director distribution can be written

$$\mathbf{n} = (\cos \phi, \sin \phi, 0) \tag{A-10}$$

where the twist angle ϕ is linearly proportional to z. A uniform electric field is applied along the z axis. As a result, the directors are tilted toward the direction of the electric field. This leads to a redistribution of the director $\mathbf{n}$ as a function of z. Letting $\phi(z)$ be the twist angle and $\theta(z)$ be the tilt angle at position z, the director can be written

$$\mathbf{n} = (\cos\theta\cos\phi, \cos\theta\sin\phi, \sin\theta) \qquad (A-11)$$

The divergence and the curl of the director $\mathbf{n}$ can be written

$$\nabla\cdot\mathbf{n} = \cos\theta\frac{d\theta}{dz} \qquad (A-12)$$

$$\nabla\times\mathbf{n} = (\sin\theta\sin\phi\,\theta' - \cos\theta\cos\phi\,\phi', -\sin\theta\cos\phi\,\theta' \\ - \cos\theta\sin\phi\,\phi', 0) \qquad (A-13)$$

where

$$\theta' = \frac{d\theta}{dz} \qquad (A-14)$$

$$\phi' = \frac{d\phi}{dz} \qquad (A-15)$$

Using Eq. (A-1), we obtain

$$U_{EL} = \frac{1}{2}k_1\cos^2\theta\left(\frac{d\theta}{dz}\right)^2 + \frac{1}{2}k_2\cos^4\theta\left(\frac{d\phi}{dz}\right)^2 \\ + \frac{1}{2}k_3\sin^2\theta\left(\frac{d\theta}{dz}\right)^2 + \frac{1}{2}k_3\sin^2\theta\cos^2\theta\left(\frac{d\phi}{dz}\right)^2 \qquad (A-16)$$

which can also be written

$$U_{EL} = \frac{1}{2}[k_1\cos^2\theta + k_3\sin^2\theta]\left(\frac{d\theta}{dz}\right)^2 + \frac{1}{2}[k_2\cos^2\theta + k_3\sin^2\theta] \\ \times\cos^2\theta\left(\frac{d\phi}{dz}\right)^2 \qquad (A-17)$$

ELECTROMAGNETIC ENERGY DENSITY

Generally speaking, the electromagnetic energy density can be written

$$U_{EM} = \tfrac{1}{2}\mathbf{E}\cdot\mathbf{D} \tag{A-18}$$

where $\mathbf{E}$ is the electric field vector and $\mathbf{D}$ is the displacement field vector. For the purpose of discussing the field-induced distortion, we consider the case of a capacitor that consists of a set of parallel electrodes with a dielectric material inside the capacitor. Specifically, we are interested in the change of the electromagnetic energy density due to the change of the dielectric constant ε as the liquid crystal director distribution changes. The discussion is divided into the following two categories:

1. *Constant Charge on the Electrodes.* In this case the electrodes are isolated from any power supply (or battery). Having a constant surface charge density σ on the surface of the electrodes, the displacement field D is a constant $(D = \sigma)$, regardless of the magnitude of the dielectric constant. The change of the energy due to a change of the dielectric constant can be written

$$\Delta U_{EM} = \frac{1}{2}\frac{D^2}{\varepsilon} - \frac{1}{2}\frac{D^2}{\varepsilon_0} \tag{A-19}$$

where ε_0 is the initial dielectric constant and ε is the final dielectric constant. In this equation, D is a constant. We note a larger dielectric constant ε leads to a lower energy density. This explains why a dielectric material is attracted into an empty capacitor.

2. *Constant Voltage on the Electrodes.* In this case the electrodes are connected to the power supply (or battery) to maintain the same voltage as the dielectric constant undergoes a change in its magnitude. Having a constant voltage, the electric field E is a constant $(E = V/d)$, regardless of the magnitude of the dielectric constant. Because of the additional charges supplied to the electrodes to maintain the same voltage, work is done by the power supply. Thus, we must take this energy into account. The net change of the electromagnetic energy density due to a change of the dielectric constant can be written

$$\Delta U_{EM} = \frac{1}{2}\varepsilon E^2 - \frac{V\,\Delta Q}{Ad} - \frac{1}{2}\varepsilon_0 E^2 \tag{A-20}$$

where A is the area of the electrodes, d is the separation between the electrodes, and ΔQ is the additional charges supplied by the power supply (or battery). It can be shown that ΔQ is given by

$$\Delta Q = A(\varepsilon - \varepsilon_0)\frac{V}{d} \qquad \text{(A-21)}$$

Using $E = V/d$ and Eqs. (A-20) and (A-21), we obtain

$$\Delta U_{EM} = \tfrac{1}{2}\varepsilon_0 E^2 - \tfrac{1}{2}\varepsilon E^2 \qquad \text{(A-22)}$$

where E is a constant and ε_0 is the initial dielectric constant. Thus, a larger final dielectric constant ε leads to a lower energy density. This explains why a dielectric material is attracted into an empty capacitor, even if the electrodes are maintained at a constant voltage.

DISCUSSION

The relationship between the three elastic constants and the vector differential operations can be displayed as follows.

Splay elastic constant k_1:   deformation with $\nabla \cdot \mathbf{n} \neq 0$

Twist elastic constant k_2:   deformation with $\mathbf{n} \cdot \nabla \times \mathbf{n} \neq 0$

Bend elastic constant k_3:   deformation with $\mathbf{n} \times \nabla \times \mathbf{n} \neq 0$

All three constants are positive, as each of these deformations can be generated individually. The magnitude of these elastic constants can be estimated as follows. Let U be the interaction energy between molecules and a be the molecular dimension. The elastic constants are of the order of U/a. Taking $U = 0.1\,\text{eV}$ and $a = 10\,\text{Å}$, we obtain k in the range of $10^{-11}\,\text{N}$. Note that the bending constant is much larger than the others, while the twist constant is small (see Table 1.2). This is consistent with the rodlike molecular model.

Appendix B

Electrooptical Distortion—Tilt Mode

In this case, we consider a nematic LC cell with it's director **n** initially parallel to the y axis. An electric field is applied along the z axis (vertical field). This is often achieved by applying an electric voltage on the electrodes. The voltage drop between the electrodes leads to an electric field parallel to the z axis. We are interested in the redistribution of the director **n** due to the presence of the electric field. Let the director **n** be written

$$\mathbf{n} = (0, \cos\theta, \sin\theta) \tag{B-1}$$

We first consider the electromagnetic field energy density due to the presence of the dielectric LC material. Initially, the director $\mathbf{n}(z)$ is uniformly parallel to the y axis. As a result of the electrostatic torque, the directors of the LC molecules are forced to align along the direction of the electric field. However, the director at the boundaries remains parallel to the rubbed direction. This leads to a redistribution of the director $\mathbf{n}(z)$ in the cell.

It is important to note that, generally speaking, the electric field vector **E** and the displacement field vector **D** are not parallel due to the anisotropy of the medium. When the electrodes are perpendicular to the z axis, the electric field **E** is parallel to the z axis as a result of the boundary condition, even if $\mathbf{n}(z)$ is not a constant. The boundary condition requires that the electric field be perpendicular to the surface of the conductors. In addition, the z component of the displacement field vector **D** remains a constant throughout the cell regardless of the distribution of the director $\mathbf{n}(z)$, specifically, $D_z = $ constant. Thus, the change of the electrostatic energy due to the reorientation of the directors can be written, according to Eq. (A-19) in Appendix A, as follows:

$$\Delta U_{EM} = \frac{1}{2} \frac{D_z^2}{(\varepsilon_\parallel \sin^2\theta + \varepsilon_\perp \cos^2\theta)} - \frac{1}{2} \frac{D_z^2}{\varepsilon_\perp} \tag{B-2}$$

where D_z ($=$ constant) is the z component of the displacement field vector. We note that the second term is a constant independent of the director orientation $\theta(z)$.

The elastic energy density in this case is given by, according to Eq. (A-5) in Appendix A

$$U_{EL} = \frac{1}{2}(k_1 \cos^2 \theta + k_3 \sin^2 \theta)\left(\frac{d\theta}{dz}\right)^2 \tag{B-3}$$

The total free energy in the cell is given by

$$U = \int_0^d (U_{EL} + \Delta U_{EM})dz \tag{B-4}$$

where U_{EL} and ΔU_{EM} are given by Eqs. (B-2) and (B-3), respectively. The integral depends on the distribution of the orientation of the director. A redistribution of the director orientation $\theta(z)$ must satisfy the following minimum free energy condition:

$$\delta U = 0 \tag{B-5}$$

subject to the boundary condition:

$$\theta(0) = \theta(d) = 0 \tag{B-6}$$

In other words, the LC molecules are aligned parallel to the rubbed direction y at the boundaries. Using techniques of calculus of variation, we obtain

$$(k_1 \cos^2 \theta + k_3 \sin^2 \theta)\left(\frac{d\theta}{dz}\right)^2 - \frac{D_z^2}{(\varepsilon_\parallel \sin^2 \theta + \varepsilon_\perp \cos^2 \theta)} = \text{constant} \tag{B-7}$$

As the director is tilted toward the direction of the applied field, the tilt angle reaches its maximum θ_{max} at the center of the cell $z = d/2$ (θ_{max} is often called the midlayer tilt angle). Thus the constant on the right side of Eq. (B-7) can be determined by assuming a maximum tilt angle at the center of the cell. This leads to

$$(k_1 \cos^2 \theta + k_3 \sin^2 \theta)\left(\frac{d\theta}{dz}\right)^2 - \frac{D_z^2}{(\varepsilon_\parallel \sin^2 \theta + \varepsilon_\perp \cos^2 \theta)}$$
$$= -\frac{D_z^2}{(\varepsilon_\parallel \sin^2 \theta_{max} + \varepsilon_\perp \cos^2 \theta_{max})} \tag{B-8}$$

This equation gives the $dz/d\theta$ as a function of θ between 0 and θ_{max}. If both D_z and θ_{max} are known, the results can then be integrated and inverted to obtain $\theta(z)$. The midlayer tilt angle θ_{max} depends on the strength of the applied field.

We now derive the relationship between θ_{max} and D_z or the applied voltage. To simplify the mathematical analysis, we define a new variable ψ as follows:

$$\sin\theta = \sin\theta_{max} \sin\psi \tag{B-9}$$

Using this new variable, we obtain

$$\varepsilon_\| \sin^2\theta + \varepsilon_\perp \cos^2\theta = \varepsilon_\perp + (\varepsilon_\| - \varepsilon_\perp)\sin^2\theta = \varepsilon_\perp(1 + \gamma\eta^2\sin^2\psi) \tag{B-10}$$

$$k_1\cos^2\theta + k_3\sin^2\theta = k_1 + (k_3 - k_1)\sin^2\theta = k_1(1 + \kappa\eta^2\sin^2\psi) \tag{B-11}$$

$$d\theta = \frac{\sqrt{\sin^2\theta_{max} - \sin^2\theta}}{\sqrt{1 - \eta^2\sin^2\psi}}d\psi \tag{B-12}$$

where

$$\eta = \sin\theta_{max} \tag{B-13}$$

$$\gamma = \frac{\varepsilon_\| - \varepsilon_\perp}{\varepsilon_\perp} \tag{B-14}$$

$$\kappa = \frac{k_3 - k_1}{k_1} \tag{B-15}$$

We note that γ is a measure of the dielectric anisotropy and κ is a measure of the elastic anisotropy. Substituting the new variable and parameters in Eqs. (B-9)–(B-15) into Eq. (B-8), we obtain

$$\frac{\sqrt{(1 + \kappa\eta^2\sin^2\psi)(1 + \gamma\eta^2)(1 + \gamma\eta^2\sin^2\psi)}}{\sqrt{1 - \eta^2\sin^2\psi}}d\psi = \frac{D_z}{\sqrt{k_1}}\frac{\sqrt{\varepsilon_\| - \varepsilon_\perp}}{\varepsilon_\perp}dz \tag{B-16}$$

where we recall that D_z is a constant. Furthermore, using the following relationship between the electric field and D_z, we obtain

$$D_z = \varepsilon_\perp(1 + \gamma\eta^2\sin^2\psi)E \tag{B-17}$$

We can rewrite Eq. (B-16) as

$$\frac{\sqrt{(1 + \kappa\eta^2 \sin^2 \psi)(1 + \gamma\eta^2)}}{\sqrt{(1 - \eta^2 \sin^2 \psi)(1 + \gamma\eta^2 \sin^2 \psi)}} d\psi = \frac{\sqrt{\varepsilon_\parallel - \varepsilon_\perp}}{\sqrt{k_1}} E\, dz \qquad \text{(B-18)}$$

where we recall that E is the electric field that is parallel to the z axis. We now integrate Eq. (B-18) on both sides. Note that $\psi = \pi/2$ corresponds to $z = d/2$. Thus, we obtain

$$\int_0^{\pi/2} \frac{\sqrt{(1 + \kappa\eta^2 \sin^2 \psi)(1 + \gamma\eta^2)}}{\sqrt{(1 - \eta^2 \sin^2 \psi)(1 + \gamma\eta^2 \sin^2 \psi)}} d\psi = \int_0^{d/2} \frac{\sqrt{\varepsilon_\parallel - \varepsilon_\perp}}{\sqrt{k_1}} E\, dz \qquad \text{(B-19)}$$

which can be further written

$$\int_0^{\pi/2} \frac{\sqrt{(1 + \kappa\eta^2 \sin^2 \psi)(1 + \gamma\eta^2)}}{\sqrt{(1 - \eta^2 \sin^2 \psi)(1 + \gamma\eta^2 \sin^2 \psi)}} d\psi = \frac{\sqrt{\varepsilon_\parallel - \varepsilon_\perp}}{\sqrt{k_1}} \frac{V}{2} \qquad \text{(B-20)}$$

where V is the applied voltage on the electrodes. Although the electric field E is a function of z, the integral over z yields the applied voltage. We note that the integrand on the left side is always greater than unity. Thus, the integral must be greater than $\pi/2$. Hence, Eq. (B-20) can be written

$$\int_0^{\pi/2} \frac{\sqrt{(1 + \kappa\eta^2 \sin^2 \psi)(1 + \gamma\eta^2)}}{\sqrt{(1 - \eta^2 \sin^2 \psi)(1 + \gamma\eta^2 \sin^2 \psi)}} d\psi = \frac{\sqrt{\varepsilon_\parallel - \varepsilon_\perp}}{\sqrt{k_1}} \frac{V}{2} \geq \frac{\pi}{2} \qquad \text{(B-21)}$$

According to this equation, there exists a critical voltage V_{c1} that is needed to create any tilt of the director in the LC medium. For applied voltage $V < V_{c1}$, the midlayer tilt angle is zero, $\theta_{max} = 0$. The critical voltage can be written

$$V_{c1} = E_{c1}d = \pi \frac{\sqrt{k_1}}{\sqrt{\varepsilon_\parallel - \varepsilon_\perp}} \qquad \text{(B-22)}$$

where E_{c1} is the critical electric field needed to create any tilt $(0 < \theta_{max})$ in the medium. We note that the critical voltage (or field) is inversely proportional to the square root of the dielectric anisotropy. Using the critical voltage V_{c1}, the relationship between θ_{max} and the applied voltage can be written

$$\int_0^{\pi/2} \frac{\sqrt{(1 + \kappa\eta^2 \sin^2 \psi)(1 + \gamma\eta^2)}}{\sqrt{(1 - \eta^2 \sin^2 \psi)(1 + \gamma\eta^2 \sin^2 \psi)}} d\psi = \frac{\pi}{2} \frac{V}{V_{c1}} \tag{B-23}$$

For a given value of η (or equivalently θ_{max}), the left side of this equation can be evaluated by numerical integration. This leads to an applied voltage needed to obtain the desired midlayer tilt angle. The z component of the displacement field vector D_z can also be obtained for each midlayer tilt angle θ_{max} by carrying out the numerical integration of Eq. (B-16). Once D_z is obtained, the distribution of the director tilt angle $\theta(z)$ can then be obtained by substituting both D_z and θ_{max} into Eq. (B-8).

Appendix C

Electrooptical Distortion—Twist Mode

In this case, we consider a nematic cell with its director **n** parallel to the x axis. An electric field is applied along the y axis (horizontal field, or in-plane field). This is often achieved by applying an electric voltage on the electrodes deposited on the inner surface of one of the glass plates. The voltage drop between the electrodes leads to an in-plane electric field parallel to the y axis. We are interested in the redistribution of the director **n** due to the presence of the in-plane electric field. Since the electric field remains in the plane of the LC cell (xy plane), the director **n** can be written

$$\mathbf{n} = (\cos \phi, \sin \phi, 0) \tag{C-1}$$

where ϕ is the twist angle measured from the x axis. According to Eq. (A-9) in Appendix A, the elastic energy density due a variation of the twist angle can be written

$$U_{EL} = \frac{1}{2}k_2 \left(\frac{d\phi}{dz}\right)^2 \tag{C-2}$$

For simplicity, we will assume that the electric field is parallel to the y axis (i.e., $E_x = E_z = 0$). In addition, we assume that the field is uniform in the xy plane. The Maxwell curl equation $\nabla \times E = 0$ leads to

$$E = E_y = \text{constant} \tag{C-3}$$

In other words, E is independent of z. The displacement field components D_x, D_y are, however, functions of z. In this case the voltage remains constant as the dielectric tensor changes as a result of the action of the field. Additional work is done by the power supply (or battery) to maintain the constant voltage. Thus, the net change in the electromagnetic energy density can be written, according to Eq. (A-22) in Appendix A

$$\Delta U_{EM} = \tfrac{1}{2}\varepsilon_\perp E^2 - \tfrac{1}{2}(\varepsilon_\parallel \sin^2 \phi + \varepsilon_\perp \cos^2 \phi)E^2 \tag{C-4}$$

where ϕ is the twist angle, which depends on z. We note that the first term is a constant independent of the twist angle.

The total free energy in the cell is given by

$$U = \int_0^d (U_{EL} + \Delta U_{EM})dz \qquad (C-5)$$

or equivalently

$$U = \int_0^{d/2} \left[k_2 \left(\frac{d\phi}{dz} \right)^2 - (\varepsilon_\parallel - \varepsilon_\perp)E^2 \sin^2 \phi \right] dz \qquad (C-6)$$

The integral depends on the distribution of the orientation of the director. A redistribution of the director orientation $\phi(z)$ must satisfy the following minimum free energy condition:

$$\delta U = 0 \qquad (C-7)$$

subject to the boundary condition:

$$\phi(0) = \phi(d) = 0 \qquad (C-8)$$

In other words, the LC molecules are aligned parallel to the rubbed direction $\mathbf{x}$ at the boundaries. Using techniques of calculus of variation, we obtain

$$k_2 \left(\frac{d\phi}{dz} \right)^2 + (\varepsilon_\parallel - \varepsilon_\perp)E^2 \sin^2 \phi = \text{constant} \qquad (C-9)$$

Letting ϕ_{max} be the midlayer twist angle at $z = d/2$ where $(d\phi/dz) = 0$, we obtain

$$k_2 \left(\frac{d\phi}{dz} \right)^2 + (\varepsilon_\parallel - \varepsilon_\perp)E^2 \sin^2 \phi = (\varepsilon_\parallel - \varepsilon_\perp)E^2 \sin^2 \phi_{max} \qquad (C-10)$$

This equation gives the $dz/d\phi$ as a function of ϕ between 0 and ϕ_{max}. If both E and ϕ_{max} are known, the results can then be integrated and inverted to obtain $\phi(z)$. The midlayer twist angle ϕ_{max} depends on the strength of the applied field E.

We now derive the relationship between ϕ_{max} and E or the applied voltage. To simplify the mathematical analysis, we define a new variable ψ as follows:

$$\sin \phi = \sin \phi_{max} \sin \psi \qquad \text{(C-11)}$$

Using this new variable, we obtain

$$(\varepsilon_{\parallel} - \varepsilon_{\perp}) \sin^2 \phi = (\varepsilon_{\parallel} - \varepsilon_{\perp}) \eta^2 \sin^2 \psi \qquad \text{(C-12)}$$

$$d\phi = \frac{\sqrt{\sin^2 \phi_{max} - \sin^2 \phi}}{\sqrt{1 - \eta^2 \sin^2 \psi}} d\psi \qquad \text{(C-13)}$$

where

$$\eta = \sin \phi_{max} \qquad \text{(C-14)}$$

Substituting the new variable and parameters in Eqs. (C-11)–(C-14) into Eq. (C-10), we obtain

$$\frac{1}{\sqrt{1 - \eta^2 \sin^2 \psi}} d\psi = \frac{\sqrt{(\varepsilon_{\parallel} - \varepsilon_{\perp})}}{\sqrt{k_2}} E \, dz \qquad \text{(C-15)}$$

where we recall that E is a constant. We now integrate this equation on both sides. Note that $\psi = \pi/2$ corresponds to $z = d/2$. Thus, we obtain

$$\int_0^{\pi/2} \frac{1}{\sqrt{1 - \eta^2 \sin^2 \psi}} d\psi = \int_0^{d/2} \frac{\sqrt{\varepsilon_{\parallel} - \varepsilon_{\perp}}}{\sqrt{k_2}} E \, dz \qquad \text{(C-16)}$$

which can be further written, since E is a constant

$$\int_0^{\pi/2} \frac{1}{\sqrt{1 - \eta^2 \sin^2 \psi}} d\psi = \frac{d}{2} E \frac{\sqrt{(\varepsilon_{\parallel} - \varepsilon_{\perp})}}{\sqrt{k_2}} \qquad \text{(C-17)}$$

We note that, again, the integrand on the left side is always greater than unity. Thus, the integral must be greater than $\pi/2$. Hence, Eq. (C-17) can be written

$$\int_0^{\pi/2} \frac{1}{\sqrt{1 - \eta^2 \sin^2 \psi}} d\psi = \frac{d}{2} E \frac{\sqrt{(\varepsilon_{\parallel} - \varepsilon_{\perp})}}{\sqrt{k_2}} \geq \frac{\pi}{2} \qquad \text{(C-18)}$$

According to this equation, there exists a critical field E_{c1} which is needed to create any twist of the director in the LC medium. For an applied electric field $E < E_{c1}$, the midlayer twist angle is zero, $\phi_{max} = 0$. The critical field can be written

$$E_{c1}d = \pi \frac{\sqrt{k_2}}{\sqrt{\varepsilon_\| - \varepsilon_\perp}} \qquad (C\text{-}19)$$

where E_{c1} is the critical in-plane electric field needed to create any twist $(0 < \phi_{max})$ in the medium. We note that the critical field is inversely proportional to the square root of the dielectric anisotropy. Using the critical field E_{c1}, the relationship between ϕ_{max} and the applied electric field can be written

$$\int_0^{\pi/2} \frac{1}{\sqrt{1 - \eta^2 \sin^2 \psi}} d\psi = \frac{\pi}{2} \frac{E}{E_{c1}} \qquad (C\text{-}20)$$

For a given value of η (or equivalently ϕ_{max}), the left side of this equation can be evaluated by numerical integration. This leads to an applied electric field needed to obtain the desired midlayer twist angle. Once E is obtained, the distribution of the director tilt angle $\phi(z)$ can then be obtained by substituting both E and ϕ_{max} into Eq. (C-10).

Appendix D

Electrooptical Distortion in TN-LC

Consider a twisted nematic LC cell with the director **n** parallel to the xy plane. The director **n** is uniformly twisted as a function of z. In other words, the initial director distribution can be written

$$\mathbf{n} = (\cos \phi, \sin \phi, 0) \tag{D-1}$$

where the twist angle ϕ is linearly proportional to z (e.g., $\phi = qz$, where q is a constant). A uniform electric field is applied along the z axis. As a result, the director is tilted toward the direction of the electric field. This leads to a redistribution of the director **n** as a function of z. Letting $\phi(z)$ be the twist angle and $\theta(z)$ be the tilt angle at position z, the director can be written

$$\mathbf{n} = (\cos \theta \cos \phi, \cos \theta \sin \phi, \sin \theta) \tag{D-2}$$

The divergence and the curl of the director **n** can be written

$$\nabla \cdot \mathbf{n} = \cos \theta \, \frac{d\theta}{dz} \tag{D-3}$$

$$\nabla \times \mathbf{n} = (\sin \theta \sin \phi \, \theta' - \cos \theta \cos \phi \, \phi', -\sin \theta \cos \phi \, \theta' - \cos \theta \sin \phi \, \phi', 0) \tag{D-4}$$

where

$$\theta' = \frac{d\theta}{dz} \tag{D-5}$$

$$\phi' = \frac{d\phi}{dz} \tag{D-6}$$

Using Eq. (A-1) in Appendix A, we obtain

$$U_{\text{EL}} = \frac{1}{2} k_1 \cos^2 \theta \left(\frac{d\theta}{dz} \right)^2 + \frac{1}{2} k_2 \cos^4 \theta \left(\frac{d\phi}{dz} \right)^2$$

$$+ \frac{1}{2} k_3 \sin^2 \theta \left(\frac{d\theta}{dz} \right)^2 + \frac{1}{2} k_3 \sin^2 \theta \cos^2 \theta \left(\frac{d\phi}{dz} \right)^2 \tag{D-7}$$

which can also be written

$$U_{EL} = \frac{1}{2} \left[k_1 \cos^2 \theta + k_3 \sin^2 \theta \right] \left(\frac{d\theta}{dz} \right)^2 + \frac{1}{2} \left[k_2 \cos^2 \theta + k_3 \sin^2 \theta \right] \cos^2 \theta \left(\frac{d\phi}{dz} \right)^2$$

$$(D-8)$$

ELECTROMAGNETIC ENERGY DENSITY

First, let us examine the components of the field vectors **E** and **D**. We assume that the LC cell is uniform in the xy plane. A voltage is applied to the electrodes that creates an electric field along the z axis. An application of the Maxwell curl equation and the boundary condition on the surface of the electrodes lead to

$$E_x = 0 \qquad \qquad (D-9)$$
$$E_y = 0 \qquad \qquad (D-10)$$

A subsequent application of the Maxwell divergence equation leads to

$$\mathbf{D} = \text{constant} \qquad \qquad (D-11)$$

A constant **D** indicates that the surface charge density remains constant during the reorientation of the directors. In this case, the directors near the electrodes remain parallel to the rubbed directions regardless of the applied field. Thus, when a voltage is applied to the electrodes, the surface charge density on the electrodes will not change as a result of the director reorientation in the middle of the cell. In other words, there is no additional induced charge at the interface between the electrodes and the LC medium due to the reorientation of the director in the middle of the cell. Having a constant surface charge density σ on the surface of the electrodes, the displacement field D is a constant ($D = \sigma$), regardless of the magnitude of the dielectric constant. The change of the electromagnetic energy density due to a change of the dielectric constant can thus be written, according to the discussion earlier in Appendix A, as follows:

$$\Delta U_{EM} = \frac{1}{2} \frac{D_z^2}{(\varepsilon_\parallel \sin^2 \theta + \varepsilon_\perp \cos^2 \theta)} - \frac{1}{2} \frac{D_z^2}{\varepsilon_\perp} \qquad (D-12)$$

where $\varepsilon_\perp$ is the initial dielectric constant. In this equation, D_z ($=$ constant) is the z component of the displacement field vector. We note that the second term is a constant independent of the director orientation $\theta(z)$. The displacement field component D_z can be written

$$D_z = (\varepsilon_\perp \cos^2\theta + \varepsilon_\parallel \sin^2\theta)E \qquad (D\text{-}13)$$

where E is the electric field in the cell. We note that E is a function of z. The total free energy in the cell is given by

$$U = \int_0^d (U_{EL} + \Delta U_{EM})dz \qquad (D\text{-}14)$$

which can be written, according to Eqs. (D-8) and (D-12), as

$$U = \frac{1}{2}\int_0^d \left(F_1\theta'^2 + F_2\phi'^2 + \frac{D_z^2}{\varepsilon_\parallel \sin^2\theta + \varepsilon_\perp \cos^2\theta} \right)dz \qquad (D\text{-}15)$$

where we have ignored the constant term in Eq. (D-12), and where F_1, F_2 are given by

$$F_1 = k_1 \cos^2\theta + k_3 \sin^2\theta \qquad (D\text{-}16)$$
$$F_2 = \left[k_2 \cos^2\theta + k_3 \sin^2\theta \right]\cos^2\theta \qquad (D\text{-}17)$$

where θ' and ϕ' are the derivatives of the tilt and twist angles with respect to z, as defined in Eqs. (D-5) and (D-6).

The orientation distribution functions $\theta(z), \phi(z)$ can be obtained by minimizing the total free energy using variational method. First, we consider an arbitrary variation of the twist angle $\delta\phi(z)$, subject to the condition that $\delta\phi(0) = \delta\phi(d) = 0$. Using Eq. (D-15), we obtain

$$\delta U = \int_0^d \left(F_2\phi'\frac{d\delta\phi}{dz} \right)dz \qquad (D\text{-}18)$$

Integrating by parts and using the boundary conditions $\delta\phi(0) = \delta\phi(d) = 0$, we obtain

$$\delta U = -\int_0^d \frac{d}{dz}(F_2\phi')\delta\phi\, dz \qquad (D\text{-}19)$$

Since $\delta\phi$ is arbitrary, $\delta U = 0$ requires that

$$\frac{d}{dz}\left\{[k_2 \cos^2\theta + k_3 \sin^2\theta]\cos^2\theta\frac{d\phi}{dz}\right\} = 0 \qquad \text{(D-20)}$$

This equation is equivalent to a statement of zero torque around the z axis. Now we consider an arbitrary variation in the tilt angle $\delta\theta(z)$, subject to the boundary condition that $\delta\phi(0) = \delta\theta(d) = 0$. For mathematical simplicity, we treat θ as an independent variable and $z = z(\theta)$. The integration in z is transformed into an integration in θ. Thus, the total free energy in Eq. (D-15) can be written

$$U = \int_0^{\theta_{max}}\left(F_1\frac{d\theta}{dz}d\theta + F_2\frac{d\phi}{dz}d\phi + \frac{D_z^2}{\varepsilon_\| \sin^2\theta + \varepsilon_\perp \cos^2\theta}\frac{dz}{d\theta}d\theta\right) \qquad \text{(D-21)}$$

where θ_{max} is the midlayer tilt angle at $z = d/2$ and F_1, F_2 are as given by Eqs. (D-16) and (D-17). The total free energy can be further simplified as

$$U = \int_0^{\theta_{max}}\left(F_3\frac{d\theta}{dz}d\theta + \frac{D_z^2}{\varepsilon_\| \sin^2\theta + \varepsilon_\perp \cos^2\theta}\frac{dz}{d\theta}d\theta\right) \qquad \text{(D-22)}$$

where F_3 is given by

$$F_3 = F_1 + F_2\left(\frac{d\phi}{d\theta}\right)^2 \qquad \text{(D-23)}$$

where F_1, F_2 are as given by Eqs. (D-16) and (D-17).

The variation in θ is now transformed into a variation $\delta z(\theta)$. Using Eq. (D-22), we obtain

$$\delta U = \int_0^{\theta_{max}}\left(-F_3\left(\frac{d\theta}{dz}\right)^2 + \frac{D_z^2}{\varepsilon_\| \sin^2\theta + \varepsilon_\perp \cos^2\theta}\right)\delta z'd\theta \qquad \text{(D-24)}$$

where z' is the derivative of $z(\theta)$ with respect to θ. Since δz is arbitrary, the minimum free energy $\delta U = 0$ requires that

$$F_3\left(\frac{d\theta}{dz}\right)^2 - \frac{D_z^2}{\varepsilon_\| \sin^2\theta + \varepsilon_\perp \cos^2\theta} = \text{constant} \qquad \text{(D-25)}$$

According to Eq. (D-20), we defined a constant of integration μ so that

$$[k_2 \cos^2 \theta + k_3 \sin^2 \theta] \cos^2 \theta \frac{d\phi}{dz} = \sqrt{k_2(\varepsilon_\parallel - \varepsilon_\perp)} \frac{D_z}{\varepsilon_\perp} \mu \qquad \text{(D-26)}$$

We note that, although μ is independent of z, it does depend on the field strength D_z. Thus, Eq. (D-25) can be written, according to Eqs. (D-16), (D-17), (D-23) and (D-26), as

$$F_1 \left(\frac{d\theta}{dz}\right)^2 + \frac{k_2(\varepsilon_\parallel - \varepsilon_\perp) D_z^2 \mu^2}{\cos^2 \theta [k_2 \cos^2 \theta + k_3 \sin^2 \theta] \varepsilon_\perp^2} - \frac{D_z^2}{\varepsilon_\parallel \sin^2 \theta + \varepsilon_\perp \cos^2 \theta} = \text{constant} \qquad \text{(D-27)}$$

For reasons of mathematical simplicity, we again write

$$\varepsilon_\parallel \sin^2 \theta + \varepsilon_\perp \cos^2 \theta = \varepsilon_\perp (1 + \gamma \sin^2 \theta) \qquad \text{(D-28)}$$

$$k_1 \cos^2 \theta + k_3 \sin^2 \theta = k_1 (1 + \kappa \sin^2 \theta) \qquad \text{(D-29)}$$

$$k_2 \cos^2 \theta + k_3 \sin^2 \theta = k_2 (1 + \alpha \sin^2 \theta) \qquad \text{(D-30)}$$

where

$$\gamma = \frac{\varepsilon_\parallel - \varepsilon_\perp}{\varepsilon_\perp} \qquad \text{(D-31)}$$

$$\kappa = \frac{k_3 - k_1}{k_1} \qquad \text{(D-32)}$$

$$\alpha = \frac{k_3 - k_2}{k_2} \qquad \text{(D-33)}$$

We note that γ is a measure of the dielectric anisotropy and κ and α are measures of the elastic anisotropy. We now define a new function $h(\theta)$ as

$$h(\theta) = \frac{\mu^2}{\cos^2 \theta [1 + \alpha \sin^2 \theta]} - \frac{1}{(1 + \gamma \sin^2 \theta)\gamma} \qquad \text{(D-34)}$$

Using this new function and the new parameters, Eq. (D-27) can be written

$$k_1(1 + \kappa \sin^2 \theta) \left(\frac{d\theta}{dz}\right)^2 + h(\theta) \frac{(\varepsilon_\parallel - \varepsilon_\perp) D_z^2}{\varepsilon_\perp^2} = \text{constant} \qquad \text{(D-35)}$$

The constant can be determined by the condition that the tilt angle θ reaches its maximum θ_{max} at midlayer $(z = d/2)$. At the midpoint, the derivative $d\theta/dz$ is zero. Thus, we obtain

$$k_1(1 + \kappa \sin^2 \theta)\left(\frac{d\theta}{dz}\right)^2 = [h(\theta_{max}) - h(\theta)]\frac{(\varepsilon_\parallel - \varepsilon_\perp)D_z^2}{\varepsilon_\perp^2} \tag{D-36}$$

where we recall that D_z is a constant. This equation can be integrated. This leads to

$$\int_0^{\theta_{max}} \frac{\sqrt{(1 + \kappa \sin^2 \theta)}}{\sqrt{h(\theta_{max}) - h(\theta)}}\, d\theta = \frac{\sqrt{\varepsilon_\parallel - \varepsilon_\perp}}{\sqrt{k_1}}\frac{D_z}{\varepsilon_\perp}\frac{d}{2} \tag{D-37}$$

This equation gives a relationship between the midlayer tilt angle θ_{max} and the displacement field D_z. Using the following relationship between the electric field and D_z

$$D_z = \varepsilon_\perp(1 + \gamma \sin^2 \theta)E \tag{D-38}$$

and taking the square root of Eq. (D-36), we obtain

$$\frac{\sqrt{(1 + \kappa \sin^2 \theta)}}{\sqrt{[h(\theta_{max}) - h(\theta)](1 + \gamma \sin^2\theta)}}\, d\theta = \frac{\sqrt{\varepsilon_\parallel - \varepsilon_\perp}}{\sqrt{k_1}}E\, dz \tag{D-39}$$

where we recall that E is the electric field that is parallel to the z axis. We now integrate this equation on both sides. Note that $\theta = \theta_{max}$ corresponds to $z = d/2$. Thus, we obtain

$$\int_0^{\theta_{max}} \frac{\sqrt{(1 + \kappa \sin^2 \theta)}}{\sqrt{[h(\theta_{max}) - h(\theta)](1 + \gamma \sin^2\theta)}}\, d\theta = \int_0^{d/2} \frac{\sqrt{\varepsilon_\parallel - \varepsilon_\perp}}{\sqrt{k_1}}E\, dz \tag{D-40}$$

which can be further written

$$\int_0^{\theta_{max}} \frac{\sqrt{(1 + \kappa \sin^2 \theta)}}{\sqrt{[h(\theta_{max}) - h(\theta)](1 + \gamma \sin^2\theta)}}\, d\theta = \frac{\sqrt{\varepsilon_\parallel - \varepsilon_\perp}}{\sqrt{k_1}}\frac{V}{2} \tag{D-41}$$

where V is the applied voltage on the electrodes. Although the electric field E is a function of z, the integral over z yields the applied voltage.

We now consider the limiting case of small distortion with θ_{max} and θ approaching zero. Using Eq. (D-34) and $\theta \ll 1$, we obtain

$$
\begin{aligned}
h(\theta) &= \mu^2(1+\theta^2)(1-\alpha\theta^2) - \frac{(1-\gamma\theta^2)}{\gamma} \\
&= \mu^2 - \frac{1}{\gamma} + [1+\mu^2(1-\alpha)]\theta^2
\end{aligned}
\tag{D-42}
$$

Thus, the integral Eq. (D-41) can be written

$$
\frac{1}{\sqrt{1+\mu^2(1-\alpha)}} \int_0^{\theta_{max}} \frac{d\theta}{\sqrt{\theta_{max}^2 - \theta^2}} = \frac{\sqrt{\varepsilon_\parallel - \varepsilon_\perp}}{\sqrt{k_1}} \frac{V_T}{2}
\tag{D-43}
$$

where V_T is the threshold voltage needed to produce an infinitesimal distortion. The integral on the left side of Eq. (D-43) is $\pi/2$. This leads to

$$
V_T = \pi \frac{\sqrt{k_1}}{\sqrt{\varepsilon_\parallel - \varepsilon_\perp}} \frac{1}{\sqrt{1+\mu^2(1-\alpha)}}
\tag{D-44}
$$

or equivalently

$$
V_T = \frac{V_{cl}}{\sqrt{1 - \left(\dfrac{k_3 - 2k_2}{k_2}\right)\mu^2}}
\tag{D-45}
$$

where V_{cl} is the critical voltage needed to produce a finite tilt angle in a parallel nematic cell:

$$
V_{cl} = \pi \frac{\sqrt{k_1}}{\sqrt{\varepsilon_\parallel - \varepsilon_\perp}}
\tag{D-46}
$$

Using Eq. (D-46) for the critical voltage for tilt in a parallel nematic cell, Eq. (D-41) can be written

$$
\frac{\pi}{2} \int_0^{\theta_{max}} \frac{\sqrt{(1+\kappa\sin^2\theta)}}{\sqrt{[h(\theta_{max}) - h(\theta)](1+\gamma\sin^2\theta)}} d\theta = \frac{V}{V_{cl}}
\tag{D-47}
$$

which gives the relationship between the applied voltage V and the midlayer tilt angle θ_{max}.

We need to evaluate the constant μ in Eqs. (4-26) and (4-45). From Eq. (D-26), we obtain

$$\frac{d\phi}{dz} = \sqrt{k_2(\varepsilon_\parallel - \varepsilon_\perp)} \frac{D_z}{\varepsilon_\perp} \frac{\mu}{[k_2 \cos^2\theta + k_3 \sin^2\theta]\cos^2\theta} \tag{D-48}$$

or equivalently

$$\frac{d\phi}{dz} = \mu\sqrt{\frac{k_1}{k_2}} \frac{\sqrt{(1 + \kappa\sin^2\theta)}}{\cos^2\theta(1 + \alpha\sin^2\theta)\sqrt{h(\theta_{max}) - h(\theta)}} \frac{d\theta}{dz} \tag{D-49}$$

where we have used Eq. (D-34) for $h(\theta)$, and Eq. (D-36). Integration of Eq. (D-49) from $z = 0$ to $z = d/2$ yields $\Phi/2$ on the left side, where Φ is the total twist angle as defined by the rubbed directions at the boundaries $z = 0$ and $z = d$. This leads to

$$\frac{\Phi}{2} = \mu\sqrt{\frac{k_1}{k_2}} \int_0^{\theta_{max}} \frac{\sqrt{(1 + \kappa\sin^2\theta)}}{\cos^2\theta(1 + \alpha\sin^2\theta)\sqrt{h(\theta_{max}) - h(\theta)}} d\theta \tag{D-50}$$

In the limit of small distortion when θ_{max} and θ approach zero, this equation becomes

$$\frac{\Phi}{2} = \mu\sqrt{\frac{k_1}{k_2}} \frac{1}{\sqrt{1 + \mu^2(1 - \alpha)}} \int_0^{\theta_{max}} \frac{d\theta}{\sqrt{\theta_{max}^2 - \theta^2}} \tag{D-51}$$

The integral on the right side is $\pi/2$. This yields

$$1 + \mu^2(1 - \alpha) = \frac{k_1}{k_2}\left(\frac{\pi}{\Phi}\right)^2\mu^2 \qquad (\theta_{max} \to 0) \tag{D-52}$$

or equivalently

$$\mu^2 = \left((1 - \alpha) + \frac{k_1}{k_2}\frac{\pi^2}{\Phi^2}\right)^{-1} \qquad (\theta_{max} \to 0) \tag{D-53}$$

Once μ is obtained, the threshold voltage can be written, according to Eqs. (D-45) and (D-53)

$$V_T = V_{c1}\left[1 + (\alpha - 1)\frac{k_2}{k_1}\frac{\Phi^2}{\pi^2}\right]^{1/2} = V_{c1}\left[1 + \frac{k_3 - 2k_2}{k_1}\frac{\Phi^2}{\pi^2}\right]^{1/2} \qquad (D-54)$$

We recall that the critical voltage V_{c1} is needed to create any tilt of the director in a parallel nematic LC cell. When a twist exist $0 < \Phi$ in the cell, the minimum voltage needed to create any distortion is V_T. Thus, in a twist nematic LC cell, for applied voltage $V < V_T$, the midlayer tilt angle is zero, $\theta_{max} = 0$.

In summary, the director distribution $\theta(z)$ and $\phi(z)$ in a TN-LC cell can be obtained by using the following steps:

1. Given a desired value of the midlayer tilt angle θ_{max}, Eq. (D-50) can be employed to find the constant μ numerically.
2. Using μ and θ_{max}, Eq. (D-47) can then be employed to obtain the voltage V needed.
3. The z component of the displacement field vector D_z can also be obtained for each midlayer tilt angle θ_{max} by carrying out the numerical integration of Eq. (D-37).
4. Once D_z is obtained, the distribution of the director tilt angle $\theta(z)$ can then be obtained by substituting both D_z and θ_{max} into Eq. (D-36) and then carrying out the numerical integration.
5. Once $\theta(z)$ is obtained, the twist angle $\phi(z)$ can be obtained by numerically integrating Eq. (D-48) or Eq. (D-49).

Appendix E

Electrooptical Distortion in STN-LC

In this appendix, we consider the general case of a supertwisted nematic LC cell with a total twist angle greater than $90°$. Consider, for example, the case of a total twist angle of $135°$. This requires that the rubbed directions of the two glass plates be angularly separated by $135°$. However, we note that the same boundary conditions can support two twist configurations, for liquid crystals without molecular chirality: a right-hand twist of $135°$ and a left-hand twist of $45°$.

When the cell is filled with LC materials without molecular chirality, it is most likely that the cell will exhibit a left-handed $45°$ twist whose twist rate is only one-third that of $135°$ twist. The elastic energy density of the $135°$ twist cell is 9 times higher than that of the $45°$ twist cell. Thus, to achieve a twist angle greater than $90°$ in practical applications, rubbed directions alone at the boundaries are not adequate. Chiral molecules with the proper handedness are often added in the nematic LC cell to ensure the desired twist (to avoid reverse twist). When the LC materials are doped with chiral molecules, a helically twisted structure exists even in the absence of the boundaries. The presence of boundaries with rubbed directions can cause a small change of the twist rate, leading to a slightly higher twist elastic energy density.

To account for the natural twist due to the presence of chiral molecules, the elastic energy density is written

$$U_{\mathrm{EL}} = \tfrac{1}{2}k_1(\nabla \cdot \mathbf{n})^2 + \tfrac{1}{2}k_2(\mathbf{n} \cdot \nabla \times \mathbf{n} + q_0)^2 + \tfrac{1}{2}k_3(\mathbf{n} \times \nabla \times \mathbf{n})^2 \qquad \text{(E-1)}$$

where $\mathbf{n}$ is a unit vector representing the director distribution in the cell, k_1, k_2, k_3 are the three principal elastic constants, and q_0 is the natural twist rate in the absence of the boundaries. We adopt a sign convention of positive q_0 for right-hand twist and negative q_0 for left-hand twist.

Consider a supertwisted nematic LC cell with the director distribution $\mathbf{n}$ parallel to the xy plane. The orientations of the LC molecules are uniformly twisted as a function of z. Thus, the director distribution $\mathbf{n}$ can be written

$$\mathbf{n} = (\cos qz, \sin qz, 0) \qquad \text{(E-2)}$$

where q is the twist rate. Here, again, a positive q represents a right-hand twist.

The divergence and the curl of the director **n** can be written

$$\nabla \cdot \mathbf{n} = 0 \tag{E-3}$$

$$\nabla \times \mathbf{n} = (-q \cos qz, -q \sin qz, 0) \tag{E-4}$$

Using Eqs. (E-1), (E-3), and (E-4), we obtain the following expression for the elastic energy density:

$$U_{EL} = \tfrac{1}{2} k_2 (q - q_0)^2 \tag{E-5}$$

We note that, in the absence of boundaries, the lowest energy configuration occurs when the actual twist rate is exactly the same as the natural twist rate. Any deviation from the natural twist rate can lead to a higher elastic energy density. Thus, when the boundaries are present with prescribed rubbed directions, the actual twist rate is often slightly different from the natural twist rate in order to comply with the boundary conditions.

ELASTIC ENERGY DENSITY

We now consider the elastic energy density due to a redistribution of the director. Again, letting $\phi(z)$ be the twist angle and $\theta(z)$ be the tilt angle at position z, the director distribution **n** can be written

$$\mathbf{n} = (\cos \theta \cos \phi, \cos \theta \sin \phi, \sin \theta) \tag{E-6}$$

The divergence and the curl of the director **n** can be written

$$\nabla \cdot \mathbf{n} = \cos \theta \, \frac{d\theta}{dz} \tag{E-7}$$

$$\nabla \times \mathbf{n} = (\sin \theta \sin \phi \, \theta' - \cos \theta \cos \phi \, \phi', -\sin \theta \cos \phi \, \theta' - \cos \theta \sin \phi \, \phi', 0) \tag{E-8}$$

where

$$\theta' = \frac{d\theta}{dz} \tag{E-9}$$

$$\phi' = \frac{d\phi}{dz} \tag{E-10}$$

As a result of the additional term q_0 to account for the natural twist due to the presence of chiral molecules, the elastic energy density can be written

$$
\begin{aligned}
U_{\text{EL}} = \frac{1}{2} k_1 \cos^2 \theta \left(\frac{d\theta}{dz}\right)^2 + \frac{1}{2} k_2 \cos^4 \theta \left(\frac{d\phi}{dz}\right)^2 - 2k_2 q_0 \cos^2 \theta \left(\frac{d\phi}{dz}\right) \\
+ \frac{1}{2} k_3 \sin^2 \theta \left(\frac{d\theta}{dz}\right)^2 + \frac{1}{2} k_3 \sin^2 \theta \cos^2 \theta \left(\frac{d\phi}{dz}\right)^2
\end{aligned}
\tag{E-11}
$$

which can also be written

$$
\begin{aligned}
U_{\text{EL}} = \frac{1}{2} \left[k_1 \cos^2 \theta + k_3 \sin^2 \theta\right] \left(\frac{d\theta}{dz}\right)^2 + \frac{1}{2} \left[k_2 \cos^2 \theta + k_3 \sin^2 \theta\right] \cos^2 \theta \left(\frac{d\phi}{dz}\right)^2 \\
- 2k_2 q_0 \cos^2 \theta \left(\frac{d\phi}{dz}\right)
\end{aligned}
\tag{E-12}
$$

where we note that the last term is proportional to the natural twist rate q_0. In the above equations, we ignore the constant term $k = q_0^2/2$.

ELECTROMAGNETIC ENERGY DENSITY

The electromagnetic energy density in this case is exactly the same as that of the TN-LC described in Appendix D. We recall that the field vectors can be written

$$
\mathbf{E} = (0, 0, E)
$$
$$
\mathbf{D} = (D_x, D_y, D_z) = \text{constant}
$$

where E is a function of z. The z component of the displacement field D_z can be written

$$
D_z = (\varepsilon_\perp \cos^2 \theta + \varepsilon_\parallel \sin^2 \theta) E
\tag{E-13}
$$

where $\varepsilon_\perp$ and $\varepsilon_\parallel$ are the principal dielectric constants.

The total free energy in the cell can be written

$$
U = \frac{1}{2} \int_0^d \left(F_1 \theta'^2 + F_2 \phi'^2 + F_3 \phi' + \frac{D_z^2}{\varepsilon_\parallel \sin^2 \theta + \varepsilon_\perp \cos^2 \theta} \right) dz
\tag{E-14}
$$

where F_1, F_2 and F_3 are given by

$$F_1 = k_1 \cos^2 \theta + k_3 \sin^2 \theta \tag{E-15}$$

$$F_2 = [k_2 \cos^2 \theta + k_3 \sin^2 \theta]\cos^2 \theta \tag{E-16}$$

$$F_3 = -2k_2 q_0 \cos^2 \theta \tag{E-17}$$

with θ' and ϕ' are the derivatives of the tilt and twist angles with respect to z, as defined in Eqs. (E-9) and (E-10). Note that $F_3 \phi'$ is the additional term due to the natural twist rate q_0.

The distribution functions $\theta(z)$, $\phi(z)$ can be obtained by minimizing the total free energy using the variational method. Following the variational procedure used in Appendix D, we obtain

$$F_2(\theta)\frac{d\phi}{dz} - k_2 q_0 \cos^2 \theta = C_1 \tag{E-18}$$

$$F_1(\theta)\left(\frac{d\theta}{dz}\right)^2 + F_2(\theta)\left(\frac{d\phi}{dz}\right)^2 - \frac{D_z^2}{\varepsilon_\parallel \sin^2 \theta + \varepsilon_\perp \cos^2 \theta} = C_2 \tag{E-19}$$

where C_1, C_2 are constants of integration and F_1, F_2 are as given by Eqs. (E-15) and (E-16). It is important to note that C_1, C_2 are independent of z, but dependent on the field strength D_z.

We now write,

$$\varepsilon_\parallel \sin^2 \theta + \varepsilon_\perp \cos^2 \theta = \varepsilon_\perp(1 + \gamma \sin^2 \theta) \tag{E-20}$$

$$k_1 \cos^2 \theta + k_3 \sin^2 \theta = k_1(1 + \alpha_1 \sin^2 \theta) \tag{E-21}$$

$$k_2 \cos^2 \theta + k_3 \sin^2 \theta = k_2(1 + \alpha_2 \sin^2 \theta) \tag{E-22}$$

where

$$\gamma = \frac{\varepsilon_\parallel - \varepsilon_\perp}{\varepsilon_\perp} \tag{E-23}$$

$$\alpha_1 = \frac{k_3 - k_1}{k_1} \tag{E-24}$$

$$\alpha_2 = \frac{k_3 - k_2}{k_2} \tag{E-25}$$

We note that γ is a measure of the dielectric anisotropy and α_1 and α_2 are measures of the elastic anisotropy. Since k_3 is often larger than k_1 and k_2, both α_1 and α_2 are positive. Here α_1 is exactly the same as κ defined in Eq. (D-32).

We assume the cell to be symmetric with respective to the midlayer plane at $z = d/2$. The constants C_1, C_2 can be determined by the condition that the tilt angle θ reach its maximum (or minimum) value θ_m at midlayer $(z = d/2)$. At the midpoint, the derivative $d\theta/dz$ is zero. Thus, we obtain

$$C_1 = F_2(\theta_m)\left(\frac{d\phi}{dz}\right)_m - k_2 q_0 \cos^2\theta_m \tag{E-26}$$

$$C_2 = F_2(\theta_m)\left(\frac{d\phi}{dz}\right)_m^2 - \frac{D_z^2}{\varepsilon_\perp(1 + \gamma\sin^2\theta_m)} \tag{E-27}$$

where we recall that D_z is a constant. We note that C_1, C_2 can be evaluated when D_z, θ_m, and $(d\phi/dz)_m$ are known. Unfortunately, both θ_m and $(d\phi/dz)_m$ are functions of the applied field D_z.

In order to find the director distribution $\theta(z)$, $\phi(z)$, we eliminate $(d\phi/dz)$ from Eqs. (E-18) and (E-19). This leads to

$$\left(\frac{d\theta}{dz}\right)^2 = \frac{1}{F_1(\theta)}\left[C_2 + \frac{D_z^2}{\varepsilon_\perp(1 + \gamma\sin^2\theta)} - \frac{(C_1 + k_2 q_0 \cos^2\theta)^2}{F_2(\theta)}\right] \tag{E-28}$$

$$\left(\frac{d\phi}{dz}\right) = \frac{C_1 + k_2 q_0 \cos^2\theta}{F_2(\theta)} \tag{E-29}$$

If C_1, C_2, and D_z are known, these two equations can be integrated numerically to yield the director distribution functions $\theta(z)$ and $\phi(z)$. In what follows we outline a procedure to obtain the director distribution functions:

1. Given an applied field strength D_z, and an initial guess of the midlayer tilt angle θ_m and the midlayer twist rate $(d\phi/dz)_m$, we obtain C_1 and C_2 from Eqs. (E-26) and (E-27).
2. Once C_1 and C_2 are known, we obtain the director distribution functions $\theta(z)$ and $\phi(z)$ by using the numerical integration of Eqs. (E-28) and (E-29). The results must satisfy the following boundary conditions:

$$\theta\left(\frac{d}{2}\right) - \theta(0) = \theta_m - \theta_0 \tag{E-30}$$

$$\phi(d) - \phi(0) = \phi_T \tag{E-31}$$

where ϕ_T is the total twist angle of the cell and θ_0 is the pretilt angle at $z = 0$.
3. If Eqs. (E-30,31) are not satisfied, we must find another guess of θ_m and $(d\phi/dz)_m$.
4. The iteration continues until Eqs. (E-30,31) are satisfied.

5. The required voltage is then obtained numerically from the following integration:

$$V = \int_0^d \frac{D_z}{\varepsilon_\perp (1 + \gamma \sin^2 \theta)} dz \qquad (E\text{-}32)$$

We now investigate the threshold voltage needed to achieve a uniform tilt $\theta(z) = \theta_0$ and a uniform twist $(d\phi/dz) = \phi_T/d$ in the cell. Substituting Eqs. (E-26) and (E-27) for C_1 and C_2 in Eq. (E-28) and after several steps of algebraic manipulations, we obtain

$$\left(\frac{d\theta}{dz}\right)^2 = \frac{(\cos^2 \theta - \cos^2 \theta_m)}{k_1(1 + \alpha_1 \sin^2 \theta)} H(\theta, \theta_m) \qquad (E\text{-}33)$$

where $H(\theta, \theta_m)$ is given by

$$H(\theta, \theta_m) = \left(\frac{d\phi}{dz}\right)_m^2 \frac{F_2(\theta_m)}{F_2(\theta)} k_2[1 + \alpha_2 - \alpha_2(\cos^2 \theta + \cos^2 \theta_m)]$$

$$- 2q_0 \left(\frac{d\phi}{dz}\right)_m \frac{F_2(\theta_m)}{F_2(\theta)} k_2 - q_0^2 k_2^2 \frac{(\cos^2 \theta - \cos^2 \theta_m)}{F_2(\theta)} \qquad (E\text{-}34)$$

$$+ \frac{\gamma D_z^2}{\varepsilon_\perp (1 + \gamma \sin^2 \theta)(1 + \gamma \sin^2 \theta_m)}$$

where we recall that θ_m is the midlayer tilt angle and $(d\phi/dz)_m$ is the midlayer twist rate at $z = d/2$. We now examine the differential equation (E-33) and the function $H(\theta, \theta_m)$. When $D_z = 0$ (no applied field), $H(\theta, \theta_m)$ is always negative for all practical cases. As $k_1(1 + \alpha_1 \sin^2 \theta)$ is positive, we conclude from Eq. (E-33) that $\theta_m \leq \theta$ for all θ. Thus, the midlayer tilt angle can be smaller than (or equal to) the pretilt angle θ_0 at the cell boundaries. When an external field is applied, $H(\theta, \theta_m)$ can become nonzero and positive, leading to $\theta \leq \theta_m$ for all θ, provided the applied field D_z is large enough. In the case when the pretilt angle is zero, θ_m is also zero when $D_z = 0$.

We now consider the limiting case when the tilt angle θ is approaching a nonzero constant throughout the cell, specifically $\theta(z) = \theta_0 \neq 0$. In this case the twist rate is uniform, according to Eq. (E-19), that is, $(d\phi/dz) = \phi_T/d$. Integrating Eq. (E-34), and taking the limit, we obtain

$$\int_{\theta_0}^{\theta_m} \frac{d\theta}{\sqrt{\cos^2 \theta - \cos^2 \theta_m}} = \frac{d}{2} \frac{1}{\sqrt{k_1[1 + \alpha_1 \sin^2 \theta_0]}}$$

$$\times \left[\left(\frac{\phi_T}{d}\right)^2 k_2(1 + \alpha_2 - 2\alpha_2 \cos^2 \theta_0) - 2q_0 k_2 \frac{\phi_T}{d} + \frac{\gamma D_z^2}{\varepsilon_\perp (1 + \gamma \sin^2 \theta_0)^2}\right]^{1/2}$$

$$(E\text{-}35)$$

It can be shown that the left side approaches zero when θ_0 approaches θ_m, provided $\theta_0 \neq 0$. The limit becomes $\pi/2$ when $\theta_0 = 0$. Thus, in the case of nonzero pretilt ($\theta_0 \neq 0$), the threshold field D_z for a uniform tilt is given by

$$\frac{\gamma D_z^2}{\varepsilon_\perp (1 + \gamma \sin^2 \theta_0)^2} = 2q_0 k_2 \frac{\phi_T}{d} - k_2 (1 + \alpha_2 - 2\alpha_2 \cos^2 \theta_0) \left(\frac{\phi_T}{d} \right)^2 \quad \text{(E-36)}$$

Using Eq. (E-32), we obtain the following expression for the threshold voltage needed to achieve a uniform tilt angle throughout the cell:

$$V_{th} = \frac{1}{\sqrt{\varepsilon_\parallel - \varepsilon_\perp}} \left[4\pi k_2 \frac{d}{p_0} \phi_T - \phi_T^2 \left[k_3 - 2(k_3 - k_2) \cos^2 \theta_0 \right] \right]^{1/2} \quad \text{(E-37)}$$

where we recall that θ_0 is the nonzero pretilt angle, $p_0 = 2\pi/q_0$ is the natural pitch and ϕ_T is the total twist angle.

For the special case of zero pretilt angle ($\theta_0 = 0$), the integral on the left side of Eq. (E-35) is $\pi/2$ as θ_m approaches $\theta_0 = 0$. The minimum voltage needed to create a small distortion in the cell is thus given by

$$V_{th} = V_{c1} \left[1 + \frac{k_3 - 2k_2}{k_1} \left(\frac{\phi_T}{\pi} \right)^2 + 4 \frac{k_2}{k_1} \frac{d}{p_0} \frac{\phi_T}{\pi} \right]^{1/2} \qquad (\theta_0 = 0) \quad \text{(E-38)}$$

where V_{c1} is the critical voltage needed to produce a finite tilt angle in a parallel nematic cell:

$$V_{c1} = \pi \frac{\sqrt{k_1}}{\sqrt{\varepsilon_\parallel - \varepsilon_\perp}} \quad \text{(E-39)}$$

It is important to remember that Eq. (E-38) is valid for the special case of zero pretilt angle ($\theta_0 = 0$). We note that Eq. (E-38) reduces to the result obtained previously for TN-LC by putting $p_0 = \infty$:

$$V_T = V_{c1} \left[1 + \frac{k_3 - 2k_2}{k_1} \frac{\phi_T^2}{\pi^2} \right]^{1/2} \quad \text{(E-40)}$$

We recall that the critical voltage V_{c1} is needed to create any tilt of the director in a parallel nematic LC.

Appendix F

Form Birefringence of Composite Media

We consider the form birefringence of a composite medium that consists of a uniform distribution of aligned ellipsoids of isotropic material with refractive index n_2 in an optically isotropic host medium of refractive index n_1.

INDUCED DIPOLE MOMENT OF AN OPTICALLY ISOTROPIC ELLIPSOID

Consider a single dielectric ellipsoid made of optically isotropic material with a dielectric susceptibility χ. The dielectric ellipsoid is placed in a uniform electric field E_0 in vacuum with the electric field vector E_0 parallel to one of the principal axes of the ellipsoid. It can be shown that a uniform polarization P is induced in the ellipsoid [1]. It can also be shown that the electric field E is uniform inside the ellipsoid. A simple relationship exists between these three vectors, which are in the same direction:

$$E = E_0 - \frac{QP}{\varepsilon_0} \tag{F-1}$$

where Q is the depolarization factor. It is important to note that E is the electric field inside the dielectric ellipsoid, whereas E_0 is the constant external electric field at infinity. Due to the presence of the ellipsoid, the electric field near the ellipsoid is no longer E_0. In addition, the polarization is related to the local electric field E by the following definition:

$$P = \varepsilon_0 \chi E \tag{F-2}$$

Using Equations (F-1) and (F-2), we obtain the induced polarization in the ellipsoid:

$$P = \frac{\varepsilon_0 \chi}{1 + Q\chi} E_0 \tag{F-3}$$

The induced dipole moment of the ellipsoid can be written

$$p = VP = V \frac{\varepsilon_0 \chi}{1 + Q\chi} E_0 \tag{F-4}$$

where V is the volume of the ellipsoid. This equation is now generalized to account for the different depolarization factors for the three different principal axes:

$$p_\alpha = V \frac{\varepsilon_0 \chi}{1 + Q_\alpha \chi} E_{0\alpha} \qquad (\alpha = a, b, c) \tag{F-5}$$

where $E_{0\alpha}$ is the component of the external electric field along the direction of the α axis and Q_α is the depolarization factor for field along the α axis.

A UNIFORM DISTRIBUTION OF DIELECTRIC ELLIPSOIDS

We now consider the effective refractive indices of the composite medium, which consists of a uniform distribution of aligned ellipsoids of isotropic material with a dielectric susceptibility χ in vacuum.

The total induced polarization can be obtained by adding all the individual induced dipole moments and then dividing by the volume of the composite medium. This leads to

$$P_\alpha = \frac{\sum p_\alpha}{V} = f \frac{\varepsilon_0 \chi}{1 + Q_\alpha \chi} E_{0\alpha} \qquad (\alpha = a, b, c) \tag{F-6}$$

where the summation $\sum$ is over all the ellipsoids and f is the fill factor $(0 \leq f < 1)$. With this induced polarization, we can now define the principal effective dielectric susceptibilities as

$$P_\alpha = \varepsilon_0 \chi'_{\alpha\alpha} E'_\alpha \qquad (\alpha = a, b, c) \tag{F-7}$$

where E'_α is the effective electric field inside a large ellipsoid of a composite medium that contains many ellipsoids. From the discussion earlier (see Eq. F-1), this effective electric field can be written

$$E'_\alpha = E_{0\alpha} - \frac{Q_\alpha P_\alpha}{\varepsilon_0} \qquad (\alpha = a, b, c) \tag{F-8}$$

Using Eq. (F-6) for the induced polarization in the composite medium P_α, we obtain

$$E'_\alpha = \left\{1 - f\frac{\varepsilon_0\chi}{1 + Q_\alpha\chi}\right\}E_{0\alpha} \qquad (\alpha = a, b, c) \qquad \text{(F-9)}$$

We now substitute Eq. (F-6) for P_α and Eq. (F-9) for E'_α into Eq. (F-7). After eliminating $E_{0\alpha}$, we obtain

$$\varepsilon_0\chi'_{\alpha\alpha} = \frac{f\varepsilon_0\chi}{1 + (1-f)Q_\alpha\chi} \qquad (\alpha = a, b, c) \qquad \text{(F-10)}$$

This equation can now be generalized for the case when the ellipsoids of refractive index are in a host medium of refractive index n_1. This is done by replacing χ with $[(n_2^2/n_1^2) - 1]$ and $\chi'_{\alpha\alpha}$ with $[(n_\alpha^2/n_1^2) - 1]$. Thus, we obtain

$$\frac{n_\alpha^2}{n_1^2} - 1 = \frac{f\left[\frac{n_2^2}{n_1^2} - 1\right]}{1 + (1-f)Q_\alpha\left[\frac{n_2^2}{n_1^2} - 1\right]} \qquad (\alpha = a, b, c) \qquad \text{(F-11)}$$

where we recall that Q_α are the depolarization factor of the ellipsoids. Detailed expressions in terms of elliptic integrals for the depolarization factors of general ellipsoids with semiaxes a,b,c can be found in Stoner [1].

REFERENCES

1. E. C. Stoner, "The demagnetization factors for ellipsoids," *Phil. Mag.* **36**, 803 (1945).

Author Index

423

Subject Index